Theory of Superconductivit

From Weak to Strong Coupling

Series in Condensed Matter Physics

Series Editors:

J M D Coey, **D R Tilley** and **D R Vij**

Other titles in the series include:

Nonlinear Dynamics and Chaos in Semiconductors
K Aoki

Modern Magnetooptics and Magnetooptical Materials
A K Zvedin and V A Kotov

Permanent Magnetism
R Skomski and J M D Coey

Series in Condensed Matter Physics

Theory of Superconductivity

From Weak to Strong Coupling

A S Alexandrov

Loughborough University, UK

CRC Press
Taylor & Francis Group
Boca Raton London New York

CRC Press is an imprint of the
Taylor & Francis Group, an **informa** business

First published 2003 by IOP Publishing Ltd

Published 2019 by CRC Press
Taylor & Francis Group
6000 Broken Sound Parkway NW, Suite 300
Boca Raton, FL 33487-2742

First issued in paperback 2019

ISBN 13: 978-0-367-45445-6 (pbk)
ISBN 13: 978-0-7503-0836-6 (hbk)

British Library Cataloguing-in-Publication Data

A catalogue record for this book is available from the British Library.

Library of Congress Cataloging-in-Publication Data are available

Cover Design: Victoria Le Billon

Typeset in LaTeX 2_ε by Text 2 Text, Torquay, Devon

Contents

Preface

The observation of high-temperature superconductivity in complex layered cuprates by Bednorz and Müller in 1986 should undoubtedly be rated as one of the greatest experimental discoveries of the last century, whereas identifying and understanding the microscopic origin of high-temperature superconductivity stands as one of the greatest theoretical challenges of this century. This book is conceived as a fairly basic introduction to the modern theory of superconductivity. It also sets out an approach to the problem of high-temperature superconductivity, based on the extension of the BCS theory to the strong-coupling regime. The book starts with the phenomenological ideas by F and H London, Ogg Jr and Shafroth-Butler-Blatt, Ginzburg and Landau, Pippard, and proceeds with the microscopic weak-coupling theory by Bardeen, Cooper and Schrieffer (BCS). The canonical Migdal–Eliashberg extension of the BCS theory to the intermediate coupling strength of electrons and phonons (or any bosons) is also discussed. Then, proceeding from the dynamic properties of a single polaron, it is shown how the BCS theory can be extended to the strong-coupling regime, where the multi-polaron problem is reduced to a charged Bose gas of bipolarons. Finally, applications of the theory to cuprates are presented in greater detail in part II along with a brief discussion of a number of alternative viewpoints. Superconductivity and, in particular, high-temperature superconductivity is the topic covered in senior graduate and postgraduate courses virtually in every physics department. Therefore the book contains introductions to the Bloch states, quantum statistics and Boltzmann kinetics, second quantization, Green's functions, and canonical transformations in the appendices to make it easy to follow for senior undergraduate and graduate students with a basic knowledge of quantum mechanics. The book could also be seen as an attempt to bring the level of university training up to the level of modern theoretical condensed matter physics.

A S Alexandrov
1 January 2003

Introduction

In 1986 Bednorz and Müller [1] discovered the onset of possible superconductivity at exceptionally high temperatures in a black ceramic material comprising four elements: lanthanum, barium, copper and oxygen. Within the next decade many more complex copper oxides (cuprates) were synthesized including the mercury-cuprate compounds which, to date, have the highest confirmed critical temperature for a superconducting transition, some $T_c = 135$ K at room pressure and approximately 160 K under high (applied) pressure. The new phenomenon initiated by Bednorz and Müller broke all constraints on the maximum T_c predicted by the conventional theory of low-temperature superconducting metals and their alloys. These discoveries could undoubtedly result in large-scale commercial applications for cheap and efficient electricity production, provided long lengths of superconducting wires operating above the liquid nitrogen temperature (ca. 80 K) can be routinely manufactured.

Following Kamerlingh–Onnes' discovery of superconductivity in elemental mercury in 1911, subsequent work revealed that many metals and alloys displayed similar superconducting properties, the transition temperature of the alloy Nb_3Ge at 23 K being the highest recorded prior to the discoveries in the high-T_c cuprates. Despite intense efforts worldwide, no adequate explanation of the superconductivity phenomenon appeared until the work by Bardeen, Cooper and Schrieffer in 1957 [2]. By that time the frictionless flow (i.e. superfluidity) of liquid ^{4}He had been discovered below a temperature of some 2.17 K. It had been known that the helium atom with its two protons, two neutrons and two electrons, was a Bose particle (boson), while its isotope ^{3}He, with two protons and only one neutron, is a fermion. An assembly of bosons obey the Bose–Einstein statistics, which allows all of them to occupy a single quantum state. Fermions obey the Pauli exclusion principle and the Fermi–Dirac statistics, which dictate that two identical particles must not occupy the same quantum state.

F London suggested in 1938 that the remarkable superfluid properties of ^{4}He were intimately linked to the Bose–Einstein 'condensation' of the entire assembly of Bose particles [3]. Nine years later, Bogoliubov [4] and Landau [5] explained how Bose statistics can lead to the frictionless flow of a liquid. The bosons in the lowest energy state within the Bose–Einstein condensate thus form a coherent macromolecule. As soon as one Bose particle in the Bose liquid meets an obstacle to its flow, for example in the form of an impurity, all the others do not allow their condensate partner to be scattered to leave the condensate. A crucial demonstration that superfluidity was linked to the Bose particles and the Bose–Einstein condensation came after experiments on liquid ^{3}He, whose atoms were fermions, which failed to show the characteristic superfluid transition within a reasonable wide temperature interval around the critical temperature for the onset of superfluidity in ^{4}He. In sharp contrast, ^{3}He becomes a superfluid only below a very low temperature of some 0.0026 K. Here we have a superfluid formed from *pairs* of two ^{3}He fermions below this temperature.

The three orders-of-magnitude difference between the critical superfluidity temperatures of ^{4}He and ^{3}He kindles the view that Bose–Einstein condensation might represent the 'smoking gun' of high-temperature superconductivity. 'Unfortunately' electrons are fermions. Therefore, it is not at all surprising that the first proposal for high-temperature superconductivity, made by Ogg Jr in 1946 [6], involved the pairing of individual electrons. If two electrons are chemically coupled together, the resulting combination is a boson with total spin $S = 0$ or 1. Thus, an ensemble of such two-electron entities can, in principle, be condensed into the Bose–Einstein superconducting condensate. This idea was further developed as a natural explanation of superconductivity by Schafroth [7] and Butler and Blatt in 1955 [8].

However, with one or two exceptions, the Ogg–Schafroth picture was condemned and practically forgotten because it neither accounted quantitatively for the critical parameters of the 'old' (i.e. low T_c) superconductors nor did it explain the microscopic nature of the attractive force which could overcome the natural Coulomb repulsion between two electrons which constitute a Bose pair. The same model which yields a rather precise estimate of the critical temperature of ^{4}He leads to an utterly unrealistic result for superconductors, namely $T_c = 10^4$ K with the atomic density of electron pairs of about 10^{22} cm^{-3}, and with the effective mass of each boson twice that of the electron, $m^{**} = 2m_e \simeq 2\times 10^{-27}$ g.

The failure of this 'bosonic' picture of individual electron pairs became fully transparent when Bardeen, Cooper and Schrieffer (BCS) [2] proposed that two electrons in a superconductor indeed formed a pair but of a very large (practically macroscopic) dimension—about 10^4 times the average inter-electron spacing. The BCS theory was derived from an early demonstration by Fröhlich [9] that conduction electrons in states near the Fermi energy could attract each other on account of their weak interaction with vibrating ions of a crystal lattice. Cooper then showed that electron pairs were stable only due to their quantum interaction with the other pairs. The ultimate BCS theory showed that in a small interval round the Fermi energy, the electrons are correlated into pairs in the momentum space. These Cooper pairs would strongly overlap in real space, in sharp contrast with the model of non-overlapping (local) pairs discussed earlier by Ogg and Schafroth. Highly successful for metals and alloys with a low T_c, the BCS theory led the vast majority of theorists to the conclusion that there could be no superconductivity above 30 K, which implied that Nb_3Ge already had the highest

T_c. While the Ogg–Schafroth phenomenology led to unrealistically high values of T_c, the BCS theory left perhaps only a limited hope for the discovery of new materials which could be superconducting at room temperatures or, at least, at liquid nitrogen temperatures.

It is now clear that the Ogg–Schafroth and BCS descriptions are actually two opposite extremes of the same problem of electron–phonon interaction. By extending the BCS theory towards the strong interaction between electrons and ion vibrations, a Bose *liquid* of tightly bound electron pairs surrounded by the lattice deformation (i.e. of so-called *small bipolarons*) was naturally predicted [10]. Further prediction was that high temperature superconductivity could exist in the crossover region of the electron–lattice interaction strength from the BCS-like to bipolaronic superconductivity [11, 12]. Compared with the early Ogg–Schafroth view, two fermions (now polarons) are bound into a bipolaron by lattice deformation. Such bipolaronic states are 'dressed' by the same lattice deformation [13] and, at first sight, they have a mass too large to be mobile. In fact, earlier studies [14, 15] considered small bipolarons as entirely localized objects. However, it has been shown later that small bipolarons are itinerant quasi-particles existing in the Bloch states at temperatures below the characteristic phonon frequency (chapter 4). As a result, the superconducting critical temperature, being proportional to the inverse mass of a bipolaron, was reduced in comparison with an 'ultra-hot' local-pair Ogg–Schafroth superconductivity but turned out to be much higher than the BCS theory prediction. Quite remarkably Bednorz and Müller noted in their original publication, and subsequently in their Nobel Prize lecture [16], that in their ground-breaking search for high-T_c superconductivity, they were stimulated and guided by the polaron model. Their expectation [16] was that if '*an electron and a surrounding lattice distortion with a high effective mass can travel through the lattice as a whole, and a strong electron–lattice coupling exists an insulator could be turned into a high temperature superconductor*'.

The book naturally divides into two parts. Part 1 describes the phenomenology of superconductivity, the microscopic BCS theory and its extension to the intermediate-coupling regime at a fairly basic level. Chapters 1–3 of this part are generally accepted themes in the conventional theory of superconductivity. Chapter 4 describes what happens to the conventional theory when the electron–phonon coupling becomes strong. Part 2 describes key physical properties of high-temperature superconductors. Chapters 5–8 also present the author's particular view of cuprates, which is not yet generally accepted.

In the course of writing the book I have profited from valuable and stimulating discussions with P W Anderson, A F Andreev, A R Bishop, J T Devreese, P P Edwards, L P Gor'kov, Yu A Firsov, J E Hirsch, V V Kabanov, P E Kornilovitch, A P Levanyuk, W Y Liang, D Mihailovic, K A Müller, J R Schrieffer, S A Trugman, G M Zhao and V N Zavaritsky. Part of the writing was done while I was on leave, from Loughborough University, as visiting

Professor at the Hewlett-Packard Laboratories in Palo Alto, California. I wish to thank R S Williams and A M Bratkovsky for arranging my visiting professorship and enlightening discussions. I am thankful to my wife Elena and to my son Maxim for their help and understanding.

PART 1

THEORY

Chapter 1

Phenomenology

At temperatures below some critical temperature T_c, many metals, alloys and doped semiconducting inorganic and organic compounds carry the electric current for infinite time without any applied electric field. Their thermodynamic and electromagnetic properties in the superconducting state (below T_c) are dramatically different from the normal state properties above T_c. Quite a few of them might be understood without any insight into the microscopic origin of superconductivity by exploiting the analogy between superconducting electrons and superfluid neutral helium atoms (liquid ^{4}He flows without any friction below $T_c \simeq 2.17$ K). Realizing this analogy, F and H London developed the successful phenomenological approach in 1935 describing the behaviour of superconductors in the external magnetic field. Ogg Jr proposed a root to high-temperature superconductivity introducing electron pairs in 1946 and Ginzburg and Landau proposed the phenomenological theory of the superconducting phase transition in 1950 providing a comprehensive understanding of the electromagnetic properties below T_c.

1.1 Normal-state Boltzmann kinetics

If ions in a metal were to be perfectly ordered and were not to vibrate around their equilibrium positions, the electric resistivity would be zero. This is due to the familiar interference of waves scattered off periodically arranged scattering centres. The electron wavefunctions in an ideal periodic potential are the Bloch waves, which obey the Schrödinger equation,

$$\left[-\frac{\nabla^2}{2m_e} + V(\boldsymbol{r})\right]\psi_{n\boldsymbol{k}}(\boldsymbol{r}) = E_{n\boldsymbol{k}}\psi_{n\boldsymbol{k}}(\boldsymbol{r}) \tag{1.1}$$

where $V(\boldsymbol{r}+\boldsymbol{l}) = V(\boldsymbol{r})$ for any lattice vector $\boldsymbol{l}$ connecting two atoms. Here and later, we use the 'theoretical' system of units, where the Planck and Boltzmann constants and light velocity are unity ($\hbar = k_B = c = 1$). The momentum and the wavevector are the same in these units ($\boldsymbol{p} = \boldsymbol{k}$). Single-particle states are classified with the wavevector $\boldsymbol{k}$ in the Brillouin zone and with the band index n, so that

$$\psi_{n\boldsymbol{k}}(\boldsymbol{r}) = u_{n\boldsymbol{k}}(\boldsymbol{r}) \exp(\mathrm{i}\boldsymbol{k} \cdot \boldsymbol{r})$$

where $u_{n\boldsymbol{k}}(\boldsymbol{r})$ is a periodic function of $\boldsymbol{r}$ and the energy is quantized into bands (appendix A). When a weak electric field $\boldsymbol{E}$ is applied, the Bloch states behave like free particles accelerated by the field in accordance with the classical Newton law

$$\frac{\mathrm{d}\boldsymbol{k}}{\mathrm{d}t} = -e\boldsymbol{E}. \tag{1.2}$$

Here e is the *magnitude* on the elementary charge. Impurities, lattice defects and ion vibrations break down the translation symmetry, therefore equation (1.2) should be modified. When the interaction of electrons with these imperfections is weak, the electric current is found from the kinetic Boltzmann equation for the distribution function $f(\boldsymbol{r}, \boldsymbol{k}, t)$ in the real $\boldsymbol{r}$ and momentum $\boldsymbol{k}$ spaces. This function can be introduced if the characteristic length of field variations is fairly long. The number of electrons in an elementary volume of this space at time t is determined by $2f(\boldsymbol{r}, \boldsymbol{k}, t)\, \mathrm{d}\boldsymbol{k}\, \mathrm{d}\boldsymbol{r}/(2\pi)^3$. In the equilibrium state, the distribution depends only on the energy $E \equiv E_{n\boldsymbol{k}}$ (appendix B):

$$f(\boldsymbol{r}, \boldsymbol{k}, t) = n_{\boldsymbol{k}} \equiv \left[1 + \exp\frac{E - \mu}{T}\right]^{-1}. \tag{1.3}$$

The Boltzmann equation for electrons in external electric, $\boldsymbol{E}$, and magnetic, $\boldsymbol{B}$, fields has the following form (appendix B):

$$\frac{\partial f}{\partial t} + \boldsymbol{v} \cdot \frac{\partial f}{\partial \boldsymbol{r}} - e(\boldsymbol{E} + \boldsymbol{v} \times \boldsymbol{B}) \cdot \frac{\partial f}{\partial \boldsymbol{k}} = \left(\frac{\partial f}{\partial t}\right)_{\mathrm{c}} \tag{1.4}$$

where $\boldsymbol{v} = \partial E_{n\boldsymbol{k}}/\partial \boldsymbol{k}$ is the group velocity. Here we consider a single band, so the band quantum number n can be dropped. By applying the Fermi–Dirac golden rule, the *collision integral* in the right-hand side for any *elastic* scattering is given by

$$\left(\frac{\partial f}{\partial t}\right)_{\mathrm{c}} = 2\pi n_{\mathrm{im}} \sum_{\boldsymbol{q}} V_{\mathrm{sc}}^2(\boldsymbol{q}) \delta(E_{\boldsymbol{k}} - E_{\boldsymbol{k}+\boldsymbol{q}})[f(\boldsymbol{r}, \boldsymbol{k} + \boldsymbol{q}, t) - f(\boldsymbol{r}, \boldsymbol{k}, t)] \tag{1.5}$$

where $V_{\mathrm{sc}}(\boldsymbol{q})$ is the Fourier component of the scattering potential and n_{im} is the density of scattering centres. The form of this integral does not depend on the particles' statistics. For weak homogeneous stationary electric and magnetic fields, the Boltzmann equation is solved by the substitution

$$f(\boldsymbol{r}, \boldsymbol{k}, t) = n_{\boldsymbol{k}} - \frac{\partial n_{\boldsymbol{k}}}{\partial E} \boldsymbol{F} \cdot \boldsymbol{v}. \tag{1.6}$$

We consider the case when the transport relaxation time defined as

$$\frac{1}{\tau(E)} \equiv 2\pi n_{\mathrm{im}} \sum_{k'} (1 - \cos\Theta) V_{\mathrm{sc}}^2(\boldsymbol{k}' - \boldsymbol{k}) \delta(E_{\boldsymbol{k}'} - E_{\boldsymbol{k}}) \tag{1.7}$$

depends only on energy E. Here Θ is the angle between $\boldsymbol{v}' = \partial E_{\boldsymbol{k}'}/\partial \boldsymbol{k}'$ and $\boldsymbol{v}$. Then keeping the terms linear in the electric field $\boldsymbol{E}$, the Boltzmann equation for the function $\boldsymbol{F}$ becomes

$$\boldsymbol{F} \cdot \boldsymbol{v} = -e\tau(E)[\boldsymbol{E} \cdot \boldsymbol{v} - \boldsymbol{F} \cdot (\boldsymbol{v} \times \boldsymbol{B} \cdot \nabla_{\boldsymbol{k}})\boldsymbol{v}]. \tag{1.8}$$

If the magnetic field is sufficiently weak ($\omega\tau \ll 1$; $\omega \propto B$ is the Larmour frequency), one can keep only the terms linear in B with the following result for the non-equilibrium part of the distribution function:

$$\boldsymbol{F} \cdot \boldsymbol{v} = -e\tau(E)[\boldsymbol{E} \cdot \boldsymbol{v} + e\tau(E)\boldsymbol{E} \cdot (\boldsymbol{v} \times \boldsymbol{B} \cdot \nabla_{\boldsymbol{k}})\boldsymbol{v}]. \tag{1.9}$$

Using the current density

$$\boldsymbol{j} = -2e \sum_{\boldsymbol{k}} \boldsymbol{v} f(\boldsymbol{r}, \boldsymbol{k}, t) = 2e \sum_{\boldsymbol{k}} \boldsymbol{v} \frac{\partial n_{\boldsymbol{k}}}{\partial E} \boldsymbol{v} \cdot \boldsymbol{F} \tag{1.10}$$

one obtains the longitudinal conductivity σ_{xx},

$$\sigma_{xx} = -2e^2 \sum_{\boldsymbol{k}} \frac{\partial n_{\boldsymbol{k}}}{\partial E} \tau(E) v_x^2 \tag{1.11}$$

and the Hall conductivity σ_{yx},

$$\sigma_{yx} = -2e^3 B \sum_{\boldsymbol{k}} \frac{\partial n_{\boldsymbol{k}}}{\partial E} \tau^2(E)(v_y^2 m_{xx}^{-1} - v_y v_x m_{yx}^{-1}). \tag{1.12}$$

Here $m_{\alpha\beta}^{-1} = \partial^2 E_{\boldsymbol{k}}/\partial k_\alpha \partial k_\beta$ is the inverse mass tensor (α, β are x, y, z).

To calculate the longitudinal conductivities (σ_{xx} and σ_{yy}) and the Hall coefficient $R_{\mathrm{H}} = -\sigma_{xy}/(B\sigma_{xx}\sigma_{yy})$, we apply the effective mass approximation (appendix A), assuming for simplicity that the inverse mass tensor is diagonal ($m_{\alpha\beta}^{-1} = \delta_{\alpha\beta}/m_\alpha$). If the transport relaxation time is independent of energy ($\tau(E) = \tau$), we obtain, integrating by parts,

$$\sigma_{xx} = \frac{ne^2\tau}{m_x} \tag{1.13}$$

and

$$\sigma_{yx} = \frac{ne^3 B\tau^2}{m_y m_x}. \tag{1.14}$$

Then the Hall coefficient is $R_H = -1/en$, which allows for an experimental determination of the carrier density $n = 2\sum_k n_k$. The Hall coefficient is negative if the carriers are electrons and positive if they are holes.

These results also follow from the classical Drude model. For example, in the case of an electric field alone, the velocity of an electron accelerated by the field at time t after a collision is $\boldsymbol{v} = \boldsymbol{v}_0 - e\boldsymbol{E}t/m$, where $\boldsymbol{v}_0$ is the velocity immediately after that collision. Since the electron emerges from a collision in a random direction, there will be no contribution from $\boldsymbol{v}_0$ to the average electron velocity, which must, therefore, be given entirely by the average of $-e\boldsymbol{E}t/m$. The average of t is the relaxation time τ and

$$\boldsymbol{j} = -ne\boldsymbol{v}_{\text{average}} = \frac{ne^2\tau}{m}\boldsymbol{E} \tag{1.15}$$

To calculate the current induced by the *time-dependent* electric field, $\boldsymbol{E}(t) = \boldsymbol{E}\mathrm{e}^{\mathrm{i}\omega t}$, one can solve the Boltzmann equation using the substitution

$$f(\boldsymbol{r}, \boldsymbol{k}, t) = n_{\boldsymbol{k}} - \frac{\partial n_{\boldsymbol{k}}}{\partial E}\boldsymbol{F} \cdot \boldsymbol{v}\mathrm{e}^{\mathrm{i}\omega t} \tag{1.16}$$

where $\boldsymbol{F}$ does not depend on time. As a result, if we replace $1/\tau$ by $-\mathrm{i}\omega + 1/\tau$, ac conductivity is obtained from dc conductivity. For example, the real part of ac longitudinal conductivity, which determines the absorption of the electromagnetic radiation, is given by

$$\mathrm{Re}\,\sigma(\nu) = \frac{ne^2\tau}{m(1 + \omega^2\tau^2)} \tag{1.17}$$

for the energy spectrum $E_{\boldsymbol{k}} = \boldsymbol{k}^2/2m$. It satisfies the sum rule

$$\int_0^\infty \mathrm{d}\omega\ \mathrm{Re}\,\sigma(\omega) = \omega_p^2/8 \tag{1.18}$$

where $\omega_p = (4\pi ne^2/m)^{1/2}$ is the plasma frequency. In our 'one-band' approximation, there is a band mass m, which comes into the expression for ω_p, rather than the free-electron mass m_{e}. In fact, using the approximation we have to cut the upper limit in the sum rule by a value which is well above the transport relaxation rate $1/\tau$ but still below the interband gap. One can prove that the band mass in ω_p of the sum rule should be replaced by the free-electron mass, if the upper limit is really infinite.

1.2 London theory

The first customary macroscopic interpretation of superconductivity as a kind of limiting case of ordinary Drude conductivity met with unresolved difficulties. Following this school of thought one was bound to search for a model of a metal which, in its most stable state, contained a *permanent current* and this without the assistance of any external field. The analogy with a ferromagnet which in its most stable state contains a permanent magnetization was recalled to support the model. The thermodynamic stability of the superconducting state, and particularly the stability of persistent currents, did not seem to allow for any other conclusion. But disfavouring this concept, a general theorem of quantum mechanics was used by Bloch, according to which the most stable state of electrons for general reasons should be without a macroscopic current. So Bloch concluded that the only theorem about superconductivity which can be proved is that any theory of superconductivity considering the phenomenon as an extreme case of ordinary conductivity is refutable.

Indeed an experiment by Meissner and Ochsenfeld [17] revealed that the phenomenon must not be interpreted as an extreme case of high metallic conductivity, in spite of the fact that it shows an electric current without an electric field. From infinite conductivity it only follows that the magnetic flux in a superconductor must be constant and, therefore, dependent upon the way in which the superconductor passes the threshold curve between the normal and superconducting states. Meissner's experiment, however, showed that the magnetic flux in a superconductor was always zero if the magnetic field was low enough. A superconductor behaves as an enormous ideal diamagnetic atom of macroscopic dimensions expelling the entire magnetic flux from its volume. The screening of an applied magnetic field is affected by volume currents instead of an atomic magnetization with a diamagnetic susceptibility

$$\chi = -\frac{1}{4\pi}. \tag{1.19}$$

F London [3] noticed that the degenerate Bose–Einstein gas provides a good basis for a phenomenological model, such as seemed needed for the superfluid state of liquid ^{4}He and for the Meissner superconducting state. For superconductivity, in particular, F and H London [18] showed that the Meissner phenomenon could be interpreted very simply by the assumption of a peculiar coupling in the momentum space, as if there were something like a condensed phase of the Bose gas (appendix B). The idea to replace Fermi statistics by Bose statistics in the theory of metals led F and H London to an equation for the current, which turned out to be microscopically exact.

The density of the electric current is known to be given in quantum mechanics by

$$\boldsymbol{j}(\boldsymbol{r}) = \frac{ie}{2m}(\psi^*\nabla\psi - \psi\nabla\psi^*) - \frac{e^2}{m}\boldsymbol{A}\psi^*\psi \tag{1.20}$$

where $\boldsymbol{A} \equiv \boldsymbol{A}(\boldsymbol{r})$ is the vector potential such that $\nabla \times \boldsymbol{A} = \boldsymbol{B}$. One should sum this expression over all electron states below the Fermi surface of the normal metal. The wavefunctions $\psi(\boldsymbol{r})$ of the electrons with a finite wavevector are considerably disturbed by the magnetic field and, therefore, the term in brackets in equation (1.20) does not vanish. However, it turns out to be of the same order of magnitude and of the opposite sign as the last term containing the vector potential. The result is only a very weak Landau diamagnetism (appendix B).

But suppose the electrons are replaced by charged bosons with mass m^{**} and charge e^*. Below the Bose–Einstein condensation temperature the number of bosons N_s, which occupy the same quantum state, is macroscopic. Their wavefunction $\phi(\boldsymbol{r})$ obeys the Schrödinger equation,

$$-\frac{1}{2m^{**}}(\nabla + \mathrm{i}e^*\boldsymbol{A})^2\phi(\boldsymbol{r}) = E\phi(\boldsymbol{r}). \tag{1.21}$$

The vector potential $\boldsymbol{A}$ can be replaced by another $\tilde{\boldsymbol{A}}$ without any change in physical observables, if

$$\tilde{\boldsymbol{A}} = \boldsymbol{A} + \nabla f \tag{1.22}$$

where $f(\boldsymbol{r})$ is an arbitrary single-valued function of coordinates. In particular, the magnetic field is the same for both potentials because $\nabla \times \nabla f = 0$. As a result, we can choose $\boldsymbol{A}$, which satisfies Maxwell's gauge $\nabla \cdot \boldsymbol{A} = 0$ by imposing the condition $\Delta f = \nabla \cdot \tilde{\boldsymbol{A}}$. In a *simply* connected superconductor (as a bulk sample with no holes), $f(\boldsymbol{r})$ can be always determined for any original choice of vector potential $\tilde{\boldsymbol{A}}$. Then, expanding equation (1.21) and the wavefunction in powers of $\boldsymbol{A}$ we obtain, in the leading zero and first order,

$$-\frac{1}{2m^{**}}\Delta\phi_0(\boldsymbol{r}) - \frac{1}{2m^{**}}\Delta\phi_1(\boldsymbol{r}) - \frac{\mathrm{i}e^*}{m^{**}}\boldsymbol{A}\nabla\phi_0(\boldsymbol{r}) = E_0\phi_1(\boldsymbol{r}) + E_1\phi_0(\boldsymbol{r}) \tag{1.23}$$

where $\phi_0(\boldsymbol{r})$ is the condensate wavefunction and E_0 is the energy in the absence of the magnetic field, while $\phi_1(\boldsymbol{r})$, E_1 are the first-order corrections to the wavefunction and to the energy, respectively. In the absence of the external field, free bosons condense into a state with zero momentum, $\boldsymbol{k} = 0$ and $E_0 = 0$ (appendix B). Hence, the unperturbed normalized condensate wavefunction is a constant ($\phi_0(\boldsymbol{r}) = 1/V^{1/2}$). The first and third terms on the left-hand side and the first term on the right-hand side of equation (1.23) vanish. The first-order correction to the energy E_1 must be proportional to $\nabla \cdot \boldsymbol{A}$ as a scalar, so that it vanishes in the Maxwell gauge. We conclude that $\phi_1(\boldsymbol{r}) = 0$ and the perturbation correction to the condensate wavefunction is proportional to the square or higher powers of the magnetic field. Thus, the bracket in equation (1.20) (paramagnetic contribution) vanishes in the first order of $\boldsymbol{A}$ and only the last (diamagnetic) term remains. All condensed bosons have the same wavefunction, $\phi_0(\mathbf{r})$, so the current density is obtained by multiplying equation (1.20) by N_s as

$$\boldsymbol{j}(\boldsymbol{r}) = -\frac{e^{*2}n_s}{m^{**}}\boldsymbol{A} \tag{1.24}$$

where $n_s = N_s/V$ is the condensate density. The non-trivial assumption that carriers in the superconductor are charged bosons leads to the simple London relation between the magnetic field and the current. We can combine the London equation with the Maxwell equation $\nabla \times \boldsymbol{B} = 4\pi \boldsymbol{j}$ to obtain an equation for the magnetic field in the superconductor:

$$\boldsymbol{B} + \lambda_{\mathrm{L}}^2 \nabla \times \nabla \times \boldsymbol{B} = 0 \tag{1.25}$$

where $\lambda_{\mathrm{L}}^2 = m^{**}/(4\pi e^{*2} n_s)$ is known as the *London penetration depth.* On a more phenomenological level, this equation is also derived by minimizing the free energy of a charged superfluid in the magnetic field,

$$F = F_s + \int \mathrm{d}\boldsymbol{r}\, \frac{m^{**} v_s^2(\boldsymbol{r})}{2} n_s + \int \mathrm{d}\boldsymbol{r}\, \frac{B^2(\boldsymbol{r})}{8\pi}. \tag{1.26}$$

Here F_s is the free energy of the unperturbed superfluid, $v_s(\boldsymbol{r})$ is the velocity of condensed carriers, so that the second term is their kinetic energy and the last term is the energy associated with the magnetic field. Replacing $v_s(\boldsymbol{r})$ by $v_s(\boldsymbol{r}) = \boldsymbol{j}(\boldsymbol{r})/(e^* n_s)$ and the current density by $\boldsymbol{j} = \nabla \times \boldsymbol{B}/4\pi$, we can rewrite the free energy:

$$F = F_s + \frac{1}{8\pi} \int \mathrm{d}\boldsymbol{r} (\boldsymbol{B}^2 + \lambda_{\mathrm{L}}^2 |\nabla \times \boldsymbol{B}|^2). \tag{1.27}$$

If $\boldsymbol{B}(\boldsymbol{r})$ changes by a small vector $\delta \boldsymbol{B}(\boldsymbol{r})$, the free energy changes by

$$\delta F = \frac{1}{4\pi} \int \mathrm{d}\boldsymbol{r} (\boldsymbol{B} + \lambda_{\mathrm{L}}^2 \nabla \times \nabla \times \boldsymbol{B}) \cdot \delta \boldsymbol{B}(\boldsymbol{r}). \tag{1.28}$$

The field, which minimizes the free energy, must, therefore, satisfy equation (1.25).

The London equation explains the Meissner–Ochsenfeld effect. In simple geometry, when a superconductor occupies half of the space with $x \geq 0$, the field $\boldsymbol{B}(x)$ is parallel to its surface and depends only on x. The London equation becomes a simple equation for the magnitude $B(x)$,

$$\lambda_{\mathrm{L}}^2 \frac{\mathrm{d}^2 B}{\mathrm{d}x^2} - B = 0 \tag{1.29}$$

with the boundary condition $B(0) = H$, where H is the external field. The solution is

$$B(x) = H \mathrm{e}^{-x/\lambda_{\mathrm{L}}}. \tag{1.30}$$

Therefore, a weak magnetic field penetrates only to a very shallow *microscopic* depth λ_{L}. Indeed, with the atomic density of carriers $n_s = 10^{22}\ \mathrm{cm}^{-3}$, $m^{**} = m_{\mathrm{e}}$ and $e^* = e$, one obtains the London penetration depth as small as $\lambda_{\mathrm{L}} = [m_{\mathrm{e}} c^2/(4\pi n_s e^2)]^{1/2} \approx 600$ Å (in ordinary units).

Figure 1.1. Flux trapped in the hole is quantized: $\Phi_B = p\Phi_0$, $p = 1, 2, 3, \ldots$.

1.3 Flux quantization

The condensate wavefunction $\phi(\boldsymbol{r})$ should be a single-valued function. This constraint leads to a quantization of the magnetic flux. Let us consider a hole in a bulk superconductor (figure 1.1) with the trapped magnetic flux

$$\Phi_B = \int \mathrm{d}\boldsymbol{s} \cdot \boldsymbol{B} \tag{1.31}$$

where the surface integral is taken over the cross section, which includes the hole.

The magnetic field does not penetrate into the bulk deeper than λ_L. Hence, we can find a contour C surrounding the hole, along which the field and the current are zero. Normalizing the condensate wavefunction as $\phi(\boldsymbol{r}) = n_s^{1/2}\exp(\mathrm{i}\Phi)$ and taking $\boldsymbol{j} = 0$ in equation (1.20), we can express the vector potential along the contour as

$$\boldsymbol{A}(\boldsymbol{r}) = -\frac{\nabla\Phi}{e^*}. \tag{1.32}$$

Then the magnetic flux becomes

$$\Phi_B = \oint_C \mathrm{d}\boldsymbol{l} \cdot \boldsymbol{A}(\boldsymbol{r}) = \frac{\delta\Phi}{e^*}. \tag{1.33}$$

Here $\delta\Phi$ is a change of the phase in the round trip along the contour. The wavefunction is single-valued if $\delta\Phi = 2\pi p$ where $p = 0, 1, 2, \ldots$. Hence, the flux is quantized ($\Phi_B = p\Phi_0$) and the flux quantum (in ordinary units)

$$\Phi_0 = \frac{\pi\hbar c}{e} = 2.07 \times 10^{-7}\ \mathrm{G\,cm}^2 \tag{1.34}$$

for $e^* = 2e$ as observed experimentally [19].

1.4 Ogg's pairs

The London equation successfully explained the Meissner–Ochsenfeld effect but could not, of course, explain superconductivity, as electrons do not obey Bose statistics. In 1946 Richard Ogg Jr [6] proposed that the London 'bosonization' of electrons could be realized due to their pairing. If two electrons are chemically coupled together, the resulting combination is a boson with the total spin $S = 0$ or 1. Ogg suggested that an ensemble of such two-electron entities could, in principle, form a superconducting Bose–Einstein condensate. The idea was motivated by his demonstration that electron pairs were a stable constituent of fairly dilute solutions of alkali metals in liquid ammonia. Sufficiently rapid cooling of the solutions to temperatures in the range from −90 to −180 °C resulted in the production of homogeneous deep-blue solids. All of the solid samples proved to be good electrical conductors. No abnormal resistance change accompanying solidification was observed, except for solutions in the concentration range characterized by the phase separation into two dilute liquid phases at sufficiently low temperatures. Extremely rapid freezing of such solutions caused an enormous decrease in measured resistance. The resistance of the liquid sample at −33 °C was some 10^4 Ω, while that of the solid at −95 °C was only 16 Ω. Ogg argued that even such a small residual resistance was due to faulty contact with platinum electrodes, and the solution in the special concentration range was actually a high-temperature superconductor up to its melting point of the order of 190 K. Other experimental studies showed the solute to be diamagnetic in the concentration range characterized by liquid–liquid phase separation. This suggested the electron constituent to be almost exclusively in the electron pair configuration. In a more dilute phase, the electrons were still predominantly paired according to Ogg but their Bose–Einstein condensation temperature was low enough due to a low concentration. In a more concentrated phase, the electron pairs became unstable, and one had essentially a liquid metal with a small temperature-independent Pauli paramagnetism (appendix B). By extremely rapid cooling, it appeared that the liquid–liquid phase separation was prevented, and that the system became frozen into the superconducting Bose–Einstein condensate. Ogg proposed that his model could also explain the previously observed superconductivity of quasi-metallic alloys and compounds.

While independent experiments in metal-ammonia solutions did not confirm Ogg's claim, his idea of real-space electron pairing was further developed as a natural explanation of superconductivity by Schafroth, Butler and Blatt [7, 8]. However, with one or two exceptions, the Ogg–Schafroth picture was condemned and practically forgotten because it neither accounted quantitatively for the critical parameters of the 'old' (i.e. low-T_c) superconductors nor did it explain the microscopic nature of the attractive force which could overcome the natural Coulomb repulsion between two electrons, which constitute the Bose pair. The microscopic BCS theory showed that in a small interval round the Fermi energy, electrons are paired in the momentum space rather than in the real space. The

Cooper pairs strongly overlap in real space, in sharp contrast with the model of non-overlapping (local) pairs, proposed by Ogg and discussed in more detail by Schafroth, Butler and Blatt.

1.5 Pippard and London superconductors

There is only one characteristic length λ_L in the London theory. A substantial step towards the microscopic theory was made by Pippard [20] and by Ginzburg and Landau [21], who introduced the second characteristic length of the superconductor, the so-called coherence length ξ. Pippard noticed that the London equation (1.24), which couples the current density at some point $\boldsymbol{r}$ with the vector potential in the same point $\boldsymbol{A}(\boldsymbol{r})$, should be modified to better fit experimentally observed penetration depths in superconducting metals and alloys. Pippard proposed a phenomenological non-local relation, which couples the current density at one point with vector potential at all neighbouring points:

$$\boldsymbol{j}(\boldsymbol{r}) = -\frac{3}{16\pi^2\lambda_L^2\xi}\int \mathrm{d}\boldsymbol{r}'\,(\boldsymbol{r}-\boldsymbol{r}')\frac{\boldsymbol{A}(\boldsymbol{r}')\cdot(\boldsymbol{r}-\boldsymbol{r}')}{|\boldsymbol{r}-\boldsymbol{r}'|^4}\mathrm{e}^{-|\boldsymbol{r}-\boldsymbol{r}'|/\xi}. \tag{1.35}$$

If the London penetration depth is large, $\lambda_L \gg \xi$, the vector potential varies slowly enough to take it out of the integral at point $\boldsymbol{r}$. In this case Pippard's equation yields the London equation. However, in the opposite case $\lambda_L \ll \xi$, the vector potential is essentially non-zero only in a thickness λ_H smaller than ξ, and the integral in equation (1.35) is reduced by the factor λ_H/ξ. We can roughly estimate the result of integration to be

$$\boldsymbol{j}(\boldsymbol{r}) \approx -\frac{\lambda_H e^{*2} n_s}{\xi m^{**}}\boldsymbol{A}(\boldsymbol{r}). \tag{1.36}$$

Applying the Maxwell equation, we obtain the exponential penetration law for the magnetic field as in the London case but with the penetration depth λ_H differing from the London penetration depth λ_L:

$$\frac{1}{\lambda_H^2} \approx \frac{4\pi\lambda_H e^{*2} n_s}{\xi m^{**}}. \tag{1.37}$$

In this Pippard limit, the true penetration depth becomes larger than the London value:

$$\lambda_H \approx \lambda_L(\xi/\lambda_L)^{1/3} > \lambda_L. \tag{1.38}$$

There is another characteristic length in 'dirty' superconductors, which is the electron mean free path l limited by the presence of impurities. Here it is natural to expect that the non-local relation between the current density and the vector potential holds within a distance of the order of l rather than ξ, if $l \ll \xi$. Thus, we need to replace ξ in the exponent of equation (1.35) by l. By doing so, Pippard assumed that the contribution to $\boldsymbol{j}(\boldsymbol{r})$ coming from distances less than the mean free path is not modified by the impurities. As a result, we obtain from Pippard and Maxwell equations in the dirty limit $\lambda_H, \xi \gg l$:

$$\lambda_{\mathrm{H}} \approx \lambda_{\mathrm{L}}(\xi/l)^{1/2}. \tag{1.39}$$

In this limit the penetration depth is proportional to the square root of the impurity concentration in agreement with experimental observations. Actually the difference between the London superconductors with $\lambda_L \gg \xi$ (type II) and the Pippard superconductors, where $\lambda_L \ll \xi$ (type I), is much deeper as follows from the Ginzburg–Landau theory.

1.6 Ginzburg–Landau theory

1.6.1 Basic equations

The condensate density n_s is homogeneous in the London theory due to a stiffness in the condensate wavefunction to small magnetic perturbations. With increasing magnetic field, the stiffness no longer holds, and the superfluid density becomes inhomogeneous, $n_s = n_s(\boldsymbol{r})$. If we normalize the condensate wavefunction by the condition $|\phi(\boldsymbol{r})|^2 = n_s(\boldsymbol{r})$, the supercurrent density is

$$\boldsymbol{j}(\boldsymbol{r}) = \frac{\mathrm{i}e^*}{2m^{**}}(\phi^*\nabla\phi - \phi\nabla\phi^*) - \frac{e^{*2}}{m^{**}}\boldsymbol{A}(\boldsymbol{r})n_s(\boldsymbol{r}). \tag{1.40}$$

To describe the superconductor in a finite magnetic field, we need an equation for the condensate density and the phase $\Phi(\boldsymbol{r})$ of the condensate wavefunction $\phi(\boldsymbol{r}) \equiv n_s(\boldsymbol{r})^{1/2}\exp[\mathrm{i}\Phi(\boldsymbol{r})]$. This equation might be rather different from the *ideal* Bose gas (equation (1.21)) because pairs are strongly correlated in real superconductors. Remarkably, Ginzburg and Landau (GL) [21] formulated the equation for $\phi(\boldsymbol{r})$, which turned out to be general enough to satisfy the microscopic theory of superconductivity. They applied the Landau theory of the second-order phase transitions assuming that $\phi(\boldsymbol{r})$ is an order parameter which distinguishes the superconducting phase, where $\phi(\mathbf{r}) \neq 0$, and the normal phase, where $\phi(\boldsymbol{r}) = 0$. In the vicinity of the transition they expanded the superfluid free energy F_s (equation (1.26)), in powers of $\phi(\boldsymbol{r})$ keeping the terms up to the fourth power:

$$F - F_n = \int \mathrm{d}\boldsymbol{r} \left\{ \alpha|\phi(\boldsymbol{r})|^2 + \frac{\beta}{2}|\phi(\boldsymbol{r})|^4 + \frac{1}{2m^{**}}|[\nabla + \mathrm{i}e^*\boldsymbol{A}(\boldsymbol{r})]\phi(\boldsymbol{r})|^2 + \frac{B^2(\boldsymbol{r})}{8\pi} \right\}. \tag{1.41}$$

Here F_n is the free energy of the normal phase in the absence of a magnetic field. There are no odd terms in this expansion. Such terms (for example, linear and cubic in $\phi(\boldsymbol{r})$) would contain an arbitrary phase Φ of the order parameter and should be excluded because F is a physical observable. In the framework of the Landau theory of phase transitions, the phenomenological coefficient α is proportional to the temperature difference $\alpha \propto (T - T_c)$ and β is temperature independent. Varying $\phi^*(\boldsymbol{r})$ by $\delta\phi^*(\boldsymbol{r})$ and integrating by parts, one obtains

$$\delta F = \int \mathrm{d}\boldsymbol{r}\, \delta\phi^*(\boldsymbol{r}) \left\{ \alpha\phi(\boldsymbol{r}) + \beta|\phi(\boldsymbol{r})|^2\phi(\boldsymbol{r}) - \frac{1}{2m^{**}}[\nabla + \mathrm{i}e^*\boldsymbol{A}(\boldsymbol{r})]^2\phi(\boldsymbol{r}) \right\} + \frac{1}{2m^{**}} \oint \delta\phi^*(\boldsymbol{r})\, \mathrm{d}\boldsymbol{s} \cdot [\nabla + \mathrm{i}e^*\boldsymbol{A}(\boldsymbol{r})]\phi(\boldsymbol{r}) \tag{1.42}$$

where the second integral is over the surface of the superconductor. In equilibrium, $\delta F = 0$ and both integrals should vanish for any $\delta\phi^*(\boldsymbol{r})$. As a result, we obtain the master equation of the theory:

$$-\frac{1}{2m^{**}}[\nabla + \mathrm{i}e^*\boldsymbol{A}(\boldsymbol{r})]^2\phi(\boldsymbol{r}) + \beta|\phi(\boldsymbol{r})|^2\phi(\boldsymbol{r}) = -\alpha\phi(\boldsymbol{r}). \tag{1.43}$$

It looks like the Schrödinger equation for the condensate wavefunction, equation (1.21), but with a nonlinear 'potential energy' proportional to $|\phi(\boldsymbol{r})|^2$ and a fixed total energy $-\alpha$. For homogeneous superconductors without a magnetic field, there are two solutions for the order parameter, $\phi(\boldsymbol{r}) = 0$, which corresponds to the normal state, and $|\phi(\boldsymbol{r})|^2 = n_s = -\alpha/\beta$, which describes the homogeneous superconducting state. The superfluid density should vanish at the transition, $T = T_c$, which is the case if $\alpha(T) \propto (T - T_c)$ and β is a constant. The coherence length is a fundamental feature of the GL theory. To determine this new length, let us consider a situation when the real order parameter changes only in one direction, $\phi(\boldsymbol{r}) = n_s^{1/2} f(x)$, and there is no magnetic field. The GL equation for a dimensionless function $f(x)$ becomes

$$-\xi(T)^2 \frac{\mathrm{d}^2 f(x)}{\mathrm{d}x^2} - f + f^3 = 0 \tag{1.44}$$

where

$$\xi(T) = \frac{1}{[2m^{**}|\alpha(T)|]^{1/2}}$$

is the natural unit of length for the variation of $\phi(\boldsymbol{r})$.

Varying $\boldsymbol{A}(\boldsymbol{r})$ by $\delta\boldsymbol{A}(\boldsymbol{r})$ in the free energy, equation (1.41), and setting $\delta F = 0$, we obtain the Maxwell equation $\nabla \times \nabla \times \boldsymbol{A} = 4\pi \boldsymbol{j}$ with the supercurrent density as in equation (1.40). The master equation and the supercurrent equation (1.40) provide a complete description of the magnetic properties of inhomogeneous superconductors for any magnetic field. To make sure that the second integral in the variation of F, equation (1.42), vanishes, the equations should be supplemented by the boundary condition at the surface:

$$\boldsymbol{n} \cdot [\nabla + \mathrm{i}e^*\boldsymbol{A}(\boldsymbol{r})]\phi(\boldsymbol{r}) = 0 \tag{1.45}$$

where $\boldsymbol{n}$ is a unit vector normal to the surface. Using this condition and the current equation (1.40), we obtain $\boldsymbol{n} \cdot \boldsymbol{j} = 0$, i.e. the normal component of the current should be zero on the surface. Hence, this boundary condition is applied not only to the interface with vacuum but also to a superconductor–insulator interface.

The London equation follows from the GL theory for a weak magnetic field because there is no contribution to $\phi(\boldsymbol{r})$ in equation (1.43) linear with respect to the vector potential in Maxwell's gauge. However, it differs from the London theory as there is another characteristic length $\xi(T)$, which is important in inhomogeneous superconductors. The ratio of the London penetration depth and the coherence length, equation (1.44), plays a crucial role in superconducting electrodynamics:

$$\kappa = \frac{\lambda_L}{\xi}. \tag{1.46}$$

It is known as the Ginzburg–Landau parameter. Substituting $\lambda_L = [m^{**}/(4\pi n_s e^{*2})]^{1/2}$, $\xi = 1/(2m^{**}|\alpha|)^{1/2}$ and $n_s = -\alpha/\beta$ we obtain the temperature independent

$$\kappa = \frac{m^{**}}{e^*}\left[\frac{\beta}{2\pi}\right]^{1/2}. \tag{1.47}$$

In 1950 the pairing hypothesis was not yet accepted, therefore Ginzburg and Landau assumed $e^* = e$ in their pioneering paper [21]. Meanwhile Ogg's phenomenology and BCS theory predict $e^* = 2e$ as a result of real or momentum space pairing, respectively.

The GL theory has its own limitations. Since the phenomenological parameters α and β were expanded near T_c in powers of $(T_c - T)$, the theory is confined to the transition region

$$\frac{|T_c - T|}{T_c} \ll 1. \tag{1.48}$$

It predicts the local London relation between the current density and the vector potential which requires $\lambda_L(T) \gg \xi(0)$, as we know from the Pippard theory. This inequality leads to another constraint:

$$\frac{|T_c - T|}{T_c} \ll \kappa^2 \tag{1.49}$$

because $\lambda_L(T) \propto (T_c - T)^{-1/2}$. This constraint is more stringent than equation (1.48) in superconductors with $\kappa \ll 1$ like Al, Sn, Hg and Pb.

There is also a general constraint on the applicability of the Landau theory of phase transitions. The expansion of F in powers of the order parameter makes sense if a statistically averaged amplitude of fluctuations $|\delta\phi|$ in the coherence volume V_c remains small compared with the order parameter itself, $|\delta\phi| \ll n_s^{1/2}$. The probability of such fluctuations is

$$w \propto \exp(-\delta F/T) \tag{1.50}$$

which gives an estimate for the fluctuation part of the free energy as $\delta F \approx T_c$. However, the free-energy fluctuation in the coherence volume $V_c(T) \approx 4\pi\xi(T)^3/3$ is some $\delta F \approx |\alpha||\delta\phi|^2 V_c$, which yields

$$|\delta\phi|^2 \approx \frac{T_c}{|\alpha| V_c}. \tag{1.51}$$

As a result, the fluctuations are small, if [22]

$$\frac{|T_c - T|}{T_c} \gg Gi \tag{1.52}$$

where

$$Gi = \left[\frac{m^{**} T_c}{n_s(0) V_c(0)^{1/3}}\right]^2. \tag{1.53}$$

In low-temperature superconductors, $V_c(0)$ is about $(E_F/T_c)^3 n^{-1}$ (chapter 2), where E_F is the Fermi energy and $n \approx n_s(0)$ is the electron density. Hence the number Gi is extremely small, $Gi \approx (T_c/E_F)^4 < 10^{-8}$ and the fluctuation region is practically absent. In novel high-temperature superconductors, a few experiments measured an extremely small coherence volume of some 100 Å^3, reduced superfluid density, $n_s(0) \approx 10^{21}$ cm^{-3}, and an enhanced effective mass of supercarriers, $m^{**} \approx 10 m_e$ (part 2). With these parameters Gi turns out to be larger than unity when $T_c \geq 30$ K. Here the GL theory does not apply in its canonical form because equations (1.52) and (1.48) are incompatible. There might be other reasons which make the expansion in powers of the order parameter impossible or make the temperature dependence of the coefficients different from that in GL theory. GL considered the superconducting transition as the second-order phase transition by taking $\alpha(T) \propto (T - T_c)$ and β as a constant near the transition. Indeed, in a homogeneous superconductor with no magnetic field, the difference of free energy densities of two phases (the so-called condensation energy, $f_{\text{cond}} = (F_n - F_s)/V$) is

$$f_{\text{cond}} \equiv -(\alpha n_s + \beta n_s^2/2) = \frac{\alpha^2(0)}{2\beta}\left(\frac{T_c - T}{T_c}\right)^2 \tag{1.54}$$

near the transition. Calculating the second temperature derivative of f_{cond} yields the difference of the specific heat $C = -T\partial^2 F/\partial T^2$ of the superconducting and normal phases which turns out to be finite at $T = T_c$:

$$\Delta C = (C_s - C_n)_{T=T_c} = \frac{\alpha^2(0)}{\beta T_c}. \tag{1.55}$$

The finite jump at T_c of the *second* derivative of the relevant thermodynamic potential justifies the assumption that the transition is a *second*-order phase transition. However, such definition of the transition order might depend on the approximation made. In particular, there is a critical region close to the transition, $|T_c - T| \leq T_c Gi$, where the specific heat deviates from the GL prediction. Landau proposed a more general definition of the second-order phase transition. While two phases with *zero-* and *finite*-order parameters coexist at T_c of the first-order transition, the ordered phase of the second-order transition should have zero-order parameter at T_c. In that sense, the second-order phase transition is a *continuous* transition. This definition does not depend on our approximations (for the theory of phase transitions see [23]). In general, the transition into a superfluid state might differ from the second-order phase transition if the order of the transition is defined by the derivative of the thermodynamic potential. For example, the transition of an ideal Bose gas into the Bose-condensed state is of the third order with a jump in the third derivative of the free energy (appendix B).

1.6.2 Surface energy and thermodynamic critical field

The GL theory allows us to understand the behaviour of superconductors in finite magnetic fields. In particular, it predicts qualitatively different properties for type I and II superconductors in sufficiently strong fields. If the external field H is fixed by external currents, the relevant thermodynamic potential describing the equilibrium state is the Gibbs energy

$$G(T, N, H) = F(T, N, B) - \frac{1}{4\pi} \int \mathrm{d}\boldsymbol{r} \boldsymbol{H} \cdot \boldsymbol{B}(\boldsymbol{r}). \qquad (1.56)$$

Let us consider the equilibrium in the external field between the normal and superconducting phases separated by an infinite plane boundary at $x = 0$, figure 1.2. If both λ_L and ξ are taken to be zero, the boundary would be sharp with no magnetic field penetrating into the superconducting phase on the right-hand side, $x \geq 0$ and with no order parameter in the normal phase on the left-hand side of the boundary, $x \leq 0$. Then the Gibbs energy per unit volume of the superconducting phase, where $B = 0$, is

$$g_s = f_s(T, N, 0) \qquad (1.57)$$

and the Gibbs energy density in the normal phase, where $B = H$, is

$$g_n = f_n(T, N, 0) + \frac{H^2}{8\pi} - \frac{H^2}{4\pi} = f_n(T, N, 0) - \frac{H^2}{8\pi}. \qquad (1.58)$$

Two phases are in equilibrium if their Gibbs energy densities are equal. This is possible only in the so-called thermodynamic critical field $H = H_c$, where

$$\frac{H_c^2}{8\pi} = f_{\mathrm{cond}} = \frac{\alpha^2}{2\beta}. \qquad (1.59)$$

The thermodynamic critical field is linear as a function of temperature near T_c:

$$H_c = \left(\frac{4\pi\alpha(0)^2}{\beta}\right)^{1/2} \left(\frac{T_c - T}{T_c}\right). \qquad (1.60)$$

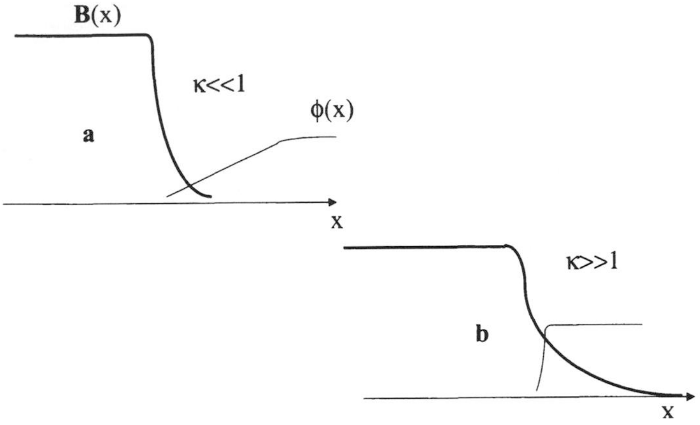

Figure 1.2. Magnetic field and order parameter near the boundary between the normal and superconducting phases of type I (*a*) and type II (*b*) superconductors.

In fact, the boundary is not sharp because both characteristic lengths are finite. To describe the effect of the finite boundary 'thickness', we introduce the surface energy of the boundary defined as the difference between the true Gibbs energy and that of a homogeneous sample in the external field $H = H_c$:

$$\sigma_s = \frac{1}{8\pi}\int_{-\infty}^{\infty} dx\,[B^2(x) - 2H_c B(x) + H_c^2] + \int_{-\infty}^{\infty} dx \left[\alpha|\phi(x)|^2 + \frac{\beta}{2}|\phi(x)|^4 + \frac{1}{2m^{**}}\left|\frac{d\phi(x)}{dx}\right|^2 + \frac{e^{*2}A^2}{2m^{**}}|\phi(x)|^2\right]. \tag{1.61}$$

The term proportional to $\boldsymbol{A} \cdot \nabla\phi(x)$ vanishes because the vector potential has no x-component ($\boldsymbol{A} = \{0, A_y, 0\}$) when the field is parallel to the boundary. We can always choose the vector potential in this form. Also using the gauge transformation (1.22), the order parameter can be made real, $\Phi = 0$, in any *simply* connected superconductor. It is convenient to introduce the dimensionless coordinate $\tilde{x} = x/\lambda_L$, the dimensionless vector potential $a = A_y/(2^{1/2}H_c\lambda_L)$ and the dimensionless magnetic field $h = B/(2^{1/2}H_c)$. Then the GL equations describing the surface energy take the following form:

$$f'' + \kappa^2[(1 - a^2)f - f^3] = 0 \tag{1.62}$$

and

$$a'' = af^2 \tag{1.63}$$

where the double prime means the second derivative with respect to $\tilde{x}$. The boundary conditions in the problem are:

$$f = 0 \qquad a' = h = 2^{-1/2} \tag{1.64}$$

for $x' = -\infty$ (in the normal phase) and

$$f = 1 \qquad a' = 0 \tag{1.65}$$

for $x' = \infty$ (in the superconducting phase). Multiplying equation (1.62) by f' and integrating it over $\tilde{x}$ yields the first integral as

$$f'^2/\kappa^2 + (1 - a^2)f^2 - f^4/2 + a'^2 = 1/2 \tag{1.66}$$

where the constant (= 1/2) on the right-hand side is found using the boundary conditions. Equation (1.66) allows us to simplify the surface energy as follows:

$$\begin{aligned}\sigma_s &= \frac{\lambda_L H_c^2}{4\pi}\int_{-\infty}^{\infty} d\tilde{x}\,\{(a' - 2^{-1/2})^2 + (a^2 - 1)f^2 + f^4/2 + f'^2/\kappa^2\}\\ &= \frac{\lambda_L H_c^2}{2\pi}\int_{-\infty}^{\infty} d\tilde{x}\,\{a'(a' - 2^{-1/2}) + f'^2/\kappa^2\}.\end{aligned} \tag{1.67}$$

Let us estimate the contributions to the surface energy of the first and second terms in the last brackets. The first term $a'(a' - 2^{-1/2})$ is zero both in the normal and superconducting phases. Its value is about -1 in the boundary region of thickness $|x'| < 1$, because $a' = h < 2^{-1/2}$ in any part of the sample. Hence, the contribution of the first term is negative and is about -1. In contrast, the contribution of the second term is positive. This term is non-zero in the region of the order of $|x'| \lesssim 1/\kappa$, where its value is about 1. Hence, its contribution to the integral is about $+1/\kappa$. We conclude that the surface energy is positive in extreme type I superconductors where $\kappa \ll 1$ but it is negative in type II superconductors where $\kappa \gg 1$.

The exact borderline between the Pippard and London superconductors is defined by the condition $\sigma_s = 0$. Integrating the last term f'^2/κ^2 in the first integral of equation (1.67) by parts and substituting f'' from the master equation (1.62) we obtain

$$\sigma_s = \frac{\lambda_L H_c^2}{4\pi}\int_{-\infty}^{\infty} d\tilde{x}\,\{(a' - 2^{-1/2})^2 - f^4/2\}. \tag{1.68}$$

Hence, the surface energy is zero if

$$a' = 2^{-1/2}(1 - f^2). \tag{1.69}$$

Now the second equation of GL theory (equation (1.63)) yields $a'' = -2^{1/2}ff' = af^2$ or $f' = -af/2^{1/2}$. Substituting this f' and a' (equation (1.69)) into the first integral of motion (equation (1.66)), we obtain

$$a^2 f^2\left(1 - \frac{1}{2\kappa^2}\right) = 0. \tag{1.70}$$

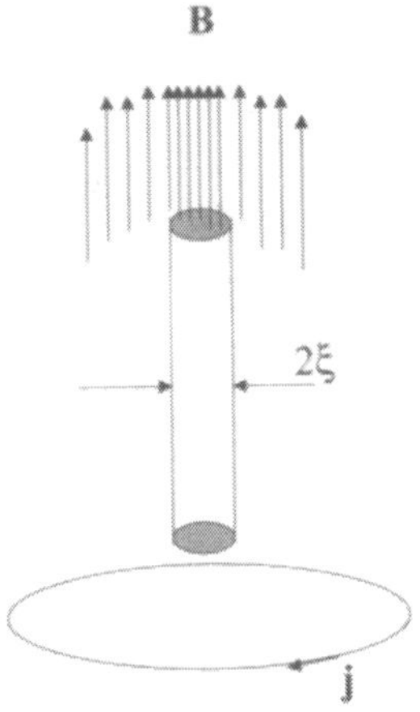

Figure 1.3. Single-vortex core screened by supercurrent.

Hence, the borderline between type I and II superconductors is found at

$$\kappa = \frac{1}{2^{1/2}}. \tag{1.71}$$

Due to the negative surface energy, type II superconductors are inhomogeneous in sufficiently strong magnetic fields. Their order parameter is modulated in space so that the normal and superconducting regions are mixed.

1.6.3 Single vortex and lower critical field

With increasing external field, the normal region in the bulk type II superconductor appears in the form of a single vortex line with the normal core of the radius about ξ surrounded by a supercurrent. Similar vortex lines were found in superfluid rotating ^{4}He and discussed theoretically by Onsager and Feynman. A generalization to superconductors is due to Abrikosov [24]. The properties of a single vortex are well described by the GL equations with proper boundary conditions. Let us assume that the vortex appears at the origin of the coordinate system and has an axial symmetry, i.e. the order parameter depends only on r of the cylindrical coordinates $\{r, \Theta, z\}$ with z parallel to the magnetic field, figure 1.3.

Then the GL equations become

$$\frac{1}{\kappa^2 \rho}\frac{\mathrm{d}}{\mathrm{d}\rho}\rho\frac{\mathrm{d}f}{\mathrm{d}\rho} - \frac{1}{f^3}\left(\frac{\mathrm{d}h}{\mathrm{d}\rho}\right)^2 - f^3 = 0 \tag{1.72}$$

and

$$\frac{1}{\rho}\frac{\mathrm{d}}{\mathrm{d}\rho}\frac{\rho}{f^2}\frac{\mathrm{d}h}{\mathrm{d}\rho} = h. \tag{1.73}$$

These equations are written in a form which introduces the dimensionless quantities $f(\rho) = n_s^{-1/2}\phi(r)$, $\rho = r\lambda_L$ and $h(\rho)$ for the order parameter, length and magnetic field, respectively. The second equation is readily derived replacing the current in equation (1.40) by $\boldsymbol{j} = \nabla \times \boldsymbol{B}/4\pi$ and taking the curl of both parts of the equation. We choose the gauge where the order parameter is real, $\Phi = 0$ and the paramagnetic term of the current in equation (1.40) is zero.

There are four boundary conditions in the single-vortex problem. Three of them are found at $\rho = \infty$. Here the superconductor is not perturbed by the magnetic field, so that $h = \mathrm{d}h/\rho = 0$ and the dimensionless order parameter is unity ($f = 1$). The fourth boundary condition is derived using the flux quantization. The total flux carried by the vortex is

$$\Phi_B = 2^{3/2}\pi\lambda_L^2 H_c \int_0^\infty \mathrm{d}\rho\, \rho h(\rho). \tag{1.74}$$

Applying equation (1.73), we obtain

$$\Phi_B = -\kappa\Phi_0 \left(\frac{\rho}{f^2}\frac{\mathrm{d}h}{\mathrm{d}\rho}\right)_{\rho=0} \tag{1.75}$$

which should be equal to $p\Phi_0$. Hence,

$$\frac{\mathrm{d}h}{\mathrm{d}\rho} = -p\frac{f^2}{\kappa\rho} \tag{1.76}$$

for $\rho = 0$, where p is a positive integer.

Let us first consider the region outside the vortex core, where $\rho \gtrsim 1/\kappa$. The order parameter should be about one in this region. Then, the second GL equation is reduced to its London form in the cylindrical coordinates:

$$\frac{\mathrm{d}^2h}{\mathrm{d}\rho^2} + \frac{1}{\rho}\frac{\mathrm{d}h}{\mathrm{d}\rho} - h = 0. \tag{1.77}$$

The solution which satisfies the boundary conditions, $h = \mathrm{d}h/\mathrm{d}\rho = 0$ for $\rho = \infty$, is

$$h = \text{constant} \times K_0(\rho). \tag{1.78}$$

Here $K_0(\rho)$ is the Hankel function of imaginary argument of zero order. It behaves as the logarithm for small ρ ($\ll 1$):

$$K_0(\rho) \approx \ln\left(\frac{2}{\rho\gamma}\right) \tag{1.79}$$

with $\gamma = 1.78$, and as the exponent for large ρ ($\gg 1$):

$$K_0(\rho) \approx \left(\frac{\pi}{2\rho}\right)^{1/2} \exp(-\rho). \tag{1.80}$$

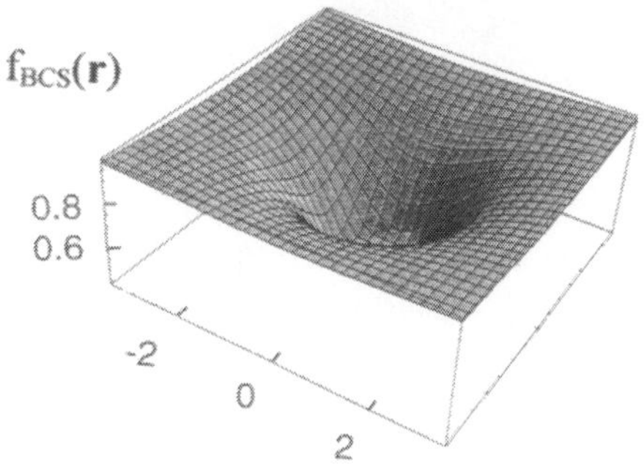

Figure 1.4. Vortex core in the BCS superconductor.

The magnetic field and current decrease exponentially in the exterior of the vortex, where $\rho > 1$. However, the magnetic field is almost constant in the interior of the vortex, $\rho \lesssim 1$. Because the magnetic field and its derivative change over the characteristic length $\rho \simeq 1$, the boundary condition (1.76) is applied for the whole interior ($0 \le \rho \lesssim 1$), not only at zero. Hence, we can use the flux quantization condition to determine the constant in equation (1.78), which is applied for $1/\kappa \le \rho < \infty$. If $\kappa \gg 1$, the two regions overlap. Therefore, $\mathrm{d}h/\mathrm{d}\rho$ calculated using equation (1.76) with $f = 1$ and equation (1.78) should be the same, which is the case if the constant $= p/\kappa$. Hence, the magnetic field around the vortex core ($r \gtrsim \xi$ in ordinary units) is

$$B(r) = \frac{\Phi_0 p}{2\pi\lambda_{\mathrm{L}}^2} K_0(r/\lambda_{\mathrm{L}}). \tag{1.81}$$

The flux quantization boundary condition allows us to 'integrate out' the magnetic field in the master equation for the interior of the vortex with the following result ($\rho \lesssim 1$):

$$\frac{1}{\rho}\frac{\mathrm{d}}{\mathrm{d}\rho}\rho\frac{\mathrm{d}f}{\mathrm{d}\rho} - \frac{p^2}{\rho^2}f - \kappa^2 f^3 = 0. \tag{1.82}$$

This equation is satisfied by a regular solution of the form $f = c_p\rho^{\mathrm{p}}$ for $\rho \to 0$. The constant c_p has to be found by numerical integration of equation (1.82). The numerical result for $p = 1$ is shown in figure 1.4 where $c_1 \simeq 1.166$.

The order parameter is significantly reduced inside the core ($\rho \ll 1/\kappa$) but it becomes almost one in the region $\rho \gg 1/\kappa$,

$$f \approx 1 - \frac{p^2}{\kappa^2\rho^2}. \tag{1.83}$$

The vortex free energy F_p is defined as the difference in the free energies of a bulk superconductor with and without a single vortex. The GL equations are reduced to the single London equation outside the vortex core where the order parameter is almost a constant, $f \approx 1$. When $\kappa \gg 1$, this region yields the main contribution to the vortex energy, while the contribution of the core is negligible. Then $\mathcal{F}_p$ is given by the London expression (1.27). It comprises the kinetic energy due to the current and the magnetic energy:

$$\mathcal{F}_p \approx \frac{1}{8\pi} \int \mathrm{d}\boldsymbol{r}\,(\boldsymbol{B}^2 + \lambda_{\mathrm{L}}^2 |\nabla \times \boldsymbol{B}|^2)$$
$$= \frac{L\lambda_{\mathrm{L}}^2 H_{\mathrm{c}}^2}{2} \int_{\kappa^{-1}}^{\infty} \mathrm{d}\rho\,\rho[h^2 + (\mathrm{d}h/\mathrm{d}\rho)^2]. \quad (1.84)$$

Here L is the vortex length along the field lines. Integrating the second term under the integral by parts, we obtain

$$\mathcal{F}_p = -\frac{\lambda_{\mathrm{L}}^2 L H_{\mathrm{c}}^2}{2} \left[h(\rho)\rho\frac{\mathrm{d}h}{\mathrm{d}\rho}\right]_{\rho=\kappa^{-1}} = \frac{L\lambda_{\mathrm{L}}^2 H_{\mathrm{c}}^2}{2} \frac{p^2}{\kappa^2} \ln\kappa. \quad (1.85)$$

The single-vortex Gibbs energy is obtained as

$$G_p = \mathcal{F}_p - \frac{1}{4\pi} \int \mathrm{d}\boldsymbol{r}\,\boldsymbol{B} \cdot \boldsymbol{H}. \quad (1.86)$$

Here the integral is proportional to the flux carried by the vortex, so that

$$G_p = \mathcal{F}_p - \frac{pL}{4\pi} \Phi_0 H. \quad (1.87)$$

$G_p > 0$ and the vortex state is unfavourable, if the external field is weak. However, if the field is strong enough so that $G_p < 0$, the vortex state becomes thermodynamically stable. The first (lower) critical field H_{c1}, where the vortex appears in bulk type II superconductors, is defined by the condition $G_p = 0$,

$$H_{\mathrm{c1}} = \frac{4\pi \mathcal{F}_p}{p\Phi_0}. \quad (1.88)$$

We see that the lowest field corresponds to a vortex carrying one flux quantum ($p = 1$),

$$H_{\mathrm{c1}} \approx H_{\mathrm{c}} \frac{\ln\kappa}{2^{1/2}\kappa}. \quad (1.89)$$

The first (lower) critical field appears to be much smaller than the thermodynamic critical field in type II superconductors with a large value of $\kappa \gg 1$. If κ is not very large, the vortex penetration and the determination of H_{c1} become more complicated.

1.6.4 Upper critical field

Bulk type I superconductors remain in the Meissner state with no field inside the sample if the external field is below H_{c}. They suddenly become normal metals when the external field is above H_c. A bulk type II superconductor also exhibits complete flux expulsion in the external field $H < H_{c1} < H_c$. However, when $H > H_{c1}$, vortices penetrate inside the sample but the flux passing through the sample remains less than its normal state value. Only for a larger magnetic field, $H \geqslant H_{c2}$ does the sample become entirely normal with no expulsion of the flux, $B = H$. The vortex state for $H_{c1} < H < H_{c2}$ is still superconducting with permanent currents in the sample. The upper (second) critical field H_{c2} is one of the fundamental characteristics of type II superconductors. One can measure H_{c2} by continuously decreasing the field from a high value. At a certain field, $H = H_{c2}$, superconducting regions begin to nucleate spontaneously, so that the resistivity and magnetization start to deviate from their normal state values. In the regions, where the nucleation occurs, superconductivity is just beginning to appear and the density of supercarriers $n_s = |\phi(\boldsymbol{r})|^2$ is small. Then approaching from the normal phase, the GL equation for the order parameter can be linearized by neglecting the cubic term:

$$\frac{1}{2m^{**}}[\nabla + \mathrm{i}e^* \boldsymbol{A}(\boldsymbol{r})]^2 \phi(\boldsymbol{r}) = \alpha \phi(\boldsymbol{r}). \tag{1.90}$$

With the same accuracy the magnetic field inside the sample does not differ from the external field, so that $\boldsymbol{A}(\boldsymbol{r}) = \{0, xH_{c2}, 0\}$. Then the master equation becomes formally identical to the Schrödinger equation for a particle of charge $2e$ in a uniform magnetic field, whose eigenvalues and eigenfunctions are well known [25]. The Hamiltonian (1.90) does not depend on coordinates y and z and the corresponding momentum components, k_y, k_z, are conserved. The eigenfunctions are found to be

$$\phi_\nu(\boldsymbol{r}) = \mathrm{e}^{\mathrm{i}(k_y y + k_z z)} \chi_n(x) \tag{1.91}$$

where $\chi(x)$ obeys the one-dimensional harmonic oscillator equation,

$$\frac{1}{2m^{**}}\chi'' - \frac{m^{**}\omega^2}{2}(x - x_{k_y})^2 \chi = \left(\alpha + \frac{k_z^2}{2m^{**}}\right)\chi. \tag{1.92}$$

Here $x_{k_y} = -k_y/(m^{**}\omega)$ is the equilibrium position, $\omega = e^* H_{c2}/m^{**}$ is the frequency of the oscillator and $\nu \equiv (k_y, k_z, n)$ are the quantum numbers. The well-behaved normalized eigenstates are found to be

$$\chi_n(x) = \left(\frac{m^{**}\omega}{\pi}\right)^{1/4} \frac{H_n[(x - x_{k_y})(m^{**}\omega)^{1/2}]}{\sqrt{2^n n!}} \mathrm{e}^{-m^{**}(x - x_{k_y})^2 \omega/2} \tag{1.93}$$

and the eigenvalues are

$$E_n = \omega(n + 1/2). \tag{1.94}$$

Here

$$H_n \xi = (-1)^n \mathrm{e}^{\xi^2} \frac{\mathrm{d}^n \mathrm{e}^{-\xi^2}}{\mathrm{d}\xi^n} \tag{1.95}$$

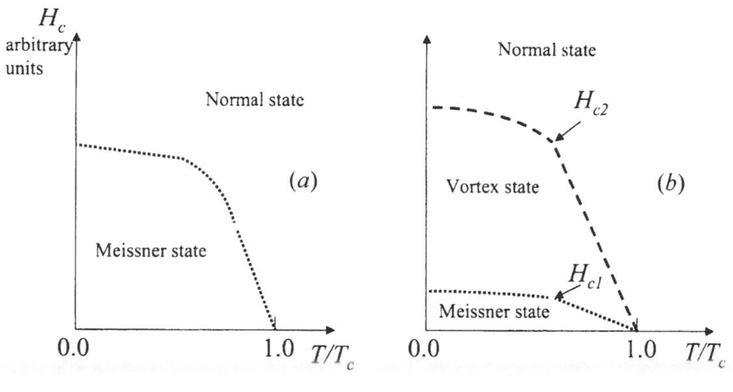

Figure 1.5. Phase diagram of bulk type I (*a*) and type II (*b*) superconductors.

are the Hermite polynomials, $n = 0, 1, 2, 3, \ldots$. The maximum value of k_y is determined by the requirement that the equilibrium position of the harmonic oscillator x_{k_y} is within the sample,

$$-\frac{m^{**}\omega}{2}L_x < k_y < \frac{m^{**}\omega}{2}L_x \tag{1.96}$$

where $L_{x,y,z}$ is the length of the sample along x, y and z, respectively. The number d of allowed values of k_y in this range is

$$d = \frac{L_y}{2\pi}m^{**}\omega L_x = \frac{L_x L_y}{\pi}eH_{c2} \tag{1.97}$$

so that every energy level is d-fold degenerate. We are interested in the highest value of $H = H_{c2}$, which is found from

$$\omega(n + 1/2) = -\left(\alpha + \frac{k_z^2}{2m^{**}}\right) \tag{1.98}$$

with $k_z = n = 0$,

$$H_{c2} = -\frac{2m^{**}\alpha}{e^*}. \tag{1.99}$$

We see that $H_{c2}(T)$ allows for a direct measurement of the superconducting coherence length $\xi(T)$, because $H_{c2} = \Phi_0/2\pi\xi^2(T)$. Near T_c, the upper critical field is linear in temperature $H_{c2}(T) \propto (T_c - T)$ in GL theory, where $\alpha \propto T - T_c$. $H_{c2}(0)$ at zero temperature is normally below the Clogston–Chandrasekhar [26] limit, which is also known as the Pauli pair-breaking limit given by $H_p \simeq 1.84T_c$ (in tesla, if T_c is in kelvin). The limit can be exceeded due to the spin–orbit coupling [27] or triplet pairing but in any case $H_{c2}(0)$ remains finite in the framework of BCS theory. Nonetheless, it might exceed the thermodynamic field by several orders of magnitude in superconductors with a large GL parameter, as follows from the relation

$$H_{c2} = 2^{1/2}\kappa H_c. \tag{1.100}$$

At the border between type II and I superconductors, where $\kappa = 2^{-1/2}$, all critical fields should be the same, $H_{c1} = H_{c2} = H_c$, in agreement with equation (1.100). The H–T phase diagram of type I and II bulk superconductors is shown in figure 1.5. The finite size of the sample and its surface might affect the magnetic field penetration and the superconducting nucleation. In particular, even type I superconductors could be in an intermediate state with normal and superconducting domains, if the field $H < H_c$, while the superconducting nucleation in type II superconductors could take place at the surface in a field $H_{c2} < H < H_{c3}$.

1.6.5 Vortex lattice

Vortex configuration in the field $H_{c1} < H < H_{c2}$ can be found by numerical integration of nonlinear GL equations. If H is only slightly less than H_{c2}, the profile of the order parameter can be established analytically [24]. In the superconducting state close to $H_{c2}(T)$ line the order parameter $\phi(x, y)$ is represented by a linear combination of degenerate solutions of the linearized GL equation with $n = k_z = 0$, equation (1.91),

$$\phi(x, y) = \sum_{k_y} c(k_y)\mathrm{e}^{\mathrm{i}k_y y} \exp\left[-\frac{(x - x_{k_y})^2}{2\xi^2}\right]. \tag{1.101}$$

We can expect the minimum of the free energy to correspond to a regular arrangement of the vortex lines. This happens because vortices repel each other. To minimize their repulsive energy, the lines will take a regular arrangement so that $\phi(x, y)$ is periodic in the x and y directions. To see how the repulsion occurs, let us consider two $p = 1$ vortices with their cores at $\boldsymbol{\rho}_1$ and $\boldsymbol{\rho}_2$ in the plane perpendicular to $\boldsymbol{B}$ in a type II superconductor with $\kappa \gg 1$. The free energy is given by the London expression:

$$\begin{aligned} F &\approx \frac{1}{8\pi}\int \mathrm{d}\boldsymbol{r}\,(\boldsymbol{B}^2 + \lambda_\mathrm{L}^2|\nabla \times \boldsymbol{B}|^2) \\ &= \frac{L\lambda_\mathrm{L}^2 H_\mathrm{c}^2}{4\pi}\int \mathrm{d}\boldsymbol{\rho}(\boldsymbol{h}^2 + |\nabla \times \boldsymbol{h}|^2) \end{aligned} \tag{1.102}$$

where the integral is taken in the plane outside the vortex cores. The magnetic field $h(\boldsymbol{\rho})$ is the superposition of the fields $h_1(\boldsymbol{\rho})$ and $h_2(\boldsymbol{\rho})$ of each vortex,

$$h(\boldsymbol{\rho}) = \frac{1}{\kappa}[K_0(|\boldsymbol{\rho} - \boldsymbol{\rho}_1|) + K_0(|\boldsymbol{\rho} - \boldsymbol{\rho}_2|)]. \tag{1.103}$$

Figure 1.6. Two parallel vortex lines repel each other.

Integrating equation (1.102) by parts and taking the limit $\kappa \to \infty$, one obtains in the leading order

$$F = 2F_1 + U_{12}. \tag{1.104}$$

The interaction energy U_{12} is given by

$$U_{12} = \frac{L\lambda_L^2 H_c^2}{2\pi} \oint \mathrm{d}\boldsymbol{l}_2 \cdot \boldsymbol{h}_1 \times \nabla \times \boldsymbol{h}_2 \tag{1.105}$$

where the contour integral is taken over the surface of the second vortex core. Integrating, one obtains the positive interaction energy

$$U_{12} = \frac{L\lambda_L^2 H_c^2}{\kappa^2} K_0(|\boldsymbol{\rho}_1 - \boldsymbol{\rho}_2|). \tag{1.106}$$

The free energy of the vortex state with two vortex lines is larger than twice the single vortex energy. Hence, two vortices repel each other. The supercurrents of the two vortices are added to each other in the outer regions but they are subtracted from each other in the region between two vortices, figure 1.6. According to the Bernoulli law, the pressure is higher where the velocity of an ideal liquid is lower. Hence, the pressure is higher in the region between two vortices, creating a repulsive force between them,

$$\boldsymbol{f}_{12} = -\nabla U_{12}. \tag{1.107}$$

The magnetic field in the core of the first vortex, created by the second vortex is $B_{12} = 2^{1/2} H_c K_0(|\boldsymbol{\rho}_1 - \boldsymbol{\rho}_2|)/\kappa$. Applying the Maxwell equation, $\nabla \times \mathbf{B}_{12} = 4\pi \boldsymbol{j}_{12}$, the repulsion force applied to the first vortex is expressed through the current $\boldsymbol{j}_{12}$ of the second vortex as

$$\boldsymbol{f}_{12} = L\Phi_0 \boldsymbol{j}_{12} \times \boldsymbol{n} \tag{1.108}$$

where $\boldsymbol{n}$ is a unit vector parallel to the field.

Let us now assume that the order parameter in the vicinity of the $H_{c2}(T)$ line is periodic with period a. The translation in the y direction by a does not change $\phi(x, y)$ if $k_y = 2\pi l/a$ in the sum (1.101) with the integer $l = 0, \pm 1, \pm 2, \ldots$. Hence, for H slightly less than H_{c2}, we have

$$\phi(x, y) = \sum_l c_l \mathrm{e}^{2\pi i l y/a} \exp\left[-\frac{[x - \pi l/(eaH_{c2})]^2}{2\xi^2}\right]. \tag{1.109}$$

In order that $\phi(x, y)$ also be periodic in x, it is necessary to impose a periodicity on the coefficients c_l. A square lattice is obtained if $c_{l+1} = c_l$. Then $\phi(x, y)$ is periodic with period $a = (2\pi)^{1/2}\xi$. A triangular lattice is obtained using the condition $c_{l+2} = c_l$ and $c_1 = ic_0$. The latter is more stable in a perfect crystal. With a further lowering of the magnetic field, the density of vortices drops and the period of the vortex lattice becomes larger than the coherence length. The vortex lattice in type II superconductors is directly observed using ferromagnetic powder sputtered on the sample surface and in neutron scattering experiments.

1.6.6 Critical current

If the current density is too high, the superconducting state is destroyed. The GL phenomenology allows us to calculate the critical current density j_c. The simplest case is the uniform current distribution across the sample. This condition is found in a thin superconducting film, when the thickness of the film is small compared with the magnetic field penetration depth and the coherence length. The magnetic field due to the current in the film is proportional to the film thickness. Hence, if the film is sufficiently thin, we can neglect the magnetic energy in the free energy density f_s,

$$f_s \approx f_n + \alpha n_s + \frac{\beta n_s^2}{2} + \frac{n_s v_s^2}{2m^{**}}. \tag{1.110}$$

Minimizing f_s as a function of the condensate density n_s, we obtain

$$n_s = -\frac{1}{\beta}\left(\alpha + \frac{v_s^2}{2m^{**}}\right). \tag{1.111}$$

Hence, the supercurrent density $j = e^* v_s n_s$ is a nonlinear function of v_s,

$$j = -\frac{e^* v_s}{\beta}\left(\alpha + \frac{v_s^2}{2m^{**}}\right) \tag{1.112}$$

with a maximum at $v_s = (2m^{**}|\alpha|/3)^{1/2}$. The maximum (i.e. critical) current density is

$$j_c = \frac{4e(m^{**})^{1/2}|\alpha/3|^{3/2}}{\beta}. \tag{1.113}$$

The *Landau criterion* of superfluidity allows us to understand j_c at a more microscopic level. Let us consider a superfluid, which flows with a constant velocity $\boldsymbol{v}$. In a homogeneous system, the elementary excitations of the liquid have well-defined momenta $\boldsymbol{k}$, so that their energy $\epsilon_{\boldsymbol{k}}$ is a function of $\boldsymbol{k}$. Because of the friction, the kinetic energy of the condensate may dissipate and the flow would gradually stop. This process creates at least one elementary excitation. In the coordinate frame moving with the liquid, the momentum and energy conservation in a 'collision' with an object (such as the retaining wall) takes the following form:

$$-M\boldsymbol{v} = -M\boldsymbol{v}' + \boldsymbol{k} \tag{1.114}$$

and

$$\frac{Mv^2}{2} = \frac{Mv'^2}{2} + \epsilon_{\boldsymbol{k}}. \tag{1.115}$$

Here M is the mass of the object and $\boldsymbol{v}'$ is the liquid velocity after the collision. Combining these two equations, one obtains

$$\epsilon_{\boldsymbol{k}} + \boldsymbol{k} \cdot \boldsymbol{v} + \frac{k^2}{2M} = 0. \tag{1.116}$$

The critical value of the velocity is obtained for $M = \infty$ as

$$v_c = \min\left[\frac{\epsilon_{\boldsymbol{k}}}{k}\right]. \tag{1.117}$$

The liquid is superfluid, if $v_c \neq 0$ and the flow is slow enough, $v \leqslant v_c$. In BCS theory (chapter 2), the excitation spectrum has a gap near the Fermi surface ($k = k_F$) and $\min \epsilon_{\boldsymbol{k}} = \Delta > 0$. Hence, there is a non-zero critical velocity in the BCS superconductor, $v_c = \Delta / k_F$. Flow with a higher velocity leads to pair breaking and a loss of superconductivity. In the ideal Bose gas $\epsilon_{\boldsymbol{k}} = k^2/2m^{**}$ and $v_c = \min[k/2m^{**}] = 0$. Hence, Bose–Einstein condensation alone is not sufficient for superfluidity. According to Bogoliubov [4], the repulsion between bosons modifies their excitation spectrum so that the repulsive Bose gas is a superfluid (chapter 4).

The critical current in the vortex state of bulk type II superconductors does not reach the pair-breaking limit. The current $\boldsymbol{j}$ in the direction perpendicular to the vortex lines creates the Lorenz force applied to the vortex core, as follows from equation (1.108) with $\boldsymbol{j}_{12} = \boldsymbol{j}$. As a result, vortices move across the current and the current inevitably flows through their normal cores. This vortex motion leads to the energy dissipation. Hence, an ideal type II superconductor has zero critical current in the vortex state, when $H_{c1} < H < H_{c2}$. However, if the external current is not very large, different defects of the crystal lattice 'pin' vortices preventing their motion. That is why the critical current of real type II superconductors depends on the sample quality. Disordered samples can carry the critical current density of about 10^7 A cm^{-2} or even higher [28].

1.7 Josephson tunnelling

Let us consider a bulk superconductor separated into two parts by a thin contact layer with different properties from those of the bulk. The layer might be an insulator, a normal metal or any *weak link* with a reduced condensate density or a small cross section. Its thickness is supposed to be small compared with the coherence length (figure 1.7). Josephson [29] predicted that a supercurrent can flow through the weak link without any voltage and it oscillates as a function of time if there is a voltage drop on the contact. Following Feynman [30], we can understand this effect in the framework of the London description of superconductivity in terms of the condensate wavefunction $\phi(x,t) = n_s^{1/2}\exp[\mathrm{i}\Phi(x,t)]$ (x is the direction perpendicular to the contact). If we suppose, that there is a supercurrent $j = (e^*/m^{**})n_s\,\mathrm{d}\Phi/\mathrm{d}x$ along x in the sample, the condensate phase should depend on x (figure 1.7). To keep the same current density along the sample, the phase gradient should be larger in the weak link, where n_s is suppressed. Hence, there is a finite phase shift $\varphi = \Phi_2 - \Phi_1$ of the condensate wavefunction on the right- (ϕ_2) and left-hand (ϕ_1) 'banks' of the contact, $x = \pm 0$. The carriers change their quantum state from $\phi_1(t)$ to $\phi_2(t)$ in the contact with a transition amplitude K, which is a property of the weak link. Hence, one can write

$$\mathrm{i}\dot{\phi}_1(t) = K\phi_2(t) \tag{1.118}$$

and

$$\mathrm{i}\dot{\phi}_2(t) = K\phi_1(t) + e^*V\phi_2(t) \tag{1.119}$$

where e^*V is a change in the electrostatic energy at the transition due to voltage V. Taking time derivatives and separating the real and imaginary parts of these equations, one obtains

$$\begin{aligned} \dot{n}_{s1} &= 2K(n_{s1}n_{s2})^{1/2}\sin\varphi \\ \dot{n}_{s2} &= -2K(n_{s1}n_{s2})^{1/2}\sin\varphi \end{aligned} \tag{1.120}$$

and

$$\dot{\varphi} = e^*V + K[(n_{s1}/n_{s2})^{1/2} - (n_{s2}/n_{s1})^{1/2}]\cos\varphi. \tag{1.121}$$

The current I through the contact is proportional to the number of carriers tunnelling per second, $I \propto \dot{n}_{s1}$, so that

$$I = I_c \sin\varphi \tag{1.122}$$

where $I_c \propto K(n_{s1}n_{s2})^{1/2}$. If both banks of the Josephson contact are made from the same superconductor, $n_{s1} = n_{s2} = n_s$ and

$$\dot{\varphi} = e^*V.$$

As a result, we obtain

$$I = I_c\sin(\varphi_0 + e^*Vt) \tag{1.123}$$

where φ_0 is the phase difference in the absence of voltage.

We conclude that some current $I < I_c$ can flow through the insulating thin layer between two bulk superconductors with no applied voltage, if φ_0 is non-zero. When the voltage is applied, the current oscillates with a frequency (in the ordinary units)

$$\omega_J = \frac{2eV}{\hbar}. \tag{1.124}$$

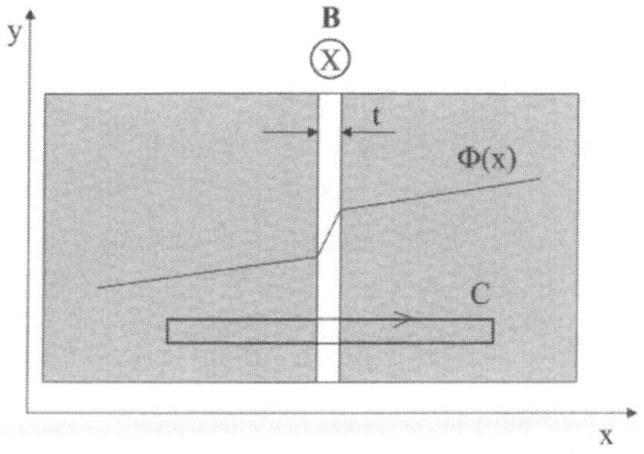

Figure 1.7. Josephson's weak link between two superconductors.

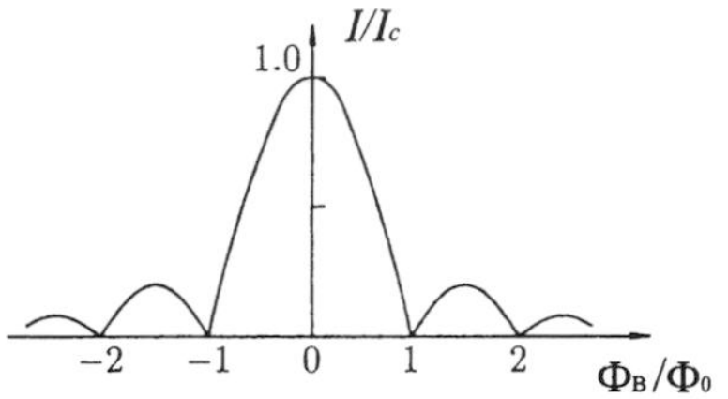

Figure 1.8. Magnetic field dependence of the Josephson current.

The Josephson current also oscillates as a function of the applied magnetic field. Indeed, let us introduce a magnetic field B into the dielectric layer parallel to the contact (figure 1.7). The magnetic field penetrates into the superconductor. The size of the region along x with the magnetic field and the supercurrent is about $w = t + 2\lambda_H$, where t is the thickness of the layer.

Using the same arguments as in section 1.3 we obtain

$$\oint_C \mathrm{d}\boldsymbol{l} \cdot \boldsymbol{A}(\boldsymbol{r}) = \frac{\delta\Phi}{e^*} \tag{1.125}$$

where the contour C crosses the contact twice as shown in figure 1.7. Here we take into account that the contributions of the transport current flowing along x are mutually cancelled. The contribution of the screening current is negligible if the length of the contour is much larger than w. The left-hand side of equation (1.125) is an elementary flux $\mathrm{d}\Phi_B = Bw\,\mathrm{d}y$ across the area of the contour, while $\delta\Phi$ is the change of the phase shift along the contact area, so that

$$\mathrm{d}\Phi_B = \frac{\varphi(y + \mathrm{d}y) - \varphi(y)}{e^*}. \tag{1.126}$$

We see that, in the presence of the magnetic field, the phase shift and Josephson current density $j = j_c \sin\varphi(y)$ are not uniform within the contact area because (from equation (1.126)) $\varphi(y) = e^* Bwy + \varphi_0$. The total current through the contact is

$$I = j_c L \int_{-L/2}^{L/2} \mathrm{d}y \, \sin(e^* Bwy + \varphi_0) = I_c(\sin\varphi_0) \left[\frac{\sin(\pi \Phi_B / \Phi_0)}{\pi \Phi_B / \Phi_0} \right] \quad (1.127)$$

where $I_c = j_c S$ and $S = L^2$ is the contact cross section. The Josephson current oscillates as a function of the magnetic flux revealing a characteristic quantum interference pattern (figure 1.8). The Josephson effect laid the groundwork for designing squids (Superconducting Quantum Interference Devices). Squids can measure magnetic fields and voltages as low as 10^{-11} G and 10^{-15} V, respectively.

Chapter 2

Weak coupling theory

The phenomenological Bose-gas picture rendered no quantitative account for the critical parameters of conventional low-temperature superconductors. Its failure stemmed from the very large diameter ξ of pairs in conventional superconductors which can be estimated using the uncertainty principle, $\xi \approx 1/\delta k$. The uncertainty in the momentum δk is estimated using the uncertainty in the kinetic energy $\delta E \simeq v_{\mathrm{F}}\delta k$, which should be of the order of T_{c}, the only characteristic energy of the superconducting state. Therefore, $\xi \approx v_{\mathrm{F}}/T_{\mathrm{c}}$ turns out to be about 1 μm for simple superconducting metals where $T_{\mathrm{c}} \simeq 1$ K and the Fermi velocity $v_{\mathrm{F}} \simeq 10^8$ cm s^{-1}. This means that pairs in conventional superconductors strongly overlap and the Ogg–Shafroth model of real space pairs cannot be applied.

An ultimately convincing theory of conventional superconductors was formulated by Bardeen, Cooper and Schrieffer [2]. In this and the next chapters we describe the essentials of BCS theory using the Bogoliubov transformation [4] in the weak-coupling regime and Green's functions for the intermediate coupling (see also the excellent books by Schrieffer [31], Abrikosov *et al* [32] and De Gennes [33]).

2.1 BCS Hamiltonian

The BCS theory of superconductivity [2] relies on Fröhlich's observation [9] that conduction electrons could attract each other due to their interaction with vibrating ions of the crystal lattice. In a more complete analysis by Bardeen and Pines [34] in which Coulomb effects and collective plasma excitations were included, the interaction between electrons and the phonon field was shown to dominate over the matrix element of the Coulomb interaction near the Fermi surface. The key point is the fact discovered by Cooper [35] that any attraction between *degenerate* electrons leads to their pairing no matter how weak the attraction is. One can understand the Cooper phenomenon by transforming the corresponding Schrödinger equation for a pair into the momentum space. Because of the Pauli principle, a pair of electrons can only 'move' along the Fermi surface in the momentum space. But it is well known that in two dimensions a bound state exists for any attraction, however weak. The Cooper solution of the two-particle problem in the presence of the Fermi sea demonstrated the instability of an attractive Fermi liquid *versus* pair formation with a possible Bose–Einstein condensation in the momentum space. Because of the Pauli principle, two electrons with parallel spins tend to be at larger distances than two electrons with antiparallel spins. Hence, following BCS, one can expect electrons to be paired into singlets rather than into triplets with the *zero* centre-of-mass momentum of the pairs. For these reasons, BCS introduced a model (truncated) Hamiltonian in which the interaction term contains only the attraction between electrons with the opposite spins and momenta,

$$H = \sum_{\boldsymbol{k},s=\uparrow,\downarrow} \xi_{\boldsymbol{k}} c^{\dagger}_{\boldsymbol{k}s} c_{\boldsymbol{k}s} + \sum_{\boldsymbol{k}} c^{\dagger}_{\boldsymbol{k}\uparrow} c^{\dagger}_{-\boldsymbol{k}\downarrow} \hat{\Delta}(\boldsymbol{k}) \tag{2.1}$$

where

$$\hat{\Delta}(\boldsymbol{k}) = \sum_{\boldsymbol{k}'} V(\boldsymbol{k}, \boldsymbol{k}') c_{-\boldsymbol{k}'\downarrow} c_{\boldsymbol{k}'\uparrow}. \tag{2.2}$$

The first term describes independent Bloch electrons, where $c_{\boldsymbol{k}s}$ is the electron annihilation operator with the momentum $\boldsymbol{k}$ and spin s, $\xi_{\boldsymbol{k}} = E_{\boldsymbol{k}} - \mu$ is the band energy dispersion referred to the chemical potential μ. It is convenient to consider an open system with a fixed chemical potential rather than with a fixed total number of particles $N_e \ggg 1$ to avoid some artificial difference between odd and even N_e. $\langle H - \mu N_e \rangle$ has a minimum in the ground state for the open system. That is why the electron energy in equation (2.1) is referred to as μ. We recall (appendix C) that annihilation and creation operators *anti*commute for fermions:

$$\{c_{\boldsymbol{k}s}, c_{\boldsymbol{k}'s'}\} \equiv c_{\boldsymbol{k}s} c_{\boldsymbol{k}'s'} + c_{\boldsymbol{k}'s'} c_{\boldsymbol{k}s} = 0 \tag{2.3}$$

and

$$\{c_{\boldsymbol{k}s}, c^{\dagger}_{\boldsymbol{k}'s'}\} \equiv c_{\boldsymbol{k}s} c^{\dagger}_{\boldsymbol{k}'s'} + c^{\dagger}_{\boldsymbol{k}'s'} c_{\boldsymbol{k}'s'} = \delta_{\boldsymbol{k}\boldsymbol{k}'} \delta_{ss'}. \tag{2.4}$$

They lower or raise the number of fermions $n_{\boldsymbol{k}s}$ in a single-particle state $|\boldsymbol{k}, s\rangle$ by one and multiply a many-particle wavefunction of non-interacting fermions by $\pm n_{\boldsymbol{k}s}$ or $\pm(1 - n_{\boldsymbol{k}s})$ as

$$\begin{aligned} c_{\boldsymbol{k}s}|n_{\boldsymbol{k}_1 s_1}, n_{\boldsymbol{k}_2 s_2}, \ldots, n_{\boldsymbol{k}s}, \ldots\rangle &= \pm\, n_{\boldsymbol{k}s}|n_{\boldsymbol{k}_1 s_1}, n_{\boldsymbol{k}_2 s_2}, \ldots, n_{\boldsymbol{k}s} - 1, \ldots\rangle \\ c^{\dagger}_{\boldsymbol{k}s}|n_{\boldsymbol{k}_1 s_1}, n_{\boldsymbol{k}_2 s_2}, \ldots, n_{\boldsymbol{k}s}, \ldots\rangle &= \pm\,(1 - n_{\boldsymbol{k}s})|n_{\boldsymbol{k}_1 s_1}, n_{\boldsymbol{k}_2 s_2}, \ldots, n_{\boldsymbol{k}s} + 1, \ldots\rangle. \end{aligned} \tag{2.5}$$

Here $+$ or $-$ depends on the evenness or oddness of the number of occupied states, respectively, which precede the state $|\boldsymbol{k}, s\rangle$ in an adopted ordering of single-particle states. Modelling the attractive potential $V(\boldsymbol{k}, \boldsymbol{k}')$, BCS took into account the retarded character of the attraction mediated by lattice vibrations. The ions of the lattice must have sufficient time to react, otherwise there would be no *overscreening* of the conventional Coulomb repulsion between electrons. In other words, the characteristic time τ_s for the scattering of one electron with the energy $\xi_{\boldsymbol{k}}$ by another one should be longer than the characteristic time of lattice relaxation which is about the inverse (Debye) frequency of vibrations $\omega_{\rm D}^{-1}$. Using the energy–time uncertainty principle, we estimate the scattering time as $\tau_s \approx |\xi_{\boldsymbol{k}} - \xi_{\boldsymbol{k}'}|^{-1}$. Hence, $V(\boldsymbol{k}, \boldsymbol{k}')$ could be negative (i.e. attractive), if $|\xi_{\boldsymbol{k}} - \xi_{\boldsymbol{k}'}| < \omega_{\rm D}$. To describe the essential physics of superconductors, BCS introduced a simple approximation for the attractive interaction:

$$V(\boldsymbol{k}, \boldsymbol{k}') = -2E_p \tag{2.6}$$

if the condition

$$|\xi_{\boldsymbol{k}}|, |\xi_{\boldsymbol{k}'}| < \omega_{\rm D} \tag{2.7}$$

is satisfied, and zero if otherwise. Here E_p is a positive energy depending on the strength of the electron–phonon coupling. Multiplying it by the density of electron states per unit cell at the Fermi level $N(E_{\rm F})$ (appendix B), we obtain a dimensionless constant

$$\lambda = 2E_p N(E_{\rm F}) \tag{2.8}$$

which conveniently characterizes the strength of the coupling. The BCS theory was originally developed for weakly coupled electrons and phonons with $\lambda \ll 1$; extended towards the intermediate coupling ($\lambda \lesssim 1$) by Eliashberg [36]; and to the strong coupling ($\lambda \gtrsim 1$) by us [10, 11].

2.2 Ground state and excitations

The operators $\hat{\Delta}$, $\hat{\Delta}^\dagger$ in the BCS Hamiltonian (2.1) annihilate and create pairs with total momentum $K = 0$. One can expect that the pairs condense in the ground state with $K = 0$ like bosons below some critical temperature $T_{\rm c}$ (appendix B). The number of condensed pairs N_s should be macroscopically large, if $T < T_{\rm c}$. The matrix elements of $\hat{\Delta}$ and $\hat{\Delta}^\dagger$ should be large as well, like the matrix elements of the annihilation and creation boson operators, $\langle N_s - 1|b_0|N_s\rangle = \langle N_s + 1|b_0^\dagger|N_s\rangle = (N_s)^{1/2} \ggg 1$ (appendix C). That is all commutators of $\hat{\Delta}$, $\hat{\Delta}^\dagger$ should be much smaller than the operators themselves because $[b_0 b_0^\dagger] = 1 \lll (N_s)^{1/2}$. Hence, we can neglect the fact that $\hat{\Delta}$ does not commute with $\hat{\Delta}^\dagger$ and $c_{\boldsymbol{k}s}^\dagger$. It is the same as replacing these operators by their expectation values in the open system,

$$\hat{\Delta}(\boldsymbol{k}) \approx \Delta_{\boldsymbol{k}} \tag{2.9}$$

where

$$\Delta_{\boldsymbol{k}} = -2E_p \Theta(\omega_{\rm D} - |\xi_{\boldsymbol{k}}|) \sum_{\boldsymbol{k}'} \Theta(\omega_{\rm D} - |\xi_{\boldsymbol{k}'}|)\langle\langle c_{-\boldsymbol{k}'\downarrow} c_{\boldsymbol{k}'\uparrow}\rangle\rangle \tag{2.10}$$

and $\Theta(x) = 1$ for positive x and zero otherwise. The wavefunction of the open system is a superposition of different eigenfunctions of the total number operator and the average equation (2.10) is not zero but macroscopically large in the superconducting state. Here the double brackets mean both quantum and statistical averages (appendix B). The substitution of equation (2.9) transforms the BCS Hamiltonian into a quadratic form with respect to the electron operators,

$$\tilde{H} = \sum_{k} [\xi_k (c^{\dagger}_{k\uparrow} c_{k,\uparrow} + c^{\dagger}_{-k\downarrow} c_{-k\downarrow}) + \Delta_k c^{\dagger}_{k\uparrow} c^{\dagger}_{-k\downarrow} + \Delta^{*}_{k} c_{-k\downarrow} c_{k\uparrow}] + \frac{|\Delta|^2}{2E_p} \tag{2.11}$$

where the last term is added to make sure that the ground-state energies of the exact BCS Hamiltonian and of the approximation (2.11) are the same. Here

$$\Delta = -2E_p \sum_{k'} \Theta(\omega_{\mathrm{D}} - |\xi_{k'}|)\langle\langle c_{-k'\downarrow} c_{k'\uparrow}\rangle\rangle \tag{2.12}$$

does not depend on $\boldsymbol{k}$. Now, following Bogoliubov [4], we can replace the electron operators by new fermion operators:

$$c_{k\uparrow} = u_k \alpha_k + v_k \beta^{\dagger}_{k} \tag{2.13}$$

$$c_{-k\downarrow} = u_k \beta_k - v_k \alpha^{\dagger}_{k}. \tag{2.14}$$

This transformation reduces the whole problem of correlated electrons to an ideal Fermi gas comprising two types of new non-interacting fermions (i.e. quasi-particles) α and β, if we choose

$$u^2_k = \frac{1}{2}\left(1 + \frac{\xi_k}{\epsilon_k}\right) \tag{2.15}$$

$$v^2_k = \frac{1}{2}\left(1 - \frac{\xi_k}{\epsilon_k}\right) \tag{2.16}$$

and

$$u_k v_k = -\frac{\Delta_k}{2\epsilon_k} \tag{2.17}$$

where

$$\epsilon_k = \sqrt{\xi^2_k + |\Delta_k|^2}. \tag{2.18}$$

Then new annihilation and creation operators anticommute like fermion operators,

$$\{\alpha_k \alpha^{\dagger}_{k'}\} = \{\beta_k \beta^{\dagger}_{k'}\} = \delta_{kk'}$$

$$\{\alpha_k \alpha_{k'}\} = \{\beta_k \beta_{k'}\} = \{\alpha_k \beta^{\dagger}_{k'}\} = \{\alpha_k \beta_{k'}\} = 0.$$

and the transformed Hamiltonian becomes diagonal,

$$\tilde{H} = E_0 + \sum_{k} \epsilon_k (\alpha^{\dagger}_{k} \alpha_k + \beta^{\dagger}_{k} \beta_k) \tag{2.19}$$

where

$$E_0 = 2\sum_k (\xi_k v_k^2 + \Delta_k u_k v_k) + \frac{|\Delta|^2}{2E_p}. \tag{2.20}$$

The *order parameter* Δ is determined from the self-consistent equation (2.12) replacing the electron operators by quasi-particle ones,

$$\Delta = E_p \sum_{k'} \frac{\Delta}{\epsilon_{k'}}(1 - 2f_{k'}) \tag{2.21}$$

where

$$f_k = \langle\langle \alpha_k^\dagger \alpha_k \rangle\rangle = \langle\langle \beta_k^\dagger \beta_k \rangle\rangle$$

is the quasi-particle distribution function. Unlike the case of bare electrons the total average number of quasi-particles is not fixed. Therefore, their chemical potential is zero in the thermal equilibrium. They do not interact, so that (appendix B)

$$f_k = \frac{1}{\exp(\epsilon_k/T) + 1}. \tag{2.22}$$

We see that there are no quasi-particles in the ground state, $f_k = 0$, at $T = 0$. Hence, E_0 in equation (2.19) is the ground-state energy of the BCS superconductor. One can replace the sum in equation (2.21) by the integral using the definition of the density of states $N(E)$ in the Bloch band (appendix A). In conventional metals, the Debye frequency is small ($\omega_D \ll \mu$) and the density of states (DOS) is practically constant in the narrow energy region $\pm\omega_D$ around the Fermi energy, $N(E) \simeq N(E_F)$. As a result, we obtain

$$\Delta = \lambda\Delta \int_0^{\omega_D} \frac{\tanh \frac{\sqrt{\xi^2+|\Delta|^2}}{2T}}{\sqrt{\xi^2 + |\Delta|^2}} \, d\xi. \tag{2.23}$$

There is a trivial solution of this equation: $\Delta = 0$. Above some T_c, this is the only solution corresponding to the normal state. However, below T_c there are two solutions: $\Delta = 0$ and a real $\Delta \neq 0$. The system prefers to be in the superconducting (condensed) state below T_c because the condensation energy E_c is negative. This energy is the ground-state energy referred to the normal state energy. At $T = 0$, we have

$$E_c = E_0 - 2\sum_{\xi_k < 0} \xi_k. \tag{2.24}$$

Using the definition of E_0 (equation (2.20)), we obtain

$$E_c = 2N(E_F) \int_0^{\omega_D} d\xi \left(\xi - \frac{\xi^2 + \Delta^2(0)/2}{\sqrt{\xi^2 + \Delta^2(0)}} \right) \tag{2.25}$$

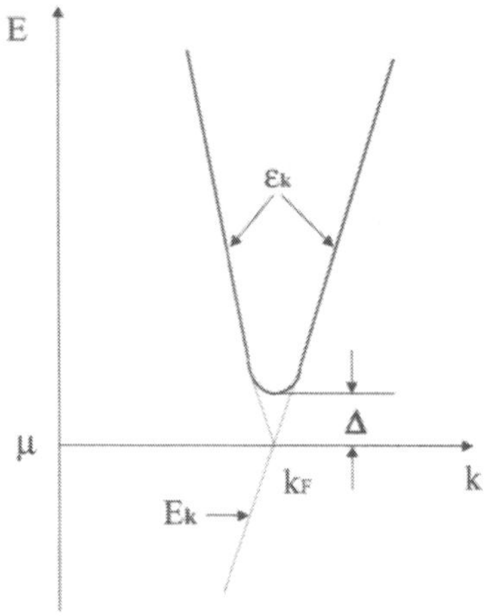

Figure 2.1. Excitation spectrum of the BCS superconductor.

where $\Delta(0)$ is the order parameter at $T = 0$. Indeed, this integral is negative:

$$E_{\mathrm{c}} \approx -\tfrac{1}{2}N(E_{\mathrm{F}})\Delta^2(0) < 0. \tag{2.26}$$

Far away from the Fermi surface, the quasi-particles, α and β, are electrons with spins 'up' and 'down', respectively, if $\xi_{\boldsymbol{k}} > 0$ and holes if $\xi_{\boldsymbol{k}} < 0$. In the vicinity of the Fermi surface, they are a mixture of both and their energy dispersion $\epsilon_{\boldsymbol{k}}$ is remarkably different from that of the non-interacting electrons and holes (see figure 2.1). The quasi-particle energy spectrum satisfies the Landau criterion (chapter 1) of superfluidity and the critical velocity $v_{\mathrm{c}} \simeq \Delta/k_{\mathrm{F}}$. The distribution of 'bare' electrons is of the form ($T = 0$)

$$n_{\boldsymbol{k}} = \langle c^{\dagger}_{\boldsymbol{k}\uparrow} c_{\boldsymbol{k}\uparrow}\rangle = v^2_{\boldsymbol{k}} \tag{2.27}$$

which has a zero step ($Z = 0$) which differs from the Fermi distribution at $T = 0$ with $Z = 1$ (figure 2.2). This is a clear manifestation of a breakdown of the Fermi-liquid description of attractive fermions at low temperatures. The 'mean-field' approximation (equation (2.9)) replacing the pair operators $\hat{\Delta}(\boldsymbol{k})$ by the *anomalous* averages $\Delta_{\boldsymbol{k}}$ is perfectly self-consistent. Indeed, let us calculate the commutator

$$[\hat{\Delta}(\boldsymbol{k}), \hat{\Delta}^{\dagger}(\boldsymbol{k})] \tag{2.28}$$

to show that its expectation value is zero. Using

$$[c_1c_2, c^{\dagger}_3c^{\dagger}_4] = \delta_{23}c_1c^{\dagger}_4 - \delta_{13}c_2c^{\dagger}_4 + \delta_{24}c^{\dagger}_3c_1 - \delta_{14}c^{\dagger}_3c_2 \tag{2.29}$$

we obtain

$$[\hat{\Delta}(\boldsymbol{k}), \hat{\Delta}^{\dagger}(\boldsymbol{k})] = 4E^2_p\Theta(\omega_{\mathrm{D}} - |\xi_{\boldsymbol{k}}|)\sum_{\boldsymbol{k}'}\Theta(\omega_{\mathrm{D}} - |\xi_{\boldsymbol{k}'}|)(1 - 2c^{\dagger}_{\boldsymbol{k}'\uparrow}c_{\boldsymbol{k}'\uparrow}). \tag{2.30}$$

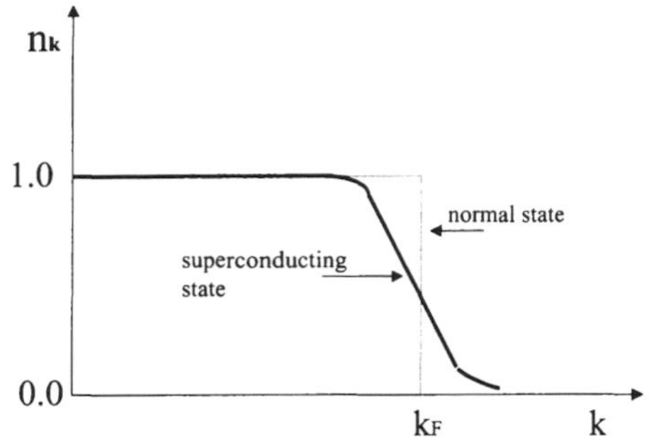

Figure 2.2. Distribution of electrons at $T = 0$.

Replacing the electron operators by the quasi-particle ones and taking the average, we obtain

$$\langle[\hat{\Delta}(\boldsymbol{k}), \hat{\Delta}^\dagger(\boldsymbol{k})]\rangle = 4E_p^2\Theta(\omega_\mathrm{D} - |\xi_{\boldsymbol{k}}|)\sum_{\boldsymbol{k}'}\Theta(\omega_\mathrm{D} - |\xi_{\boldsymbol{k}'}|)(u_{\boldsymbol{k}'}^2 - v_{\boldsymbol{k}'}^2)(1-2f_{\boldsymbol{k}'}) = 0 \tag{2.31}$$

because the function under the integral is odd with respect to $\xi_{\boldsymbol{k}'}$.

The ground state of the BCS Hamiltonian Φ_0 is a vacuum with respect to quasi-particles. Quasi-particles are fermions and their vacuum state is obtained from the electron vacuum $|0\rangle$ by applying the quasi-particle annihilation operators for every momentum $\boldsymbol{k}$:

$$\Phi_0 = A\prod_{\boldsymbol{k}}\alpha_{\boldsymbol{k}}\beta_{\boldsymbol{k}}|0\rangle. \tag{2.32}$$

Indeed, every single particle state $\boldsymbol{k}$ in Φ_0 is free from quasi-particles, so that applying their annihilation operator to Φ_0, we obtain zero. The eigenfunction should be normalized, which is the case if

$$A = \langle 0|\prod_{\boldsymbol{k}}\beta_{\boldsymbol{k}}^\dagger\alpha_{\boldsymbol{k}}^\dagger\alpha_{\boldsymbol{k}}\beta_{\boldsymbol{k}}|0\rangle^{-1/2}. \tag{2.33}$$

In terms of the electron operators, we have

$$\alpha_{\boldsymbol{k}} = u_{\boldsymbol{k}}c_{\boldsymbol{k}\uparrow} - v_{\boldsymbol{k}}c_{-\boldsymbol{k}\downarrow}^\dagger$$
$$\beta_{\boldsymbol{k}} = v_{\boldsymbol{k}}c_{\boldsymbol{k}\uparrow}^\dagger + u_{\boldsymbol{k}}c_{-\boldsymbol{k}\downarrow}$$

and

$$\Phi_0 = A\prod_{\boldsymbol{k}}(u_{\boldsymbol{k}}v_{\boldsymbol{k}} - v_{\boldsymbol{k}}^2c_{-\boldsymbol{k}\downarrow}^\dagger c_{\boldsymbol{k}\uparrow}^\dagger)|0\rangle \tag{2.34}$$

$$A = \left(\prod_{\boldsymbol{k}} v_{\boldsymbol{k}}\right)^{-1}. \tag{2.35}$$

Thus the ground-state wavefunction is

$$\Phi_0 = \prod_k (u_k - v_k c^\dagger_{-k\downarrow} c^\dagger_{k\uparrow})|0\rangle. \tag{2.36}$$

It is a superposition of the eigenfunctions of the total number operator $\hat{N} \equiv \sum_{ks} c^\dagger_{ks} c_{ks}$.

2.3 Meissner–Ochsenfeld effect

BCS reduced the many-body problem to a non-interacting gas of quasi-particles allowing for an analytical description of thermodynamic and kinetic properties of conventional superconductors. Let us first discuss the BCS theory of the magnetic flux expulsion. We apply the perturbation theory for a linear interaction of electrons with the vector potential $\boldsymbol{A}(\boldsymbol{r})$ taking the BCS Hamiltonian as a zero-order one,

$$H_{\text{int}} = -\frac{e}{m} \sum_{\boldsymbol{k},\boldsymbol{q},s} (\boldsymbol{k} \cdot \boldsymbol{A}_{\boldsymbol{q}}) c^\dagger_{\boldsymbol{k}+\boldsymbol{q}s} c_{\boldsymbol{k}s} \tag{2.37}$$

where $\boldsymbol{A}_{\boldsymbol{q}}$ is the Fourier component of $\boldsymbol{A}(\boldsymbol{r})$. This form of the interaction follows from the velocity operator $[-\mathrm{i}\nabla - e\boldsymbol{A}(\boldsymbol{r})]/m$ in the effective mass (m) approximation for the band energy dispersion and the gauge, where $\nabla \cdot \boldsymbol{A}(\boldsymbol{r}) = 0$. The field distribution in the sample is determined by the average of the current density operator, which follows as the symmetrized form of the velocity operator in the second quantization

$$\hat{\boldsymbol{j}}(\boldsymbol{r}) = \hat{\boldsymbol{j}}_p(\boldsymbol{r}) + \hat{\boldsymbol{j}}_d(\boldsymbol{r}) \tag{2.38}$$

where

$$\hat{\boldsymbol{j}}_p(\boldsymbol{r}) = -\frac{e}{2m} \sum_{\boldsymbol{k},\boldsymbol{q},s} c^\dagger_{\boldsymbol{k}+\boldsymbol{q}s} c_{\boldsymbol{k}s} (2\boldsymbol{k} + \boldsymbol{q}) \mathrm{e}^{-\mathrm{i}\boldsymbol{q}\cdot\boldsymbol{r}} \tag{2.39}$$

is the paramagnetic part and

$$\hat{\boldsymbol{j}}_d(\boldsymbol{r}) = -\frac{e^2}{m} \sum_{\boldsymbol{k},\boldsymbol{q},s} c^\dagger_{\boldsymbol{k}+\boldsymbol{q}s} c_{\boldsymbol{k}s} \boldsymbol{A}(\boldsymbol{r}) \mathrm{e}^{-\mathrm{i}\boldsymbol{q}\cdot\boldsymbol{r}} \tag{2.40}$$

the diamagnetic part. The perturbed many-particle state $\tilde{\Phi}$ in the first order in $\boldsymbol{A}$ is given by

$$\tilde{\Phi} = |n_\alpha, n_\beta\rangle + \sum_{n'_\alpha, n'_\beta} |n'_\alpha, n'_\beta\rangle \frac{\langle n'_\alpha, n'_\beta | H_{\text{int}} | n_\alpha, n_\beta\rangle}{E_{n_\alpha n_\beta} - E_{n'_\alpha n'_\beta}} \tag{2.41}$$

where $|n_\alpha, n_\beta\rangle$ and $E_{n_\alpha n_\beta}$ are the eigenstates and eigenvalues of the zero-order Hamiltonian $\tilde{H}$, respectively. Applying the Bogoliubov transformation (equations (2.13) and (2.14)) for $H_{\rm int}$ yields the average current density at $T = 0$:

$$j_p(\boldsymbol{r}) = \frac{2e^2}{m^2} \sum_{\boldsymbol{k},\boldsymbol{q}} \frac{u_{\boldsymbol{k}+\boldsymbol{q}} v_{\boldsymbol{k}} (2\boldsymbol{k} + \boldsymbol{q})(\boldsymbol{k} \cdot \boldsymbol{A}_{\boldsymbol{q}})}{\epsilon_{\boldsymbol{k}+\boldsymbol{q}} + \epsilon_{\boldsymbol{k}}} \times \exp(-\mathrm{i}\boldsymbol{q} \cdot \boldsymbol{r})(u_{\boldsymbol{k}+\boldsymbol{q}} v_{\boldsymbol{k}} - u_{\boldsymbol{k}} v_{\boldsymbol{k}+\boldsymbol{q}}) \tag{2.42}$$

$$j_d = -\frac{ne^2}{m} \boldsymbol{A}(\boldsymbol{r}). \tag{2.43}$$

Let us assume that the magnetic field varies over the characteristic length $\lambda_{\rm H}$, which is large compared with the coherence length ξ. In this case one can take the limit $\boldsymbol{q} \to 0$ in equation (2.42). In this limit $u_{\boldsymbol{k}+\boldsymbol{q}} v_{\boldsymbol{k}} - u_{\boldsymbol{k}} v_{\boldsymbol{k}+\boldsymbol{q}} = 0$, while the denominator remains finite: $\epsilon_{\boldsymbol{k}+\boldsymbol{q}} + \epsilon_{\boldsymbol{k}} > 2\Delta(0)$. Therefore, the paramagnetic contribution vanishes and we obtain the London equation (chapter 1),

$$j(\boldsymbol{r}) = -\frac{ne^2}{m} \boldsymbol{A}(\boldsymbol{r}). \tag{2.44}$$

In the opposite limit ($\lambda_{\rm H} < \xi$), the Pippard non-local theory of the flux expulsion follows from equations (2.42) and (2.43). In the normal state the denominator in equation (2.42) turns out to be zero at the Fermi level and the paramagnetic current appears to be finite. Actually, one can show that it cancels the diamagnetic part so that the normal state current is zero in a permanent magnetic field.

2.4 BCS gap, critical temperature and single-electron tunnelling

The BCS theory introduces the order parameter Δ (equation (2.21)) which is also a gap in the quasi-particle spectrum (figure 2.1) for a homogeneous system. The value of the gap at $T = 0$ should be of the order of $T_{\rm c}$. In fact, BCS theory predicts a universal ratio $2\Delta(0)/T_{\rm c} \simeq 3.5$ as follows from the master equation (2.21). At $T = 0$, the non-trivial solution is determined from

$$\frac{1}{\lambda} = \int_0^{\omega_{\rm D}} \frac{\mathrm{d}\xi}{\sqrt{\xi^2 + \Delta^2(0)}}. \tag{2.45}$$

Integration yields

$$\Delta(0) \simeq 2\omega_{\rm D} \exp\left(-\frac{1}{\lambda}\right) \tag{2.46}$$

for $\lambda \ll 1$, the limit to which the theory is applied. This is a remarkable result which demonstrates the instability of the Fermi liquid for *any value* of the attraction λ in agreement with the Cooper two-particle solution. The exponent in equation (2.46) cannot be expanded in a series of λ. Thus the superconducting ground state cannot be derived by using the perturbation theory with respect to the pairing potential up to any order.

At $T = T_c$, the gap should be zero so that T_c is determined by

$$\frac{1}{\lambda} = \int_0^{\omega_D} \frac{d\xi \, \tanh(\xi/2T_c)}{\xi}. \tag{2.47}$$

Integrating by parts and replacing the upper limit by infinity, we have

$$T_c = \frac{2e^C \omega_D}{\pi} \exp\left(-\frac{1}{\lambda}\right) \tag{2.48}$$

where $C \simeq 0.577$ is the Euler constant. The numerical coefficient in equation (2.48) is ≈ 1.14, so that

$$\frac{2\Delta(0)}{T_c} \simeq 3.5. \tag{2.49}$$

To calculate the temperature dependence of $\Delta(T)$ at low temperatures, $T \ll T_c$ we rewrite the master equation using its zero-temperature form (equation (2.45)):

$$\ln \frac{\Delta(T)}{\Delta(0)} = -2 \int_0^{\infty} \frac{d\xi}{\sqrt{\xi^2 + \Delta^2(T)}} f(\xi). \tag{2.50}$$

Here we have replaced the upper limit in the integral for infinity which is justified because the quasi-particle distribution function falls down exponentially at low temperatures,

$$f(\xi) \approx \exp\left[-\frac{\sqrt{\xi^2 + \Delta^2(T)}}{T}\right]. \tag{2.51}$$

Replacing the integration over ξ by the integration over the energy $\epsilon = \sqrt{\xi^2 + \Delta^2(T)}$ yields

$$\ln \frac{\Delta(T)}{\Delta(0)} = -2 \int_{\Delta(T)}^{\infty} d\epsilon \frac{e^{-\epsilon/T}}{\sqrt{\epsilon^2 - \Delta^2(T)}}. \tag{2.52}$$

The remaining integral is exponentially small at low temperatures, so that we can replace $\Delta(T)$ by $\Delta(0)$ on the right-hand side and expand ln in powers of $\Delta_1(T) = \Delta(T) - \Delta(0)$ on the left-hand side with the following result:

$$\Delta_1(T) = -\sqrt{2\pi T \Delta(0)} \exp\left(-\frac{\Delta(0)}{T}\right). \tag{2.53}$$

The temperature correction to the gap appears to be exponentially small at low temperatures. In the vicinity of T_c, where $(T_c - T)/T_c \ll 1$, the whole gap is small compared with the temperature. However, a direct expansion in powers of Δ cannot be applied in equation (2.23), because every term of such an expansion would render a divergent integral. Instead we use

$$\frac{\tanh x}{x} = \sum_{n=-\infty}^{\infty} \frac{1}{x^2 + [\pi(n + 1/2)]^2} \tag{2.54}$$

so that

$$\frac{1}{\lambda} = 2T \sum_n \int_0^{\omega_{\mathrm{D}}} \frac{\mathrm{d}\xi}{\xi^2 + \omega_n^2 + \Delta^2(T)} \tag{2.55}$$

where $\omega_n = \pi T(2n + 1)$ are the so-called Matsubara frequencies, $n = 0, \pm 1, \pm 2, \ldots$. The last equation can be expanded in powers of $\Delta(T)$ as

$$\ln \frac{T_{\mathrm{c}}}{T} = 2\Delta^2(T) T_{\mathrm{c}} \sum_n \int_0^{\infty} \frac{\mathrm{d}\xi}{(\xi^2 + \omega_n^2)^2}. \tag{2.56}$$

Calculating the integral over ξ and the sum over n (appendix B, equation (B.27)) we obtain

$$\Delta(T) = \pi T_{\mathrm{c}} \left[\frac{8}{7\zeta(3)}\right]^{1/2} \sqrt{1 - \frac{T}{T_{\mathrm{c}}}} \approx 3.06 T_{\mathrm{c}} \sqrt{1 - \frac{T}{T_{\mathrm{c}}}}. \tag{2.57}$$

There is a discontinuity in the temperature derivative of $\Delta(T)$ at T_{c} which leads to a jump of the specific heat (section 2.6).

The gap can be measured directly in tunnelling experiments where one applies voltage V to a thin dielectric layer between the normal metal and the superconductor (figure 2.3) [37]. The current running through the dielectric is proportional to the number of electrons tunnelling under the barrier per second. The electron, tunnelling from the normal metal, becomes a quasi-particle in the superconductor. Applying the Fermi–Dirac golden rule, we obtain the transition rate as

$$I(V) \propto \sum_{\boldsymbol{k}(\xi_{\boldsymbol{k}}<0), \boldsymbol{k}'} T_{\boldsymbol{k}\boldsymbol{k}'}^2 \delta(\epsilon_{\boldsymbol{k}'} - \xi_{\boldsymbol{k}} - eV). \tag{2.58}$$

The matrix element $T_{\boldsymbol{k}\boldsymbol{k}'}$ is almost independent of the momentum $\boldsymbol{k}$ in the normal metal and of $\boldsymbol{k}'$ in the superconductor, if the voltage is not very high $eV \sim \Delta \ll \mu$. The δ-function in equation (2.58) takes into account the difference eV in the normal and superconducting chemical potentials. Replacing the sum by the integral we obtain

$$I(V) \propto \int_{-\infty}^{eV} \mathrm{d}\xi \int_{-\infty}^{+\infty} \mathrm{d}\xi' \, \delta\left[\sqrt{\xi'^2 + \Delta^2(T)} - \xi\right]. \tag{2.59}$$

Then the conductance $\sigma = \mathrm{d}I/\mathrm{d}V$ is found:

$$\sigma \propto \int_{-\infty}^{+\infty} \mathrm{d}\xi' \, \delta\left[\sqrt{\xi'^2 + \Delta^2(T)} - eV\right]. \tag{2.60}$$

Calculating the remaining integral, we obtain

$$\frac{\sigma}{\sigma_{\mathrm{N}}} = \frac{eV}{\sqrt{(eV)^2 - \Delta^2(T)}} \tag{2.61}$$

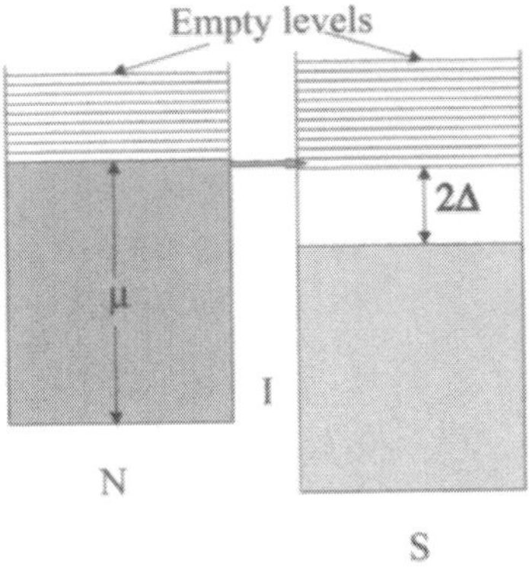

Figure 2.3. Tunnelling from the normal metal (N) to the superconductor (S) through a dielectric barrier. Shaded areas correspond to occupied states.

where σ_N is the normal state conductance of the barrier above T_c.

There is no current if $|eV| < \Delta$ because there are no states inside the gap in the superconductor (figure 2.3). Just above the threshold, $eV = \Delta$, the conductance has a maximum because the quasi-particle density of states $\rho(\epsilon)$ diverges at $\epsilon = \Delta$,

$$\rho(\epsilon) \equiv \frac{\partial |\xi|}{\partial \epsilon} = \frac{\epsilon}{\sqrt{\epsilon^2 - \Delta^2}}. \tag{2.62}$$

The typical experimental ratio σ/σ_N as a function of the voltage and the temperature follows the BCS prediction (2.61) rather well in conventional superconductors.

2.5 Isotope effect

The origin of the electron–electron attraction in superconductors can be tested by isotope substitution, when an ion mass M is varied without any change of the electronic configuration of the ion. There are two parameters in the BCS expression for T_c (equation (2.48)) which depend on the mechanism of the interaction. The characteristic phonon frequency ω_D is proportional to $1/\sqrt{M}$ as a frequency of any harmonic oscillator. However, the coupling constant λ is independent of the ion mass (section 3.3). Hence, the isotope exponent is found as

$$\alpha = -\frac{\mathrm{d}\ln T_c}{\mathrm{d}\ln M} = 0.5. \tag{2.63}$$

In fact, the isotope exponent α could be lower than 0.5 in a BCS superconductor because of the Coulomb repulsion and the anharmonicity of phonons. But, in any case, the finite value of α measured experimentally proves that phonons are involved in the pairing mechanism. The isotope effect has been observed in many conventional superconductors and in high-temperature superconductors (Part II).

2.6 Heat capacity

Only the electrons near the Fermi surface can absorb heat in a metal because of the Pauli principle. The number of these electrons is proportional to temperature. Therefore, their specific heat C_e in the normal state is linear as the function of temperature (appendix B). The temperature dependence of C_e changes drastically in the superconducting state due to the gap. The quasi-particle energies depend on the temperature in the self-consistent BCS approximation which should be taken into account in the calculations of temperature derivatives of the thermodynamic potential. We apply the definition of the specific heat as $C_e = T\,\mathrm{d}S/\mathrm{d}T$, where the quasi-particle entropy is defined as

$$S = -\langle \ln P(U_Q, N) \rangle. \tag{2.64}$$

Here $P(U_Q, N) = Z^{-1}\mathrm{e}^{-\beta U_Q}$ is the statistical probability of finding the Fermi gas of N quasi-particles with energy U_Q (appendix B),

$$U_Q = \sum_{\boldsymbol{k}} \epsilon_{\boldsymbol{k}} (n_{\boldsymbol{k}\alpha} + n_{\boldsymbol{k}\beta}) \tag{2.65}$$

where $n_{\boldsymbol{k}\alpha,\beta} = 0, 1$ are the quasi-particle occupation numbers; and

$$Z = \prod_{\boldsymbol{k}} [1 + \mathrm{e}^{\beta\epsilon_{\boldsymbol{k}}}]^2 \tag{2.66}$$

is the quasi-particle grand partition function. Here we take into account the fact that the quasi-particle chemical potential is zero. Calculating the statistical average in equation (2.64), we obtain the entropy of the ideal Fermi gas of quasi-particles as

$$S = -2\sum_{\boldsymbol{k}} [f_{\boldsymbol{k}} \ln f_{\boldsymbol{k}} + (1 - f_{\boldsymbol{k}}) \ln(1 - f_{\boldsymbol{k}})] \tag{2.67}$$

where the distribution function $f_{\boldsymbol{k}} = \langle n_{\boldsymbol{k}\alpha} \rangle = \langle n_{\boldsymbol{k}\beta} \rangle$ is defined in equation (2.22). The temperature derivative of the distribution function includes the derivative of $\epsilon_{\boldsymbol{k}}$ as

$$\frac{\mathrm{d}f_{\boldsymbol{k}}}{\mathrm{d}T} = f_{\boldsymbol{k}}(1 - f_{\boldsymbol{k}}) \left(\frac{\epsilon_{\boldsymbol{k}}}{T^2} - \frac{1}{T}\frac{\mathrm{d}\epsilon_{\boldsymbol{k}}}{\mathrm{d}T} \right)$$

so that

$$C_e = 2\sum_{\boldsymbol{k}} f_{\boldsymbol{k}}(1 - f_{\boldsymbol{k}}) \left(\frac{\epsilon_{\boldsymbol{k}}^2}{T^2} - \frac{\epsilon_{\boldsymbol{k}}}{T}\frac{\mathrm{d}\epsilon_{\boldsymbol{k}}}{\mathrm{d}T} \right). \tag{2.68}$$

At low temperatures, the number of quasi-particles is exponentially small $f_{\boldsymbol{k}} \propto e^{-\Delta(0)/T}$ and so is the specific heat (equation (2.68)). Above T_c, $\epsilon_{\boldsymbol{k}} = |\xi|$ and we obtain $C_e = C_N$, where

$$C_N = \frac{N(E_F)}{T^2} \int_0^\infty d\xi \, \frac{\xi^2}{\cosh^2(\xi/2T)} = \frac{2\pi^2}{3} N(E_F) T \tag{2.69}$$

as expected for the ideal Fermi gas (appendix B). However, just below T_c, the second term in the brackets of equation (2.68) appears to be finite:

$$\frac{\epsilon_{\boldsymbol{k}}}{T} \frac{d\epsilon_{\boldsymbol{k}}}{dT} = \frac{\Delta}{T} \frac{d\Delta}{dT} = -\frac{\pi^2}{2} \left(\frac{8}{7\zeta(3)} \right) \tag{2.70}$$

and the specific heat has a discontinuity,

$$C_e = C_N + \frac{8\pi^2}{7\zeta(3)} N(E_F) T_c \tag{2.71}$$

if $T = T_c - 0$. Here $\zeta(3) \simeq 1.202$. The relative value of the jump is

$$\frac{C_e(T_c - 0) - C_e(T_c + 0)}{C_e(T_c + 0)} = \frac{12}{7\zeta(3)} \simeq 1.43 \tag{2.72}$$

which agrees with the value measured in many conventional superconductors. The phase transition turns out to be second order as in the GL phenomenology (chapter 1).

2.7 Sound attenuation

The interaction of ultrasound waves with electrons is described by the following Hamiltonian:

$$H_{\text{int}} = V(q) e^{i\nu t} \sum_{\boldsymbol{k},s} c^\dagger_{\boldsymbol{k}s} c_{\boldsymbol{k}-\boldsymbol{q}s} + H.c. \tag{2.73}$$

where $q = \nu/s$ and ν are the wavevector and frequency of the sound, respectively (s is the sound velocity), $V(q)$ is proportional to the sound amplitude and $H.c.$ is the Hermitian conjugate.

Applying the Bogoliubov transformation, we obtain four terms in the interaction (2.73). Two of them correspond to the annihilation and creation of two different quasi-particles and the other two correspond to their scattering. The sound frequency is low ($\nu \ll \Delta \simeq T_c$) and only the scattering terms are relevant:

$$H_{\text{int}} \propto \sum_{\boldsymbol{k}} M^{(-)}_{\boldsymbol{k}\boldsymbol{q}} (\alpha^\dagger_{\boldsymbol{k}} \alpha_{\boldsymbol{k}-\boldsymbol{q}} + \beta^\dagger_{\boldsymbol{k}} \beta_{\boldsymbol{k}-\boldsymbol{q}}) \tag{2.74}$$

where

$$M^{(-)}_{\boldsymbol{k}\boldsymbol{q}} = u_{\boldsymbol{k}} u_{\boldsymbol{k}-\boldsymbol{q}} - v_{\boldsymbol{k}} v_{\boldsymbol{k}-\boldsymbol{q}} \tag{2.75}$$

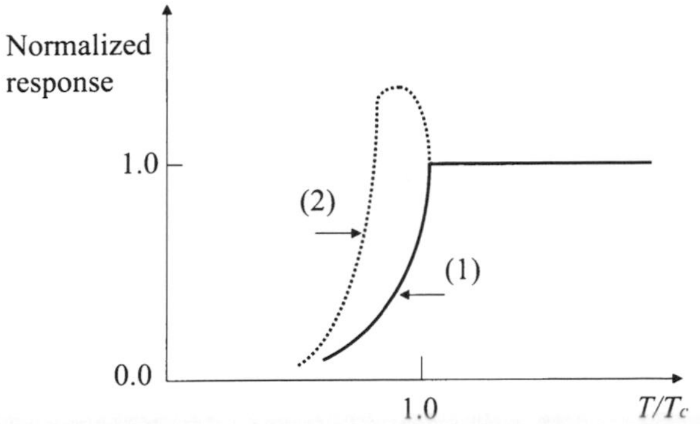

Figure 2.4. Temperature dependence of sound attenuation and thermal conductivity (1) compared with the nuclear spin relaxation rate (2).

is a so-called coherence factor.

The rate of the sound absorption is given by the Fermi–Dirac golden rule as

$$W_{\text{abs}} \propto \sum_{k} [M_{kq}^{(-)}]^2 f_{k-q}(1 - f_k)\delta(\epsilon_{k-q} - \epsilon_k + \nu) \tag{2.76}$$

and the emission rate is

$$W_{\text{emi}} \propto \sum_{k} [M_{kq}^{(-)}]^2 f_{k-q}(1 - f_k)\delta(\epsilon_{k-q} - \epsilon_k - \nu). \tag{2.77}$$

The sound attenuation Γ, which is the difference between these two rates, is given by

$$\Gamma \propto \sum_{k} [M_{kq}^{(-)}]^2 (f_k - f_{k-q})\delta(\epsilon_{k-q} - \epsilon_k - \nu). \tag{2.78}$$

If $\nu \ll T, \Delta$, we have $f_k - f_{k-q} \approx \nu \partial f/\partial\epsilon$. Replacing the integration over k and over the angle between $\boldsymbol{k}$ and $\boldsymbol{q}$ by the integration over $\xi = \xi_k$ and $\xi' = \xi_{k-q}$, respectively, we obtain

$$\Gamma \propto -\int \mathrm{d}\xi \int \mathrm{d}\xi' \, [u(\xi)u(\xi') - v(\xi)v(\xi')]^2 \frac{\partial f}{\partial\epsilon}\delta(\epsilon - \epsilon') \tag{2.79}$$

where $\epsilon = \sqrt{\xi^2 + \Delta^2}$ and $\epsilon' = \sqrt{\xi'^2 + \Delta^2}$. Only transitions with $\xi = \xi'$ contribute to the integral over ξ', so that

$$\Gamma \propto -\int_{\Delta}^{\infty} \mathrm{d}\epsilon \left(\frac{\partial\xi}{\partial\epsilon}\right)^2 [u^2(\xi) - v^2(\xi)]^2 \frac{\partial f}{\partial\epsilon}. \tag{2.80}$$

Here we can see that the large density of quasi-particle states $\partial|\xi|/\partial\epsilon = \epsilon/\sqrt{\epsilon^2 - \Delta^2}$ is cancelled by the small coherence factor in the integral in (2.80) as

$$\Gamma \propto -\int_{\Delta}^{\infty} \mathrm{d}\epsilon \, \frac{\epsilon^2 - \Delta^2}{\epsilon^2 - \Delta^2}\frac{\partial f}{\partial\epsilon} = \frac{1}{\exp(\Delta/T) + 1}. \tag{2.81}$$

The sound attenuation in the superconducting state depends exponentially on temperature (figure 2.4). Using its ratio to the normal-state attenuation,

$$\frac{\Gamma_s}{\Gamma_n} = \frac{2}{\exp(\Delta(T)/T) + 1} \tag{2.82}$$

one can measure the temperature dependence of the BCS gap.

2.8 Nuclear spin relaxation rate

The measurement of the linewidth of the nuclear magnetic resonance (NMR) is another powerful method of determining $\Delta(T)$. The linewidth depends on the inverse time of the relaxation of nuclear magnetic moment due to the spin-flip scattering of carriers off nuclei, $1/T_1$. This scattering is described by the hyperfine interaction of the nucleus with the electron spin:

$$H_{\text{int}} \propto \sum_{k,k'} c^{\dagger}_{k'\downarrow} c_{k\uparrow} + H.c. \tag{2.83}$$

The NMR frequency is very small and the spin-flip scattering is practically elastic. That is why only the scattering of quasi-particles contribute to $1/T_1$ as in the case of sound attenuation,

$$H_{\text{int}} \propto \sum_{k} M^{(+)}_{kk'} (\beta^{\dagger}_{k'} \alpha_k + \alpha^{\dagger}_{k} \beta_{k'}). \tag{2.84}$$

However, here the coherence factor

$$M^{(+)}_{kk'} = u_k u_{k'} + v_k v_{k'} \tag{2.85}$$

is different. Applying the Fermi–Dirac golden rule, we obtain

$$1/T_1 \propto \sum_{k,k'} [M^{(+)}_{kk'}]^2 (f_k - f_{k'}) \delta(\epsilon_{k'} - \epsilon_k - \nu) \tag{2.86}$$

$\nu \to 0$. Replacing the sums by the integrals yields a divergent integral,

$$1/T_1 \propto -T \int_{\Delta}^{\infty} \mathrm{d}\epsilon \, \frac{\epsilon^2 + \Delta^2}{\epsilon^2 - \Delta^2} \frac{\partial f}{\partial \epsilon}. \tag{2.87}$$

In fact, the divergency is cut by some damping of excitations τ^{-1}, for example due to the inelastic electron–phonon scattering. Above T_c, where $\Delta = 0$, the relaxation rate is proportional to T. This linear temperature dependence of $1/T_1$ in a normal metal is known as the Korringa law. It has the same origin as the linear specific heat. Both are due to the Pauli principle. Only electrons in a narrow energy region around the Fermi surface can exchange their spin with nuclei and absorb heat. Well below T_c, the NMR relaxation rate is exponentially small because of the gap:

$$1/T_1 \propto \mathrm{e}^{-\Delta/T} \ln(\Delta\tau) \tag{2.88}$$

as in sound attenuation. However, just below T_c, it has a maximum (see figure 2.4). The maximum in $1/T_1$ is one of the most interesting and important features of BCS theory. This result is distinctly different from the result for sound attenuation. They differ because the coherence factors are different:

$$M^{(+,-)} = (uu' \pm vv'). \tag{2.89}$$

A simple energy-gap form of a *two-fluid* model could account for the drop in the sound attenuation but not for the rapid rise of $1/T_1$ just below T_c. The observation by Hebel and Slichter [38] of the peak in $1/T_1$ was one of the first in the body of evidence for the detailed nature of the pairing correlations in BCS superconductors.

2.9 Thermal conductivity

Important information about the excitation spectrum of the superconducting state can also be obtained from thermal conductivity [39]. Quasi-particles contribute to heat transfer in a superconductor. Their contribution $\boldsymbol{Q}$ to the heat flow is determined by the deviation $\tilde{f}$ of the distribution function from the equilibrium,

$$\boldsymbol{Q} = \sum_k \tilde{f} \boldsymbol{v} \epsilon_k \tag{2.90}$$

where $\boldsymbol{v} = \partial\epsilon_k/\partial\boldsymbol{k}$ is the quasi-particle group velocity. The distribution function obeys the Boltzmann equation, which for a small deviation from the equilibrium has the form:

$$\boldsymbol{v} \cdot \frac{\partial f}{\partial \boldsymbol{r}} - \frac{\partial \epsilon_k}{\partial \boldsymbol{r}} \cdot \frac{\partial f}{\partial \boldsymbol{k}} = -\frac{\tilde{f}}{\tau_{\mathrm{tr}}^{\mathrm{s}}} \tag{2.91}$$

where the transport relaxation rate for elastic scattering is obtained using the Fermi–Dirac golden rule:

$$\frac{1}{\tau_{\mathrm{tr}}^{\mathrm{s}}} = \frac{|\xi|}{\epsilon}\frac{1}{\tau_{\mathrm{tr}}^{\mathrm{n}}}. \tag{2.92}$$

In the superconducting state, the transport relaxation rate is diminished by a factor $|\xi|/\epsilon < 1$ compared with that in the normal state $(1/\tau_{\mathrm{tr}}^{\mathrm{n}})$ because of the square of the coherence factor $M^{(-)}$ in the probability of scattering. The second term on the left-hand side of the Boltzmann equation accounts for the driving force, which acts on a quasi-particle due to the temperature dependence of its energy, $\epsilon = \sqrt{\xi^2 + |\Delta(T(\boldsymbol{r})|^2}$. The solution of equation (2.91) is

$$\tilde{f} = \tau_{\mathrm{tr}}^{\mathrm{n}} \frac{\epsilon^2}{T|\xi|} \frac{\partial f}{\partial \epsilon} \boldsymbol{v} \cdot \nabla T. \tag{2.93}$$

Substituting it into equation (2.90) yields the thermal conductivity $K = |\boldsymbol{Q}|/|\nabla T|$ as

$$K_s \propto -\frac{1}{T}\int_{\Delta}^{\infty} \mathrm{d}\epsilon\, \epsilon^2 \frac{\partial f}{\partial \epsilon}. \tag{2.94}$$

Hence, the ratio to the normal-state thermal conductivity is

$$\frac{K_s}{K_n} = \frac{\int_{\Delta/2T}^{\infty} \mathrm{d}x\, x^2/\cosh^2(x)}{\int_0^{\infty} \mathrm{d}x\, x^2/\cosh^2(x)}. \tag{2.95}$$

The normalized thermal conductivity (equation (2.95)) drops exponentially with the temperature lowering below T_c just like sound attenuation (figure 2.4).

2.10 Unconventional Cooper pairing

In BCS theory electrons are paired with the opposite momenta on the Fermi surface and the opposite spins. The simple approximation for the pairing potential (equation (2.6)) leads to the momentum-independent gap $\Delta_{\boldsymbol{k}} = \Delta$, which is uniform along the Fermi surface. This is an s-wave order parameter because the pair wavefunction is isotropic in real space. It depends only on the distance between two correlated electrons. BCS theory describes an anisotropic Cooper pairing with non-zero orbital momentum of pairs $l \neq 0$ as well, if the potential allows for such pairing [40]. In the general case, the mean-field Hamiltonian is

$$\tilde{H} = \sum_{\boldsymbol{k},s=\uparrow,\downarrow} \left[\xi_{\boldsymbol{k}} c_{\boldsymbol{k}s}^{\dagger} c_{\boldsymbol{k}s} + \frac{1}{2}\sum_{s'}(\Delta_{\boldsymbol{k}}^{ss'} c_{\boldsymbol{k}s}^{\dagger} c_{-\boldsymbol{k}s'}^{\dagger} + H.c.)\right] \tag{2.96}$$

where we omit the c-number term, which does not contain fermionic operators. The order parameter should be antisymmetric:

$$\Delta_{\boldsymbol{k}}^{ss'} = -\Delta_{-\boldsymbol{k}}^{ss'} \tag{2.97}$$

according to the Pauli exclusion principle. Applying the Bogoliubov transformation, the master equation for any pairing potential $V(\boldsymbol{k},\boldsymbol{k}')$ takes the following form:

$$\Delta_{\boldsymbol{k}}^{ss'} = -\sum_{\boldsymbol{k}'} V(\boldsymbol{k},\boldsymbol{k}')\frac{\Delta_{\boldsymbol{k}'}^{ss'}}{2\epsilon_{\boldsymbol{k}'}}(1 - 2f_{\boldsymbol{k}'}). \tag{2.98}$$

Let us assume that $V(\boldsymbol{k},\boldsymbol{k}')$ depends on the value of the momentum transfer $q = |\boldsymbol{k} - \boldsymbol{k}'|$, which is the case for an isotropic system. In BCS theory, the states near the Fermi surface contribute to the sum in equation (2.97) and $q \approx 2^{1/2}k_{\mathrm{F}}\sqrt{1-\cos\Theta'}$, where Θ' is the angle between $\boldsymbol{k}$ and $\boldsymbol{k}'$. Then all functions in the master equation depend on the angles of $\boldsymbol{k}$ and $\boldsymbol{k}'$ with the absolute values of the momenta equal to the Fermi momentum, $|\boldsymbol{k}| = |\boldsymbol{k}'| = k_{\mathrm{F}}$. We can expand the order parameter and the potential in the polynomial series:

$$\Delta_{\boldsymbol{k}}^{ss'} = \sum_{l=0}^{\infty} \Delta(l) P_l(\cos\Theta) \tag{2.99}$$

$$V(q) = \sum_{l=0}^{\infty} V(l) P_l(\cos\Theta')$$

where $P_l(\cos\Theta)$ are the Legendre polynomials,

$$P_l(x) = \frac{1}{2^l l!} \frac{\mathrm{d}^l}{\mathrm{d}x^l}(x^2 - 1)^l \tag{2.100}$$

which are orthogonal and normalized as

$$\int_{-1}^{1} \mathrm{d}x\, P_l(x) P_{l'}(x) = \frac{2\delta_{ll'}}{2l+1}. \tag{2.101}$$

This expansion is instrumental due to the summation theorem:

$$P_l(\cos\Theta') = P_l(\cos\Theta) P_l(\cos\Theta'') + 2\sum_{m=1}^{l} \frac{(l-m)!}{(l+m)!} P_l^m(\cos\Theta) P_l^m(\cos\Theta'') \cos[m(\phi - \phi'')] \tag{2.102}$$

where $P_l^m(x)$ are associated Legendre functions defined as

$$P_l^m(\cos\Theta) = \sin^m\Theta \frac{\mathrm{d}^m P_l(\cos\Theta)}{(\mathrm{d}\cos\Theta)^m}$$

and Θ' is the angle between any two directions determined by the spherical angles (Θ, ϕ) and (Θ'', ϕ''), respectively. Substitution of the series into the master equation yields a system of coupled nonlinear equations for the amplitudes $\Delta(l)$. These equations are linearized and decoupled at $T = T_{\mathrm{c}}$. Using the orthogonality of the Legendre polynomials and the summation theorem (equation (2.102)), we obtain, for an isotropic system with parabolic energy dispersion,

$$\Delta(l) = -\frac{V(l)N(E_{\mathrm{F}})}{2l+1}\Delta(l) \int_0^{\omega_{\mathrm{D}}} \frac{\mathrm{d}\xi\, \tanh(\xi/2T_{\mathrm{c}})}{\xi}. \tag{2.103}$$

Then the critical temperature

$$T_{\mathrm{c}} = \frac{2e^C \omega_{\mathrm{D}}}{\pi} \exp\left(-\frac{1}{\lambda_{l^*}}\right) \tag{2.104}$$

where $\lambda_{l^*} \equiv |V(l^*)|N(E_F)/(2l^* + 1)$ for negative $V(l^*)$. The solution of equation (2.103) is trivial: $\Delta(l) = 0$ for all l except $l = l^*$ with a maximum value of λ_{l^*} and negative $V(l^*)$. Hence, if the Fourier transform of the pairing potential has an attractive component with $l \neq 0$, the system condenses into the state with non-zero orbital momentum of the pair wavefunction. The single-particle gap, $\Delta_{\boldsymbol{k}} \propto P_l(\cos\Theta)$, has nodes on the Fermi surface if $l \neq 0$. It is odd for $l = 1, 3, 5, \ldots$ and even for even l, $\Delta_{-\boldsymbol{k}} = (-1)^l \Delta_{\boldsymbol{k}}$. The order parameter, $\Delta_{\boldsymbol{k}}$, is proportional to the Fourier transform of the pair wavefunction. The change in sign of $\boldsymbol{k}$ corresponds to the permutation of real-space coordinates of the two fermions of the pair. According to the Pauli exclusion principle, the total wavefunction of the pair should change its sign after the permutation of both the orbital and spin coordinates of two fermions. Hence, the spin component of the paired state with an odd orbital momentum should be symmetric under the permutation of the spin coordinates of two particles, while the spin component of even orbital states should be antisymmetric. Therefore, the even l states $(s, d, \ldots)$ are singlets (total spin S is zero) and the odd l states $(p, f, \ldots)$ are triplets $(S = 1)$.

2.11 Bogoliubov equations

The Bogoliubov transformation of the BCS Hamiltonian can be readily generalized for inhomogeneous superconductors. Let us introduce field operators (appendix C)

$$\Psi_s(\boldsymbol{r}) = \sum_{\boldsymbol{k}} c_{\boldsymbol{k}s} \exp(\mathrm{i}\boldsymbol{k} \cdot \boldsymbol{r}). \tag{2.105}$$

Here and further on, we take the volume of the system as $V = 1$. Then the BCS Hamiltonian can be written as follows:

$$\tilde{H} = \int \mathrm{d}\boldsymbol{r} \left[\sum_s \Psi_s^\dagger(\boldsymbol{r})\hat{h}(\boldsymbol{r})\Psi_s(\boldsymbol{r}) + \Delta(\boldsymbol{r})\Psi_\uparrow^\dagger(\boldsymbol{r})\Psi_\downarrow^\dagger(\boldsymbol{r}) + \Delta^*(\boldsymbol{r})\Psi_\downarrow(\boldsymbol{r})\Psi_\uparrow(\boldsymbol{r}) \right] \tag{2.106}$$

where

$$\hat{h}(\boldsymbol{r}) = -\frac{[\nabla + \mathrm{i}e\boldsymbol{A}(\boldsymbol{r})]^2}{2m} + U(\boldsymbol{r}) - \mu \tag{2.107}$$

is the one-electron Hamiltonian in the external magnetic $(\boldsymbol{A}(\boldsymbol{r}))$ and electric $(U(\boldsymbol{r}))$ fields. We apply the effective mass (m) approximation for the band dispersion (see appendix A) and drop c-number terms in the total energy. The coordinate-dependent order parameter is given by

$$\Delta(\boldsymbol{r}) = -2E_p\langle\langle\Psi_\downarrow(\boldsymbol{r})\Psi_\uparrow(\boldsymbol{r})\rangle\rangle. \tag{2.108}$$

Superfluid properties of inhomogeneous superconductors can be studied by the use of the Bogoliubov equations, fully taking into account the interaction of quasi-particles with the condensate. To derive these equations, we introduce the time-dependent Heisenberg operators (appendix D) as

$$\psi_s(\boldsymbol{r}, t) = e^{i\tilde{H}t}\Psi_s(\boldsymbol{r})e^{-i\tilde{H}t}. \tag{2.109}$$

The equations of motion for these operators are readily derived:

$$\begin{aligned} i\frac{\partial \psi_\uparrow(\boldsymbol{r}, t)}{\partial t} &= -[\tilde{H}, \psi_\uparrow(\boldsymbol{r}, t)] \\ &= \hat{h}(\boldsymbol{r})\psi_\uparrow(\boldsymbol{r}, t) + \Delta(\boldsymbol{r})\psi_\downarrow^\dagger(\boldsymbol{r}, t) \end{aligned} \tag{2.110}$$

and

$$\begin{aligned} -i\frac{\partial \psi_\downarrow^\dagger(\boldsymbol{r}, t)}{\partial t} &= [\tilde{H}, \psi_\downarrow^\dagger(\boldsymbol{r}, t)] \\ &= \hat{h}^*(\boldsymbol{r})\psi_\downarrow^\dagger(\boldsymbol{r}, t) - \Delta^*(\boldsymbol{r})\psi_\uparrow(\boldsymbol{r}, t). \end{aligned} \tag{2.111}$$

Here we have applied commutation relations for the field operators:

$$\begin{aligned} \{\Psi_s(\boldsymbol{r}), \Psi_{s'}^\dagger(\boldsymbol{r}')\} &= \sum_{\boldsymbol{k},\boldsymbol{k}'}\{c_{\boldsymbol{k}s}, c_{\boldsymbol{k}'s'}^\dagger\}\exp[i\boldsymbol{k}\cdot\boldsymbol{r} - i\boldsymbol{k}'\cdot\boldsymbol{r}') \\ &= \delta_{ss'}\sum_{\boldsymbol{k}}\exp[i\boldsymbol{k}\cdot(\boldsymbol{r} - \boldsymbol{r}')] = \delta_{ss'}\delta(\boldsymbol{r} - \boldsymbol{r}') \end{aligned}$$

and

$$\begin{aligned} [\Psi_\uparrow^\dagger(\boldsymbol{r}')\Psi_\downarrow^\dagger(\boldsymbol{r}'), \Psi_\uparrow(\boldsymbol{r})] &= -\delta(\boldsymbol{r} - \boldsymbol{r}')\Psi_\downarrow^\dagger(\boldsymbol{r}) \\ [\Psi_\uparrow^\dagger(\boldsymbol{r}')\Psi_\uparrow(\boldsymbol{r}'), \Psi_\uparrow(\boldsymbol{r})] &= -\delta(\boldsymbol{r} - \boldsymbol{r}')\Psi_\uparrow^\dagger(\boldsymbol{r}). \end{aligned} \tag{2.112}$$

The linear Bogoliubov transformation of ψ-operators has the form

$$\psi_\uparrow(\boldsymbol{r}, t) = \sum_n [u_n(\boldsymbol{r}, t)\alpha_n + v_n^*(\boldsymbol{r}, t)\beta_n^\dagger] \tag{2.113}$$

$$\psi_\downarrow(\boldsymbol{r}, t) = \sum_n [u_n(\boldsymbol{r}, t)\beta_n - v_n^*(\boldsymbol{r}, t)\alpha_n^\dagger] \tag{2.114}$$

where α_n and $\alpha_n^\dagger$ are the fermion operators, which annihilate and create quasi-particles in a quantum state n. Using this transformation in equations (2.110) and (2.111), we obtain two coupled Schrödinger equations for the wavefunctions $u(\boldsymbol{r}, t)$ and $v(\boldsymbol{r}, t)$:

$$\begin{aligned} i\frac{d}{dt}u(\boldsymbol{r}, t) &= \hat{h}(\boldsymbol{r})u(\boldsymbol{r}, t) - \Delta(\boldsymbol{r})v(\boldsymbol{r}, t), \\ -i\frac{d}{dt}v(\boldsymbol{r}, t) &= \hat{h}^*(\boldsymbol{r})v(\boldsymbol{r}, t) + \Delta^*(\boldsymbol{r})u(\boldsymbol{r}, t). \end{aligned} \tag{2.115}$$

There is also the sum rule,

$$\sum_n [u_n(\boldsymbol{r},t)u_n^*(\boldsymbol{r}',t) + v_n(\boldsymbol{r},t)v_n^*(\boldsymbol{r}',t)] = \delta(\boldsymbol{r}-\boldsymbol{r}') \tag{2.116}$$

which retains the Fermi commutation relations for all operators. When the magnetic and electric fields are stationary, these equations are reduced to the steady-state ones:

$$\epsilon_n u_n(\boldsymbol{r}) = \left\{-\frac{[\boldsymbol{\nabla} + \mathrm{i}e\boldsymbol{A}(\boldsymbol{r})]^2}{2m} + U(\boldsymbol{r}) - \mu\right\} u_n(\boldsymbol{r}) - \Delta(\boldsymbol{r})v_n(\boldsymbol{r}) \tag{2.117}$$

$$\epsilon_n v_n(\boldsymbol{r}) = -\left\{-\frac{[\boldsymbol{\nabla} - \mathrm{i}e\boldsymbol{A}(\boldsymbol{r})]^2}{2m} + U(\boldsymbol{r}) - \mu\right\} v_n(\boldsymbol{r}) - \Delta^*(\boldsymbol{r})u_n(\boldsymbol{r}).$$

Using the same transformation, the quantum and statistical averages in equations (2.108) yield the self-consistent equation for the order parameter:

$$\Delta(\boldsymbol{r}) = -2E_p \sum_n u_n(\boldsymbol{r})v_n^*(\boldsymbol{r})(1-2f_n) \tag{2.118}$$

where $f_n = \langle\langle \alpha_n^\dagger \alpha_n \rangle\rangle = \langle\langle \beta_n^\dagger \beta_n \rangle\rangle = (1+\exp\epsilon_n/T)^{-1}$ is the equilibrium quasi-particle distribution in the quantum states n with energy ϵ_n. We can readily solve the set of Bogoliubov equations (2.117) in the homogeneous case, when $\boldsymbol{A}(\boldsymbol{r}) = U(\boldsymbol{r}) = 0$. In this case the excitation wavefunctions are plane waves,

$$u_{\boldsymbol{k}}(\boldsymbol{r}) = u_{\boldsymbol{k}} \mathrm{e}^{\mathrm{i}\boldsymbol{k}\cdot\boldsymbol{r}} \tag{2.119}$$

$$v_{\boldsymbol{k}}(\boldsymbol{r}) = v_{\boldsymbol{k}} \mathrm{e}^{\mathrm{i}\boldsymbol{k}\cdot\boldsymbol{r}} \tag{2.120}$$

and the order parameter $\Delta(\boldsymbol{r})$ is $\boldsymbol{r}$-independent, $\Delta(\boldsymbol{r}) = \Delta$. Substituting equations (2.119) and (2.120) into equations (2.117), we obtain

$$\epsilon_{\boldsymbol{k}} u_{\boldsymbol{k}} = \xi_{\boldsymbol{k}} u_{\boldsymbol{k}} - \Delta v_{\boldsymbol{k}} \tag{2.121}$$

$$\epsilon_{\boldsymbol{k}} v_{\boldsymbol{k}} = -\xi_{\boldsymbol{k}} v_{\boldsymbol{k}} - \Delta^* u_{\boldsymbol{k}} \tag{2.122}$$

and, from equation (2.116),

$$|u_{\boldsymbol{k}}|^2 + |v_{\boldsymbol{k}}|^2 = 1. \tag{2.123}$$

As a result, we find

$$|u_{\boldsymbol{k}}|^2 = \frac{1}{2}\left(1 + \frac{\xi_{\boldsymbol{k}}}{\epsilon_{\boldsymbol{k}}}\right) \tag{2.124}$$

$$|v_{\boldsymbol{k}}|^2 = \frac{1}{2}\left(1 - \frac{\xi_{\boldsymbol{k}}}{\epsilon_{\boldsymbol{k}}}\right) \tag{2.125}$$

$$u_{\boldsymbol{k}} v_{\boldsymbol{k}}^* = -\frac{\Delta}{2\epsilon_{\boldsymbol{k}}}. \tag{2.126}$$

The elementary excitation energy is

$$\epsilon_k = \sqrt{\xi_k^2 + |\Delta|^2} \tag{2.127}$$

as it should be (see equation (2.18)), where the gap is found using the master equation,

$$\Delta = 2E_p \sum_k \frac{\Delta}{2\epsilon_k}(1 - 2f_k). \tag{2.128}$$

2.12 Landau criterion and gapless superconductivity

The Bogoliubov equations are coupled nonlinear integra-differential equations. They are mathematically transparent but the analytical solution is possible only in a few simple cases. As an example, we consider a uniform flow with the superfluid velocity $\boldsymbol{v}_s$. This state is described by an oscillating complex order parameter

$$\Delta(\boldsymbol{r}) = \Delta \exp(2\mathrm{i}\boldsymbol{q} \cdot \boldsymbol{r}) \tag{2.129}$$

where $2\boldsymbol{q} = 2m\boldsymbol{v}_s$ is the centre-of-mass momentum of the Cooper pair, and Δ is the real amplitude. Let us examine how the flow destroys superconductivity. If the flow is uniform, the system remains translation invariant and the momentum $\boldsymbol{k}$ is a quantum number. The solution has the form

$$\begin{aligned} u_n(\boldsymbol{r}) &= u_k \exp[\mathrm{i}(\boldsymbol{k} + \boldsymbol{q}) \cdot \boldsymbol{r}] \\ v_n(\boldsymbol{r}) &= v_k \exp[\mathrm{i}(\boldsymbol{k} - \boldsymbol{q}) \cdot \boldsymbol{r}] \end{aligned} \tag{2.130}$$

where the coefficients u_k and v_k are found from

$$\begin{aligned} (E_k - \xi_{k+q})u_k + \Delta v_k &= 0 \\ (E_k + \xi_{k-q})v_k + \Delta u_k &= 0. \end{aligned} \tag{2.131}$$

Taking into account the normalization condition $u_k^2 + v_k^2 = 1$, we obtain

$$\begin{aligned} u_k^2, v_k^2 &= \frac{1}{2}\left(1 \pm \frac{\xi_{k+q} + \xi_{k-q}}{2\epsilon_k}\right) \\ u_k v_k &= -\frac{\Delta}{2\epsilon_k} \end{aligned} \tag{2.132}$$

and the excitation spectrum

$$E_k = \frac{\xi_{k+q} - \xi_{k-q}}{2} + \sqrt{\frac{(\xi_{k+q} + \xi_{k-q})^2}{4} + \Delta^2}. \tag{2.133}$$

If we apply the Landau criterion (1.117) to the spectrum equation (2.133), the critical velocity will be $v_c = \Delta/k_F$. Hence, we can keep only the terms of the first order in q as long as $q \lesssim m v_c \ll k_F$ in equation (2.133). Then the excitation spectrum becomes

$$E_{\boldsymbol{k}} \approx \epsilon_{\boldsymbol{k}} + \boldsymbol{k} \cdot \boldsymbol{v}_s \tag{2.134}$$

and the equation for the order parameter is

$$\Delta = 2E_p \sum_{\boldsymbol{k}} \frac{\Delta}{2\epsilon_{\boldsymbol{k}}} [1 - 2f(\epsilon_{\boldsymbol{k}} - \boldsymbol{k} \cdot \boldsymbol{v}_s)]. \tag{2.135}$$

The excitation energy (equation (2.133)) is negative for some directions of $\boldsymbol{k}$ and quasi-particles appear even at zero temperature, when the superfluid velocity exceeds v_c. The Landau criterion tells us that superconductivity should disappear at $v_s > v_c$. Let us see how it happens by solving the BCS equation (2.135) in two and three dimensions. For $T = 0$, we have

$$f(\epsilon_{\boldsymbol{k}} - \boldsymbol{k} \cdot \boldsymbol{v}_s) = \theta(-\epsilon_{\boldsymbol{k}} + k v_s \cos\varphi)$$

where φ is the angle between $\boldsymbol{v}_s$ and $\boldsymbol{k}$. If $v_s < v_c$, there are no quasi-particles ($f_{\boldsymbol{k}} = 0$) and

$$\Delta = \Delta(0). \tag{2.136}$$

Here $\Delta(0)$ is the BCS gap (equation (2.46)) in the absence of the flow. However, if $v_s > v_c$, there is an interval of $|\varphi| \leqslant \arccos[\epsilon_{\boldsymbol{k}}/(k_F v_s)]$, where $f(\epsilon_{\boldsymbol{k}} - \boldsymbol{k} \cdot \boldsymbol{v}_s) = 1$ and the gap should be different from $\Delta(0)$. Integrating over the angle, we obtain

$$\Delta \ln \frac{\Delta(0)}{\Delta} = \frac{2\Delta}{\pi} \int_{\Delta}^{k_F v_s} d\epsilon \, \frac{\arccos(\epsilon/k_F v_s)}{\sqrt{\epsilon^2 - \Delta^2}}. \tag{2.137}$$

in two dimensions. Calculating the remaining integral, we arrive at

$$\Delta \ln \frac{\Delta(0)}{\Delta} = \Delta \ln \frac{k_F v_s}{\Delta}. \tag{2.138}$$

This equation has only the trivial solution $\Delta = 0$. Hence, the Landau criterion cannot be compromised in two-dimensional s-wave BCS superconductors, that is superconductivity disappears precisely at $v_s \geqslant v_c$. The superflow in three-dimensional superconductors is different [41]. In this case, integrating over the angle yields

$$\Delta \ln \frac{\Delta(0)}{\Delta} = \Delta \int_{\Delta}^{k_F v_s} d\epsilon \, \frac{1 - (\epsilon/k_F v_s)}{\sqrt{\epsilon^2 - \Delta^2}}. \tag{2.139}$$

There is a non-trivial solution $\Delta \neq 0$ even if $v_s > v_c$, which is found from

$$\ln \frac{k_F v_s}{\Delta(0)} = \sqrt{1 - \left(\frac{\Delta}{k_F v_s}\right)^2} - \ln\left[1 + \sqrt{1 - \left(\frac{\Delta}{k_F v_s}\right)^2}\,\right]. \tag{2.140}$$

It disappears only at $v_s \geqslant e v_c/2$, where $e = 2.718$. The excitation spectrum (equation (2.133)) has no gap but the order parameter is still non-zero if $v_c < v_s <$

$ev_c/2$. The superconductivity in this region is *gapless*. Normal excitations fill the negative part of their energy spectrum in $\boldsymbol{k}$-space resulting in two components (normal and superfluid), as at finite temperatures. The quasi-particles do not fully destroy the order parameter, they renormalize it while $v_s < ev_c/2$. The two-fluid situation at $T = 0$ is possible due to the Fermi statistics of quasi-particles in the BCS superconductor because the Pauli exclusion principle limits their density. When quasi-particles are bosons (chapter 4), the kinetic energy of a moving condensate entirely dissipates into quasi-particles as soon as $v_s \geqslant v_c$.

There are other examples of the gapless superconductivity like unconventional Cooper pairs (section 2.10) and 'dirty' superconductors with magnetic impurities [42].

2.13 Andreev reflection

The Bogoliubov equations are particularly instrumental in handling the interface between normal and superconducting metals (NS interface) and between different superconductors (SS interface). They allow us to calculate the I–V characteristics of tunnelling structures in the same fashion as in the conventional single-particle tunnelling problem in quantum mechanics. If there is a potential barrier due to a dielectric layer, NS conductance can be readily calculated using the Fermi–Dirac golden rule (section 2.4). It shows the gap structure. In the absence of the barrier, a new phenomenon is observed, in which an incoming electron from the normal side of the normal/superconducting contact is reflected as a hole along the same trajectory [46]. The *Andreev reflection* results in an *increase* in the tunnelling conductance in the voltage range $|eV| \lesssim \Delta$ in sharp contrast to its suppression in the case of the barrier.

A simple theory of NS tunnelling in a metallic (no-barrier) regime [47] follows from the one-dimensional Bogoliubov equations, which can be written in the matrix form:

$$E\psi(x) = \begin{pmatrix} -(1/2m)\,\mathrm{d}^2/\mathrm{d}x^2 - \mu(x) & \Delta(x) \\ \Delta(x) & (1/2m)\,\mathrm{d}^2/\mathrm{d}x^2 + \mu(x) \end{pmatrix} \psi(x). \tag{2.141}$$

The gap $\Delta(x)$ and the chemical potential $\mu(x)$ depend on the coordinate x perpendicular to the contact area. In the normal state ($x \leqslant 0$, $\Delta(x) = 0$, $\mu(x) = \mu_n$), equation (2.141) is the free-particle Schrödinger equation (first row) or its time-reversed version (second row). The two-component wavefunction of the normal metal is given by

$$\psi_n(x < 0) = \begin{pmatrix} 1 \\ 0 \end{pmatrix} \mathrm{e}^{\mathrm{i}q^+x} + b \begin{pmatrix} 1 \\ 0 \end{pmatrix} \mathrm{e}^{-\mathrm{i}q^+x} + a \begin{pmatrix} 0 \\ 1 \end{pmatrix} \mathrm{e}^{-\mathrm{i}q^-x} \tag{2.142}$$

where momenta associated with the energy E are $q^\pm = [2m(\mu_n \pm E)]^{1/2}$. Here the first and second terms describe the incident and reflected electron plane waves, respectively. The third term describes the reflected (Andreev) hole. The hole appears when the electron with the momentum $q \approx k_F$ tunnels from a normal metal into a superconducting condensate together with its 'partner' having the opposite momentum $-q$. This simultaneous two-electron tunnelling results in the hole excitation on the hole branch of the excitation energy spectrum near $-k_F$ in the normal metal (figure 2.5). In the superconductor ($x \geqslant 0$, $\Delta(x) = \Delta$, $\mu(x) = \mu_s$) the incoming electron can produce only outgoing particles (i.e. with the positive group velocity $\partial\epsilon_{\boldsymbol{k}}/\partial k$) (figure 2.5). The solution in this region is given by

$$\psi_s(x > 0) = c \begin{pmatrix} 1 \\ \dfrac{\Delta}{E+\xi} \end{pmatrix} \mathrm{e}^{\mathrm{i}k^+x} + d \begin{pmatrix} 1 \\ \dfrac{\Delta}{E-\xi} \end{pmatrix} \mathrm{e}^{-\mathrm{i}k^-x} \tag{2.143}$$

Here the momenta associated with the energy E are $k^\pm = [2m(\mu_s \pm \xi)]^{1/2}$, where $\xi = \sqrt{E^2 - \Delta^2}$. The energy of the incident electron is defined in the whole positive region ($E \geqslant 0$), so that ξ is not necessary real.

The coefficients a, b, c, d are determined from the boundary conditions, which are the continuity of $\psi(x)$ and its first derivative at $x = 0$, as in the conventional single-particle tunnelling problem. Applying the boundary conditions, we obtain

$$1 + b = c + d \tag{2.144}$$

$$a = c\frac{\Delta}{E+\xi} + d\frac{\Delta}{E-\xi} \tag{2.145}$$

$$q^+(1-b) = ck^+ - dk^- \tag{2.146}$$

and

$$q^- a = ck^+ \frac{\Delta}{E+\xi} - dk^- \frac{\Delta}{E-\xi}. \tag{2.147}$$

For simplicity, we now take $q^\pm \approx k^\pm \approx k_F$ because the Fermi energy is huge compared with the gap in the BCS superconductors. Then we find $b = d = 0$, which, physically speaking, means that all reflection is the Andreev reflection and all transmission occurs without branch crossing.

The transmission coefficient, which determines the conductance, is given by

$$T(E) = 1 + |a|^2 - |b|^2. \tag{2.148}$$

With $b = d = 0$, we obtain $c = 1$ and

$$a = \frac{\Delta}{E+\xi} \tag{2.149}$$

from equations (2.144) and (2.145), respectively. If the incident energy (or the voltage $V = E/e$) is larger than the gap ($E \geqslant \Delta$), ξ is real and the transmission is obtained as

$$T(E) = \frac{2E}{E + \sqrt{E^2 - \Delta^2}}. \tag{2.150}$$

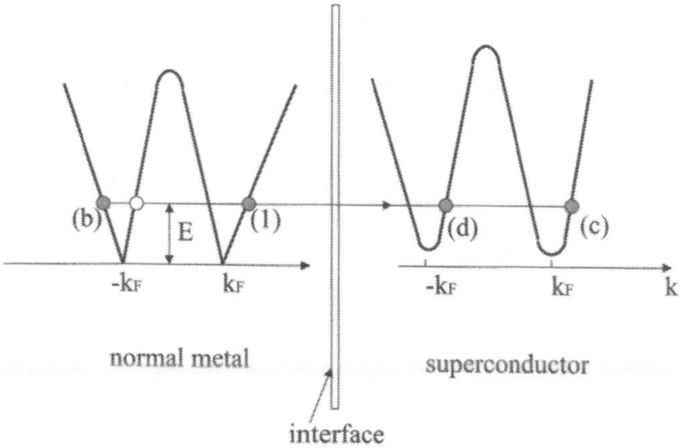

Figure 2.5. Schematic plot of excitation energies *versus* k at an NS interface. The open circle denotes the Andreev hole, the closed circles denote an incident at (1), reflected (b) electrons, and transmitted (c), (d) quasi-particles.

It tends to one in the high energy limit ($E \gg \Delta$). Remarkably, when the energy is below the gap ($E \leqslant \Delta$), the transmission is doubled compared with the normal state limit:

$$T(E) = 1 + \frac{\Delta^2}{(E+\xi)(E+\xi^*)} = 2. \tag{2.151}$$

Hence, the Andreev reflection is observed as an enhancement of the tunnelling conductance of NS metallic contacts in the gap region $V \leqslant \Delta/e$. The phenomenon serves as a powerful tool in the gap determination (part 2).

2.14 Green's function formulation of the BCS theory, $T = 0$

There is yet another elegant formalism introduced by Gor'kov [43], which allows for an economic derivation of BCS results and an extension of the theory to the intermediate coupling regime (chapter 3). Using the Heisenberg operators, equation (2.109), let us define the time-dependent 'normal' $G(\boldsymbol{r}, \boldsymbol{r}', t)$ and 'anomalous' $F^{\pm}(\boldsymbol{r}, \boldsymbol{r}', t)$ one-particle Green's functions (GF) as

$$\begin{aligned} \mathrm{i}G(\boldsymbol{r}, \boldsymbol{r}', t) &= \Theta(t)\langle\psi_s(\boldsymbol{r}, t)\psi_s^\dagger(\boldsymbol{r}', 0)\rangle - \Theta(-t)\langle\psi_s^\dagger(\boldsymbol{r}', 0)\psi_s(\boldsymbol{r}, t)\rangle \\ \mathrm{i}F^+(\boldsymbol{r}, \boldsymbol{r}', t) &= \Theta(t)\langle\psi_\downarrow^\dagger(\boldsymbol{r}, t)\psi_\uparrow^\dagger(\boldsymbol{r}', 0)\rangle - \Theta(-t)\langle\psi_\uparrow^\dagger(\boldsymbol{r}', 0)\psi_\downarrow^\dagger(\boldsymbol{r}, t)\rangle \\ \mathrm{i}F(\boldsymbol{r}, \boldsymbol{r}', t) &= \Theta(t)\langle\psi_\downarrow(\boldsymbol{r}, t)\psi_\uparrow(\boldsymbol{r}', 0)\rangle - \Theta(-t)\langle\psi_\uparrow(\boldsymbol{r}', 0)\psi_\downarrow(\boldsymbol{r}, t)\rangle. \end{aligned} \tag{2.152}$$

Here the quantum averages are calculated in the ground state of the system at zero temperature. The operators $\psi_\uparrow(\boldsymbol{r}, t)$ and $\psi_\downarrow(\boldsymbol{r}', t')$ anticommute if $t' = t$. Hence, $F(\boldsymbol{r}, \boldsymbol{r}', 0)$ and $F^+(\boldsymbol{r}, \boldsymbol{r}', 0)$ are connected by the relation

$$F^*(\boldsymbol{r}, \boldsymbol{r}', 0) = -F^+(\boldsymbol{r}, \boldsymbol{r}', 0). \tag{2.153}$$

The first time derivatives of the GFs are calculated using the equations of motion (2.110) and (2.111). Taking into account that $\mathrm{d}\Theta(t)/\mathrm{d}t = \delta(t)$, we obtain a system of two coupled equations:

$$\mathrm{i}\frac{\partial G(\boldsymbol{r},\boldsymbol{r}',t)}{\partial t} = \delta(t)\delta(\boldsymbol{r}-\boldsymbol{r}') + \hat{h}(\boldsymbol{r})G(\boldsymbol{r},\boldsymbol{r}',t) + \Delta(\boldsymbol{r})F^{+}(\boldsymbol{r},\boldsymbol{r}',t) \tag{2.154}$$

and

$$\mathrm{i}\frac{\partial F^{+}(\boldsymbol{r},\boldsymbol{r}',t)}{\partial t} = -\hat{h}^{*}(\boldsymbol{r})F^{+}(\boldsymbol{r},\boldsymbol{r}',t) + \Delta^{*}(\boldsymbol{r})G(\boldsymbol{r},\boldsymbol{r}',t). \tag{2.155}$$

The order parameter is expressed in terms of the anomalous GF as

$$\Delta(\boldsymbol{r}) = -2\mathrm{i}E_p[F^{+}(\boldsymbol{r},\boldsymbol{r},+0)]^{*}. \tag{2.156}$$

We can readily solve these self-consistent equations in the absence of an external field. In a homogeneous superconductor, $G(\boldsymbol{r},\boldsymbol{r}',t) = G(\boldsymbol{r}-\boldsymbol{r}',t)$, $F^{+}(\boldsymbol{r},\boldsymbol{r}',t) = F^{+}(\boldsymbol{r}-\boldsymbol{r}',t)$ and $\Delta(\boldsymbol{r})$ is a real constant. Applying the Fourier transforms

$$G(\boldsymbol{r}-\boldsymbol{r}',t) = \frac{1}{2\pi}\sum_{\boldsymbol{k}}\int_{-\infty}^{\infty}\mathrm{d}\omega\, G(\boldsymbol{k},\omega)\exp(\mathrm{i}\boldsymbol{k}\cdot\boldsymbol{r}-\mathrm{i}\omega t) \tag{2.157}$$

$$F^{+}(\boldsymbol{r}-\boldsymbol{r}',t) = \frac{1}{2\pi}\sum_{\boldsymbol{k}}\int_{-\infty}^{\infty}\mathrm{d}\omega\, F^{+}(\boldsymbol{k},\omega)\exp(\mathrm{i}\boldsymbol{k}\cdot\boldsymbol{r}-\mathrm{i}\omega t)$$

we obtain

$$\begin{aligned}(\omega-\xi_{\boldsymbol{k}})G(\boldsymbol{k},\omega) - \Delta F^{+}(\boldsymbol{k},\omega) &= 1\\ (\omega+\xi_{\boldsymbol{k}})F^{+}(\boldsymbol{k},\omega) - \Delta G(\boldsymbol{k},\omega) &= 0.\end{aligned} \tag{2.158}$$

Then the Fourier components are found to be

$$\begin{aligned}G(\boldsymbol{k},\omega) &= \frac{\omega+\xi_{\boldsymbol{k}}}{\omega^2-\epsilon_{\boldsymbol{k}}^2} = \frac{u_{\boldsymbol{k}}^2}{\omega-\epsilon_{\boldsymbol{k}}} + \frac{v_{\boldsymbol{k}}^2}{\omega+\epsilon_{\boldsymbol{k}}}\\ F^{+}(\boldsymbol{k},\omega) &= \frac{\Delta}{\omega^2-\epsilon_{\boldsymbol{k}}^2}.\end{aligned} \tag{2.159}$$

These expressions are well defined for any ω but not for $\omega = \pm\epsilon_{\boldsymbol{k}}$. They cannot be used for integration with respect to ω because their simple poles are just on the real axis of ω. We have to define a way of bypassing the poles in evaluating the integral. The solution of the first-order differential equations (2.154) and (2.155) are not well defined because we have not applied any initial condition so far. The easiest way to apply the condition is to consider the normal state limit of equation (2.159) with $\Delta = 0$,

$$\begin{aligned}G^{(0)}(\boldsymbol{k},\omega) &= \frac{\Theta(\xi_{\boldsymbol{k}})}{\omega-\xi_{\boldsymbol{k}}} + \frac{\Theta(-\xi_{\boldsymbol{k}})}{\omega-\xi_{\boldsymbol{k}}}\\ [F^{+}(\boldsymbol{k},\omega)]^{(0)} &= 0.\end{aligned} \tag{2.160}$$

Integrating the Fourier transform $G^{(0)}(\boldsymbol{k}, \omega)$ should yield the normal state GF, which can be directly calculated using its definition. The free-electron Heisenberg operators are found as

$$\psi_s(\boldsymbol{r}, t) = \sum_{\boldsymbol{k}} c_{\boldsymbol{k}s} \exp[\mathrm{i}(\boldsymbol{k} \cdot \boldsymbol{r} - \xi_{\boldsymbol{k}} t)]$$

so that

$$\begin{aligned} \mathrm{i}G^{(0)}(\boldsymbol{r} - \boldsymbol{r}', t) = \Theta(t) \sum_{\boldsymbol{k},\boldsymbol{k}'} \langle c_{\boldsymbol{k}s} c^{\dagger}_{\boldsymbol{k}'s} \rangle \exp[\mathrm{i}(\boldsymbol{k} \cdot \boldsymbol{r} - \boldsymbol{k}' \cdot \boldsymbol{r}') - \mathrm{i}\xi_{\boldsymbol{k}} t] \\ - \Theta(-t) \langle c^{\dagger}_{\boldsymbol{k}'s} c_{\boldsymbol{k}s} \rangle \exp[\mathrm{i}(\boldsymbol{k} \cdot \boldsymbol{r} - \boldsymbol{k}' \cdot \boldsymbol{r}') - i\xi_{\boldsymbol{k}} t] \end{aligned} \quad (2.161)$$

where $\langle c^{\dagger}_{\boldsymbol{k}'s} c_{\boldsymbol{k}s} \rangle = \delta_{\boldsymbol{k}\boldsymbol{k}'} \Theta(-\xi_{\boldsymbol{k}})$ is the Fermi–Dirac distribution at $T = 0$ and $\langle c_{\boldsymbol{k}s} c^{\dagger}_{\boldsymbol{k}'s} \rangle = \delta_{\boldsymbol{k}\boldsymbol{k}'} \Theta(\xi_{\boldsymbol{k}})$. Hence, we have

$$\mathrm{i}G^{(0)}(\boldsymbol{r} - \boldsymbol{r}', t) = \sum_{\boldsymbol{k}} \exp[\mathrm{i}\boldsymbol{k} \cdot (\boldsymbol{r} - \boldsymbol{r}') - \mathrm{i}\xi_{\boldsymbol{k}} t] \{\Theta(t)\Theta(\xi_{\boldsymbol{k}}) - \Theta(-t)\Theta(-\xi_{\boldsymbol{k}})\}. \quad (2.162)$$

We obtain the same result using the Fourier transform equation (2.160), if we choose the following way of bypassing the poles:

$$G^{(0)}(\boldsymbol{k}, \omega) = \frac{\Theta(\xi_{\boldsymbol{k}})}{\omega - \xi_{\boldsymbol{k}} + \mathrm{i}\delta} + \frac{\Theta(-\xi_{\boldsymbol{k}})}{\omega - \xi_{\boldsymbol{k}} - \mathrm{i}\delta} \quad (2.163)$$

where $\delta = +0$ is an infinitesimal positive constant. Indeed let us calculate the integral

$$\begin{aligned} \int_{-\infty}^{\infty} \mathrm{d}\omega\, G^{(0)}(\boldsymbol{k}, \omega) \exp(-\mathrm{i}\omega t) = \Theta(\xi_{\boldsymbol{k}}) \int_{-\infty}^{\infty} \mathrm{d}\omega \frac{\mathrm{e}^{-\mathrm{i}\omega t}}{\omega - \xi_{\boldsymbol{k}} + \mathrm{i}\delta} \\ + \Theta(-\xi_{\boldsymbol{k}}) \int_{-\infty}^{\infty} \mathrm{d}\omega \frac{\mathrm{e}^{-\mathrm{i}\omega t}}{\omega - \xi_{\boldsymbol{k}} - \mathrm{i}\delta}. \end{aligned} \quad (2.164)$$

When t is positive the contour in both integrals of the right-hand side should be chosen in the lower half-plane of the complex variable. The pole in the first term is below the real axis, while the pole in the second term is found in the upper half-plane. The first integral yields $-2\pi \mathrm{i}\Theta(t)\Theta(\xi_{\boldsymbol{k}})$ and the second integral is zero. When t is negative, the first integral is zero, while the second one is $2\pi \mathrm{i}\Theta(-t)\Theta(-\xi_{\boldsymbol{k}})$. As a result, we recover the ideal Fermi-gas GF (equation (2.162)). Hence, the superconducting GFs, which provide a correct normal state limit, are

$$G(\boldsymbol{k}, \omega) = \frac{u_{\boldsymbol{k}}^2}{\omega - \epsilon_{\boldsymbol{k}} + \mathrm{i}\delta} + \frac{v_{\boldsymbol{k}}^2}{\omega + \epsilon_{\boldsymbol{k}} - \mathrm{i}\delta} \quad (2.165)$$

$$F^{+}(\boldsymbol{k}, \omega) = u_{\boldsymbol{k}} v_{\boldsymbol{k}} \left[\frac{1}{\omega + \epsilon_{\boldsymbol{k}} - \mathrm{i}\delta} - \frac{1}{\omega - \epsilon_{\boldsymbol{k}} + \mathrm{i}\delta} \right].$$

We note that the poles of the Fourier transform of GF yields the excitation spectrum of the superconductor.

Let us show that these GFs also provide the BCS results for the electron distribution function and the gap. The electron density is calculated as

$$n_{\mathrm{e}} = -\mathrm{i}G(\boldsymbol{r}, \boldsymbol{r}, -0) = -\frac{\mathrm{i}}{2\pi}\sum_{\boldsymbol{k}}\left[\int_{-\infty}^{\infty} \mathrm{d}\omega\, \mathrm{e}^{-\mathrm{i}\omega t} G(\boldsymbol{k}, \omega)\right]_{t\to -0}. \tag{2.166}$$

Here t is negative and the contour should be taken in the upper half-plane. Only the second term of the normal GF, equation (2.165), contributes with the following result:

$$n_{\mathrm{e}} = \sum_{\boldsymbol{k}} v_{\boldsymbol{k}}^2 \tag{2.167}$$

so that the electron distribution function is $v_{\boldsymbol{k}}^2$, as it should be (see equation (2.27)). The gap is found to be

$$\Delta = -2\mathrm{i}E_p[F^+(\boldsymbol{r}, \boldsymbol{r}, +0)]^* = -\frac{2\mathrm{i}E_p}{2\pi}\sum_{\boldsymbol{k}}\int_{-\infty}^{\infty} \mathrm{d}\omega\, \mathrm{e}^{\mathrm{i}\omega t}\{F^+(\boldsymbol{k}, \omega)\}^* \tag{2.168}$$

where $t = +0$. Calculating the integral with the Fourier transform equation (2.165), we obtain the BCS equation at $T = 0$,

$$\Delta = -2E_p \sum_{\boldsymbol{k}} u_{\boldsymbol{k}} v_{\boldsymbol{k}}. \tag{2.169}$$

2.15 Green's functions of the BCS superconductor at finite temperatures

At finite temperatures, the BCS theory can be formulated with the 'temperature' GFs (appendix D). Following Matsubara [44], we replace time t in the definition of the Heisenberg operators by a 'thermodynamic time' $\tau = \mathrm{i}t$. Then the *temperature* GF is defined as

$$\mathcal{G}(\boldsymbol{r}, \boldsymbol{r}', \tau_1, \tau_2) = -\langle\langle T_\tau \psi_s(\boldsymbol{r}, \tau_1)\psi_s^\dagger(\boldsymbol{r}', \tau_2)\rangle\rangle \tag{2.170}$$

where $\psi_s(\boldsymbol{r}, \tau) = \exp(H\tau)\Psi_s(\boldsymbol{r})\exp(-H\tau)$ and Gor'kov's temperature GF is

$$\mathcal{F}^+(\boldsymbol{r}, \boldsymbol{r}', \tau_1, \tau_2) = -\langle\langle T_\tau \psi_\downarrow^\dagger(\boldsymbol{r}, \tau_1)\psi_\uparrow^\dagger(\boldsymbol{r}', \tau_2)\rangle\rangle.$$

Here the thermodynamic 'times', τ_1, τ_2, are real and positive, varying in the interval $0 < \tau_1, \tau_2 < 1/T$. The double angular brackets correspond to quantum as well as statistical averages of any operator $\hat{A}$ with the Gibbs distribution (appendix B),

$$\langle\langle \hat{A}\rangle\rangle = \sum_\nu \mathrm{e}^{(\Omega - E_\nu)/T}\langle\nu|\hat{A}|\nu\rangle \equiv \mathrm{Tr}\{\mathrm{e}^{(\Omega-\tilde{H})/T}\hat{A}\} \tag{2.171}$$

where Ω is the thermodynamic potential and $|\nu\rangle$ are the eigenstates of $\tilde{H}$ with the eigenvalues E_ν. The operation T_τ performs the 'time' ordering according to the following definition:

$$T_\tau \psi_s(\boldsymbol{r}, \tau_1)\psi^\dagger_{s'}(\boldsymbol{r}', \tau_2) \equiv \Theta(\tau_1 - \tau_2)\psi_s(\boldsymbol{r}, \tau_1)\psi^\dagger_{s'}(\boldsymbol{r}', \tau_2) \\ - \Theta(\tau_2 - \tau_1)\psi^\dagger_{s'}(\boldsymbol{r}', \tau_2)\psi_s(\boldsymbol{r}, \tau_1). \quad (2.172)$$

Actually the temperature GFs depend on the difference $\tau = \tau_1 - \tau_2$ because of the trace in their definition. Indeed let us consider $\tau_1 < \tau_2$, so that

$$\mathcal{G}(\boldsymbol{r}, \boldsymbol{r}', \tau_1, \tau_2) = \exp(\Omega/T)\,\mathrm{Tr}\{\mathrm{e}^{-\tilde{H}/T}\mathrm{e}^{\tau_2\tilde{H}}\Psi^\dagger_s(\boldsymbol{r}')\mathrm{e}^{\tau\tilde{H}}\Psi_s(\boldsymbol{r})\mathrm{e}^{-\tau_1\tilde{H}}\} \\ = \exp(\Omega/T)\,\mathrm{Tr}\{\mathrm{e}^{-(\tau+1/T)\tilde{H}}\Psi^\dagger_s(\boldsymbol{r}')\mathrm{e}^{\tau\tilde{H}}\Psi_s(\boldsymbol{r})\} \quad (2.173)$$

where we have performed the cyclic permutation under the trace. Differing from positive τ_1 and τ_2, the variable τ is defined in the domain

$$-1/T \leqslant \tau \leqslant 1/T. \quad (2.174)$$

There is a connection between GFs with negative and positive τ. Following the definition in equation (2.170), we find, for $\tau > 0$, that

$$\mathcal{G}(\boldsymbol{r}, \boldsymbol{r}', \tau) = -\exp(\Omega/T)\,\mathrm{Tr}\{\mathrm{e}^{(\tau-1/T)\tilde{H}}\Psi_s(\boldsymbol{r})\mathrm{e}^{-\tau\tilde{H}}\Psi^\dagger_s(\boldsymbol{r}')\}. \quad (2.175)$$

Here replacing the positive τ by a negative $\tilde{\tau}$ as $\tau = \tilde{\tau} + 1/T$, we obtain ($\tilde{\tau} < 0$)

$$\mathcal{G}(\boldsymbol{r}, \boldsymbol{r}', \tilde{\tau} + 1/T) = -\exp(\Omega/T)\,\mathrm{Tr}\{\mathrm{e}^{\tilde{\tau}\tilde{H}}\Psi_s(\boldsymbol{r})\mathrm{e}^{-(\tilde{\tau}+1/T)\tilde{H}}\Psi^\dagger_s(\boldsymbol{r}')\} \\ = -\mathcal{G}(\boldsymbol{r}, \boldsymbol{r}', \tilde{\tau}). \quad (2.176)$$

The Fourier transform theory states that if a function $F(\tau)$ is defined over the interval $-1/T \leqslant \tau \leqslant 1/T$, then its Fourier expansion is

$$F(\tau) = T\sum_{n=-\infty}^{\infty} f_n \exp(-\mathrm{i}\pi nT\tau) \quad (2.177)$$

where $n = 0, \pm 1, \pm 2, \ldots$, and

$$f_n = \tfrac{1}{2}\int_{-1/T}^{1/T} \mathrm{d}\tau\, F(\tau)\exp(\mathrm{i}\pi nT\tau). \quad (2.178)$$

In our case, $F(\tau) = -F(\tau + 1/T)$ for a negative τ and

$$f_n = \tfrac{1}{2}(1 - \mathrm{e}^{\mathrm{i}\pi n})\int_0^{1/T} \mathrm{d}\tau\, F(\tau)\exp(\mathrm{i}\pi nT\tau). \quad (2.179)$$

Hence the Fourier components f_n of the fermionic GFs are non-zero only for *odd n*. As a result we can expand the temperature GFs of a homogeneous superconductor into a Fourier *series* as follows:

$$\mathcal{G}(\boldsymbol{r}, \boldsymbol{r}', \tau) = T \sum_{\omega_n} \mathcal{G}(\boldsymbol{k}, \omega_n) \exp[\mathrm{i}\boldsymbol{k} \cdot (\boldsymbol{r} - \boldsymbol{r}') - \mathrm{i}\omega_n \tau] \tag{2.180}$$

$$\mathcal{F}^+(\boldsymbol{r}, \boldsymbol{r}', \tau) = T \sum_{\omega_n} \mathcal{F}^+(\boldsymbol{k}, \omega_n) \exp[\mathrm{i}\boldsymbol{k} \cdot (\boldsymbol{r} - \boldsymbol{r}') - \mathrm{i}\omega_n \tau] \tag{2.181}$$

where the discrete Matsubara frequencies are $\omega_n = \pi T(2n + 1)$, $n = 0, \pm 1, \pm 2, \ldots$. Differentiating the Matsubara operators with respect to τ, we obtain the equations of motion,

$$-\frac{\partial \psi_\uparrow(\boldsymbol{r}, \tau)}{\partial \tau} = \hat{h}(\boldsymbol{r})\psi_\uparrow(\boldsymbol{r}, \tau) + \Delta(\boldsymbol{r})\psi_\downarrow^\dagger(\boldsymbol{r}, \tau) \tag{2.182}$$

$$\frac{\partial \psi_\downarrow^\dagger(\boldsymbol{r}, \tau)}{\partial \tau} = \hat{h}^*(\boldsymbol{r})\psi_\downarrow^\dagger(\boldsymbol{r}, \tau) - \Delta^*(\boldsymbol{r})\psi_\uparrow(\boldsymbol{r}, \tau)$$

and the equations for temperature GFs,

$$-\frac{\partial \mathcal{G}(\boldsymbol{r}, \boldsymbol{r}', \tau)}{\partial \tau} = \delta(\tau)\delta(\boldsymbol{r} - \boldsymbol{r}') + \hat{h}(\boldsymbol{r})\mathcal{G}(\boldsymbol{r}, \boldsymbol{r}', \tau) + \Delta(\boldsymbol{r})\mathcal{F}^+(\boldsymbol{r}, \boldsymbol{r}', \tau), \tag{2.183}$$

$$\frac{\partial \mathcal{F}^+(\boldsymbol{r}, \boldsymbol{r}', \tau)}{\partial \tau} = \hat{h}^*(\boldsymbol{r})\mathcal{F}^+(\boldsymbol{r}, \boldsymbol{r}', \tau) - \Delta^*(\boldsymbol{r})\mathcal{G}(\boldsymbol{r}, \boldsymbol{r}', \tau). \tag{2.184}$$

The order parameter is given by

$$\Delta^*(\boldsymbol{r}) = -2E_p\mathcal{F}^+(\boldsymbol{r}, \boldsymbol{r}, 0). \tag{2.185}$$

In the absence of an external field, the Fourier transforms of these equations are:

$$(\mathrm{i}\omega_n - \xi_{\boldsymbol{k}})\mathcal{G}(\boldsymbol{k}, \omega_n) - \Delta\mathcal{F}^+(\boldsymbol{k}, \omega_n) = 1 \tag{2.186}$$

$$(\mathrm{i}\omega_n + \xi_{\boldsymbol{k}})\mathcal{F}^+(\boldsymbol{k}, \omega_n) - \Delta^*\mathcal{G}(\boldsymbol{k}, \omega_n) = 0$$

and

$$\Delta^* = -2E_pT \sum_{\boldsymbol{k}} \sum_{\omega_n} \mathcal{F}^+(\boldsymbol{k}, \omega_n). \tag{2.187}$$

The solution is

$$\mathcal{G}(\boldsymbol{k}, \omega_n) = -\frac{\mathrm{i}\omega_n + \xi_{\boldsymbol{k}}}{\omega_n^2 + \epsilon_{\boldsymbol{k}}^2} \tag{2.188}$$

$$\mathcal{F}^+(\boldsymbol{k}, \omega_n) = -\frac{\Delta^*}{\omega_n^2 + \epsilon_{\boldsymbol{k}}^2}$$

and the equation for the order parameter takes the following form:

$$\Delta = 2E_pT \sum_{k} \sum_{\omega_n} \frac{\Delta}{\omega_n^2 + \epsilon_k^2}. \tag{2.189}$$

The sum over frequencies in equation (2.189) is calculated using the expansion of $\tanh(x)$ (equation (2.54)). As a result, we obtain the BCS equation for the order parameter:

$$1 = \frac{\lambda}{2} \int \frac{d\xi}{\sqrt{\xi^2 + \Delta(T)^2}} \tanh \frac{\sqrt{\xi^2 + \Delta(T)^2}}{2T} \tag{2.190}$$

where the integral should be cut at $|\xi| \leqslant \omega_D$.

There is no direct physical meaning of poles of the Fourier transforms of temperature GFs, which are found on the imaginary frequency axis. Nonetheless, they allow for a direct calculation of the thermodynamic properties of the system, for example of the gap (equation (2.190)). The kinetic properties are expressed in terms of real-time GFs. In fact, the Matsubara GFs lead directly to real-time GFs. A one-particle real-time GF is defined at finite temperatures as

$$G(k, t) = -i\langle\langle T_t c_{ks}(t) c_{ks}^{\dagger} \rangle\rangle \tag{2.191}$$

with the real time t. There are also retarded G^R and advanced G^A GFs:

$$G^R(k, t) = -i\Theta(t)\langle\langle\{c_k(t)c_k^{\dagger}\}\rangle\rangle \tag{2.192}$$

$$G^A(k, t) = i\Theta(-t)\langle\langle\{c_k(t)c_k^{\dagger}\}\rangle\rangle. \tag{2.193}$$

Their Fourier components are analytical in the upper or lower half-plane of ω, respectively. There is a simple connection between the Fourier components of G and $G^{R,A}$ (appendix D):

$$G^{R,A}(k, \omega) = \operatorname{Re} G(k, \omega) \pm i \coth\left(\frac{\omega}{2T}\right) \operatorname{Im} G(k, \omega) \tag{2.194}$$

and between those of $G^{R,A}$ and $\mathcal{G}$,

$$G^R(k, i\omega_n) = \mathcal{G}(k, \omega_n) \tag{2.195}$$

for $\omega_n > 0$ and

$$\mathcal{G}(k, -\omega_n) = \mathcal{G}^*(k, \omega_n). \tag{2.196}$$

In our case the temperature GF is

$$\mathcal{G}(k, \omega_n) = \frac{u_k^2}{i\omega_n - \epsilon_k} + \frac{v_k^2}{i\omega_n + \epsilon_k} \tag{2.197}$$

where $u_k^2, v_k^2 = (\epsilon_k \pm \xi_k)/2\epsilon_k$ and $\epsilon_k = \sqrt{\xi_k^2 + \Delta^2}$. The analytical continuation of this expression to the upper half-plane yields

$$G^R(k, \omega) = \frac{u_k^2}{\omega - \epsilon_k + i\delta} + \frac{v_k^2}{\omega + \epsilon_k + i\delta}. \tag{2.198}$$

Then using equation (2.194), we obtain

$$G(\boldsymbol{k},\omega) = P\left(\frac{u_{\boldsymbol{k}}^2}{\omega-\epsilon_{\boldsymbol{k}}}+\frac{v_{\boldsymbol{k}}^2}{\omega+\epsilon_{\boldsymbol{k}}}\right) - \mathrm{i}\pi\tanh\left(\frac{\omega}{2T}\right)[u_{\boldsymbol{k}}^2\delta(\omega-\epsilon_{\boldsymbol{k}})+v_{\boldsymbol{k}}^2\delta(\omega+\epsilon_{\boldsymbol{k}})] \tag{2.199}$$

where we have applied the relation $1/(\omega+\mathrm{i}\delta) = P(1/\omega) - \mathrm{i}\pi\delta(\omega)$. Here the first term is understood as the principal value of the integral when integrating with respect to ω. For zero temperature, $\tanh(\omega/2T) = \mathrm{sign}(\omega)$ and

$$G(\boldsymbol{k},\omega) = \frac{u_{\boldsymbol{k}}^2}{\omega-\epsilon_{\boldsymbol{k}}+\mathrm{i}\delta}+\frac{v_{\boldsymbol{k}}^2}{\omega+\epsilon_{\boldsymbol{k}}-\mathrm{i}\delta} \tag{2.200}$$

as it should (see equation (2.165)).

2.16 Microscopic derivation of the Ginzburg–Landau equations

Using the Green's function formalism near T_c, Gor'kov [45] derived the Ginzburg–Landau equations from BCS theory. Let us consider the BCS superconductor in a stationary magnetic field with the vector potential $\boldsymbol{A}(\boldsymbol{r})$. The Fourier transform of equations (2.183) and (2.184) with respect to the thermodynamic time yields

$$\left\{\mathrm{i}\omega_n + \frac{[\nabla+\mathrm{i}e\boldsymbol{A}(\boldsymbol{r})]^2}{2m}+\mu\right\}\mathcal{G}_{\omega_n}(\boldsymbol{r},\boldsymbol{r}') = \delta(\boldsymbol{r}-\boldsymbol{r}') + \Delta(\boldsymbol{r})\mathcal{F}^+_{\omega_n}(\boldsymbol{r},\boldsymbol{r}'),$$

$$\left\{-\mathrm{i}\omega_n + \frac{[\nabla-\mathrm{i}e\boldsymbol{A}(\boldsymbol{r})]^2}{2m}+\mu\right\}\mathcal{F}^+_{\omega_n}(\boldsymbol{r},\boldsymbol{r}') = -\Delta^*(\boldsymbol{r})\mathcal{G}_{\omega_n}(\boldsymbol{r},\boldsymbol{r}') \tag{2.201}$$

where

$$\mathcal{G}_{\omega_n}(\boldsymbol{r},\boldsymbol{r}') = \int_0^{1/T} \mathrm{d}\tau\, \mathcal{G}(\boldsymbol{r},\boldsymbol{r}',\tau)\exp(\mathrm{i}\pi nT\tau)$$

$$\mathcal{F}^+_{\omega_n}(\boldsymbol{r},\boldsymbol{r}') = \int_0^{1/T} \mathrm{d}\tau\, \mathcal{F}^+(\boldsymbol{r},\boldsymbol{r}',\tau)\exp(\mathrm{i}\pi nT\tau)$$

and

$$\Delta^*(\boldsymbol{r}) = -2E_pT\sum_{\omega_n}\mathcal{F}^+_{\omega_n}(\boldsymbol{r},\boldsymbol{r}).$$

If the temperature is close to T_c, the order parameter is small. Then we can expand the GFs in powers of $\Delta(\boldsymbol{r})$. The zero-order GF is a temperature GF of an ideal Fermi gas, $\mathcal{G}^{(n)}_{\omega_n}(\boldsymbol{r},\boldsymbol{r}')$ in the magnetic field, which satisfies the following equation:

$$\left\{\mathrm{i}\omega_n + \frac{[\nabla+\mathrm{i}e\boldsymbol{A}(\boldsymbol{r})]^2}{2m}+\mu\right\}\mathcal{G}^{(n)}_{\omega_n}(\boldsymbol{r},\boldsymbol{r}') = \delta(\boldsymbol{r}-\boldsymbol{r}'). \tag{2.202}$$

If there is no magnetic field, $\mathcal{G}^{(n)}_{\omega_n}(\boldsymbol{r}, \boldsymbol{r}') = \mathcal{G}^{(0)}_{\omega_n}(\boldsymbol{r} - \boldsymbol{r}')$ and its Fourier transform is found as

$$\mathcal{G}^{(0)}(\boldsymbol{k}, \omega_n) = \frac{1}{\mathrm{i}\omega_n - \xi_{\boldsymbol{k}}}. \tag{2.203}$$

Transforming back to real space we obtain

$$\mathcal{G}^{(0)}_{\omega_n}(\boldsymbol{\rho}) = \frac{1}{(2\pi)^3} \int \mathrm{d}\boldsymbol{k} \, \frac{\mathrm{e}^{\mathrm{i}\boldsymbol{k}\cdot\boldsymbol{\rho}}}{\mathrm{i}\omega_n - \xi_{\boldsymbol{k}}} \tag{2.204}$$

where $\boldsymbol{\rho} = \boldsymbol{r} - \boldsymbol{r}'$. Calculating the integral over the angles in the polar spherical coordinates yields

$$\mathcal{G}^{(0)}_{\omega_n}(\rho) = \frac{1}{(2\pi)^2 \mathrm{i}\rho} \int_0^\infty k \, \mathrm{d}k \, \frac{\mathrm{e}^{\mathrm{i}k\rho} - \mathrm{e}^{-\mathrm{i}k\rho}}{\mathrm{i}\omega_n - \xi}. \tag{2.205}$$

We can replace k by $k = k_{\mathrm{F}} + \xi/v_{\mathrm{F}}$ because only the states near the Fermi surface contribute to the integral in equation (2.205), when ω_n is of the order of $T_{\mathrm{c}} \ll E_{\mathrm{F}}$. The result is

$$\mathcal{G}^{(0)}_{\omega_n}(\rho) = \frac{m}{(2\pi)^2 \mathrm{i}\rho} \int_{-\infty}^{\infty} \mathrm{d}\xi \, \frac{\mathrm{e}^{\mathrm{i}k_{\mathrm{F}}\rho} \mathrm{e}^{\mathrm{i}\xi\rho/v_{\mathrm{F}}} - \mathrm{e}^{-\mathrm{i}k\rho} \mathrm{e}^{-\mathrm{i}\xi\rho/v_{\mathrm{F}}}}{\mathrm{i}\omega_n - \xi}. \tag{2.206}$$

Using the contour in the upper or lower half-plane, we obtain

$$\mathcal{G}^{(0)}_{\omega_n}(\rho) = -\frac{m}{2\pi\rho} \exp[\mathrm{i} \, \mathrm{sign}(\omega_n) k_{\mathrm{F}}\rho - |\omega_n|\rho/v_{\mathrm{F}}]. \tag{2.207}$$

The Fourier component of the normal state GF falls exponentially with the characteristic length

$$\rho \approx \xi_0 \tag{2.208}$$

where $\xi_0 = v_{\mathrm{F}}/(2\pi T_{\mathrm{c}})$ is the zero temperature coherence length. GFs in a magnetic field are not translation invariant and the exact calculation of the normal $\mathcal{G}^{(n)}_{\omega_n}(\boldsymbol{r}, \boldsymbol{r}')$ using equation (2.202) is a challenging problem. But we can assume that the spatial variations of the vector potential are small over the characteristic distance ξ_0. If we neglect these variations, then the field operator in the magnetic field can be readily expressed via the free-electron operator. It satisfies the equation of motion

$$\frac{\partial \psi_s(\boldsymbol{r}, \tau)}{\partial \tau} = \left\{ \frac{[\nabla + \mathrm{i}e\boldsymbol{A}(\boldsymbol{r})]^2}{2m} + \mu \right\} \psi_s(\boldsymbol{r}, \tau). \tag{2.209}$$

If $\boldsymbol{A}(\boldsymbol{r})$ is a constant then the solution is

$$\psi_s(\boldsymbol{r}, \tau) = \sum_{\boldsymbol{k}} c_{\boldsymbol{k}s} \exp[\mathrm{i}(\boldsymbol{k} - e\boldsymbol{A}) \cdot \boldsymbol{r} - \xi_{\boldsymbol{k}}\tau)] \tag{2.210}$$

which is verified by direct substitution into equation (2.209). Hence, if the vector potential varies slowly, the GF differs from the zero-field GF only by the phase:

$$\mathcal{G}^{(n)}_{\omega_n}(\boldsymbol{r},\boldsymbol{r}') = \mathrm{e}^{-\mathrm{i}e\boldsymbol{A}(\boldsymbol{r})\cdot\boldsymbol{\rho}}\mathcal{G}^{(0)}_{\omega_n}(\rho). \tag{2.211}$$

This quasi-classical approximation with respect to the magnetic field is applied if the phase change over the distance ξ_0 is small. $|\boldsymbol{A}(\boldsymbol{r})|$ varies by $B\xi_0$ over this distance, and the phase changes by $eB\xi_0^2$. The maximum field is (see section 1.6.4)

$$B \leqslant H_{c2} \approx \frac{\Phi_0}{2\pi\xi_0^2}\frac{T_c - T}{T_c} \tag{2.212}$$

so that the quasi-classical approximation is applied, if

$$\frac{T_c - T}{T_c} \ll 1. \tag{2.213}$$

It is the very same region where the Ginzburg–Landau theory is applied. Using the normal state GF in the magnetic field (equation (2.211)), we can transform Gor'kov equations (2.201) into the integral form:

$$\mathcal{G}_{\omega_n}(\boldsymbol{r},\boldsymbol{r}') = \mathcal{G}^{(n)}_{\omega_n}(\boldsymbol{r},\boldsymbol{r}') + \int \mathrm{d}\boldsymbol{x}\mathcal{G}^{(n)}_{\omega_n}(\boldsymbol{r},\boldsymbol{x})\Delta(\boldsymbol{x})\mathcal{F}^{+}_{\omega_n}(\boldsymbol{x},\boldsymbol{r}') \tag{2.214}$$

$$\mathcal{F}^{+}_{\omega_n}(\boldsymbol{r},\boldsymbol{r}') = -\int \mathrm{d}\boldsymbol{x}\,\mathcal{G}^{(n)}_{-\omega_n}(\boldsymbol{x},\boldsymbol{r})\Delta^*(\boldsymbol{x})\mathcal{G}_{\omega_n}(\boldsymbol{x},\boldsymbol{r}').$$

Now we expand $\mathcal{G}_{\omega_n}(\boldsymbol{r},\boldsymbol{r}')$ to the terms of the second order and $\mathcal{F}_{\omega_n}(\boldsymbol{r},\boldsymbol{r}')$ to the terms of the third order in $\Delta(\boldsymbol{r})$ inclusive:

$$\begin{aligned}
\mathcal{G}_{\omega_n}(\boldsymbol{r},\boldsymbol{r}') = {} & \mathcal{G}^{(n)}_{\omega_n}(\boldsymbol{r},\boldsymbol{r}') - \int \mathrm{d}\boldsymbol{x}\int \mathrm{d}\boldsymbol{y}\,\mathcal{G}^{(n)}_{\omega_n}(\boldsymbol{r},\boldsymbol{x})\Delta(\boldsymbol{x}) \\
& \times \mathcal{G}^{(n)}_{-\omega_n}(\boldsymbol{y},\boldsymbol{x})\Delta^*(\boldsymbol{y})\mathcal{G}^{(n)}_{\omega_n}(\boldsymbol{y},\boldsymbol{r}') \\
\mathcal{F}^{+}_{\omega_n}(\boldsymbol{r},\boldsymbol{r}') = {} & -\int \mathrm{d}\boldsymbol{x}\mathcal{G}^{(n)}_{-\omega_n}(\boldsymbol{x},\boldsymbol{r})\Delta^*(\boldsymbol{x})\mathcal{G}^{(n)}_{\omega_n}(\boldsymbol{x},\boldsymbol{r}') \\
& + \int \mathrm{d}\boldsymbol{x}\int \mathrm{d}\boldsymbol{y}\int \mathrm{d}\boldsymbol{z}\,\mathcal{G}^{(n)}_{-\omega_n}(\boldsymbol{x},\boldsymbol{r})\Delta^*(\boldsymbol{x})\mathcal{G}^{(n)}_{\omega_n}(\boldsymbol{x},\boldsymbol{y})\Delta(\boldsymbol{y}) \\
& \times \mathcal{G}^{(n)}_{-\omega_n}(\boldsymbol{z},\boldsymbol{y})\Delta^*(\boldsymbol{z})\mathcal{G}^{(n)}_{\omega_n}(\boldsymbol{z},\boldsymbol{r}').
\end{aligned} \tag{2.215}$$

Using this expansion, we obtain the following equation for the order parameter:

$$\begin{aligned}
\Delta^*(\boldsymbol{r}) = {} & 2E_pT\sum_{\omega_n}\int \mathrm{d}\boldsymbol{x}\,\mathcal{G}^{(n)}_{-\omega_n}(\boldsymbol{x},\boldsymbol{r})\Delta^*(\boldsymbol{x})\mathcal{G}^{(n)}_{\omega_n}(\boldsymbol{x},\boldsymbol{r}) \\
& - 2E_pT\sum_{\omega_n}\int \mathrm{d}\boldsymbol{x}\int \mathrm{d}\boldsymbol{y}\int \mathrm{d}\boldsymbol{z}\,\mathcal{G}^{(n)}_{-\omega_n}(\boldsymbol{x},\boldsymbol{r})\Delta^*(\boldsymbol{x})\mathcal{G}^{(n)}_{\omega_n}(\boldsymbol{x},\boldsymbol{y})\Delta(\boldsymbol{y}) \\
& \times \mathcal{G}^{(n)}_{-\omega_n}(\boldsymbol{z},\boldsymbol{y})\Delta^*(\boldsymbol{z})\mathcal{G}^{(n)}_{\omega_n}(\boldsymbol{z},\boldsymbol{r}).
\end{aligned} \tag{2.216}$$

Let us consider the kernel in the first term

$$K(\boldsymbol{x}-\boldsymbol{r}) = T\sum_{\omega_n} \mathcal{G}^{(n)}_{-\omega_n}(\boldsymbol{x},\boldsymbol{r})\mathcal{G}^{(n)}_{\omega_n}(\boldsymbol{x},\boldsymbol{r}) = K^{(0)}(\boldsymbol{x}-\boldsymbol{r})\mathrm{e}^{-2\mathrm{i}e\boldsymbol{A}(\boldsymbol{r})\cdot(\boldsymbol{x}-\boldsymbol{r})} \quad (2.217)$$

where

$$\begin{aligned} K^{(0)}(\boldsymbol{x}-\boldsymbol{r}) &= T\frac{m^2}{(2\pi\rho)^2}\sum_{\omega_n}\exp[-2|\omega_n|\rho/v_{\mathrm{F}}] \\ &= \frac{m^2 T}{(2\pi\rho)^2\sinh(2\pi T\rho/v_{\mathrm{F}})} \end{aligned} \quad (2.218)$$

and $\rho = |\boldsymbol{x}-\boldsymbol{r}|$. The kernel $K^{(0)}(\rho)$ has a characteristic radius about ξ_0, while $\Delta(\boldsymbol{x})$ changes over a much larger distance of the order of $\xi(T) \gg \xi_0$ near T_{c}. Therefore in the integral over $\boldsymbol{x}$ we can expand all quantities near the point $\boldsymbol{x} = \boldsymbol{r}$ up to the second order in ρ inclusive:

$$\begin{aligned} &\int \mathrm{d}\boldsymbol{x}\, K^{(0)}(\boldsymbol{x}-\boldsymbol{r})\mathrm{e}^{-2\mathrm{i}e\boldsymbol{A}(\boldsymbol{r})\cdot(\boldsymbol{x}-\boldsymbol{r})}\Delta^*(\boldsymbol{x}) \\ &\quad\approx \Delta^*(\boldsymbol{r})\int \mathrm{d}\boldsymbol{\rho}\, K^{(0)}(\rho) + \int \mathrm{d}\boldsymbol{\rho}\, K^{(0)}(\rho)\boldsymbol{\rho}\cdot[\nabla_{\boldsymbol{r}} - 2\mathrm{i}e\boldsymbol{A}(\boldsymbol{r})]\Delta^*(\boldsymbol{r}) \\ &\quad\quad + \frac{1}{2}\sum_{i,j=1}^{3}\int \mathrm{d}\boldsymbol{\rho}\, K^{(0)}(\rho)\rho_i\rho_j\left(\frac{\partial}{\partial r_i} - 2\mathrm{i}eA_i\right)\left(\frac{\partial}{\partial r_j} - 2\mathrm{i}eA_j\right)\Delta^*(\boldsymbol{r}). \end{aligned} \quad (2.219)$$

Here the second term is zero because the function under the integral is odd. The remaining integrals are:

$$\begin{aligned} \int \mathrm{d}\boldsymbol{\rho}\, K^{(0)}(\rho) &= T\sum_{\omega_n}\int \mathrm{d}\boldsymbol{\rho}\, \mathcal{G}^{(0)}_{-\omega_n}(\boldsymbol{\rho})\mathcal{G}^{(0)}_{\omega_n}(\boldsymbol{\rho}) \\ &= T\sum_{\boldsymbol{k}}\sum_{\omega_n}\frac{1}{\omega_n^2+\xi_{\boldsymbol{k}}^2} = N(E_{\mathrm{F}})\int_0^{\omega_{\mathrm{D}}}\mathrm{d}\xi\,\frac{\tanh(\xi/2T)}{\xi} \end{aligned} \quad (2.220)$$

and

$$\begin{aligned} \int \mathrm{d}\boldsymbol{\rho}\, K^{(0)}(\rho)\rho_i\rho_j &= \frac{\delta_{ij}}{3}\int \mathrm{d}\boldsymbol{\rho}\, K^{(0)}(\rho)\rho^2 \\ &= \frac{\delta_{ij}}{3}\frac{v_{\mathrm{F}}^2}{4(\pi T)^2}N(E_{\mathrm{F}})\int_0^{\infty}\mathrm{d}x\,\frac{x^2}{\sinh x} = \frac{\delta_{ij}}{3}\frac{7\zeta(3)v_{\mathrm{F}}^2}{8(\pi T)^2}N(E_{\mathrm{F}}). \end{aligned} \quad (2.221)$$

We cut the divergent integral in equation (2.220) at $|\xi| = \omega_{\mathrm{D}}$ as usual and introduce the DOS at the Fermi level: $N(E_{\mathrm{F}}) = mk_{\mathrm{F}}/(2\pi^2)$. The second term on the right-hand side of equation (2.216) is cubic in $\Delta(\boldsymbol{r})$. Here we can neglect the space variations of $\Delta(\boldsymbol{x})$ and the phase due to the vector potential under the integrals. Then applying the Fourier transform of the zero-field ideal gas GF, we obtain the following expression for this term:

$$-2E_pT\Delta^*(\boldsymbol{r})|\Delta(\boldsymbol{r})|^2\sum_{\boldsymbol{k}}\sum_{\omega_n}\frac{1}{(\omega_n^2+\xi_{\boldsymbol{k}}^2)^2}=-\lambda\frac{7\zeta(3)}{8(\pi T)^2}\Delta^*(\boldsymbol{r})|\Delta(\boldsymbol{r})|^2. \tag{2.222}$$

We note that T_{c} is determined from the linearized BCS equation:

$$1=\lambda\int_0^{\omega_{\mathrm{D}}}\mathrm{d}\xi\,\frac{\tanh(\xi/2T_{\mathrm{c}})}{\xi}.$$

Then close to T_{c}, we can expand it as

$$\int_0^{\omega_{\mathrm{D}}}\mathrm{d}\xi\,\frac{\tanh(\xi/2T)}{\xi}\approx\frac{1}{\lambda}+\frac{T_{\mathrm{c}}-T}{T_{\mathrm{c}}} \tag{2.223}$$

where we use the integral, equation (B.28) from the appendix. Introducing the 'condensate wavefunction' $\phi(\boldsymbol{r})$ as

$$\phi(\boldsymbol{r})=\Delta(\boldsymbol{r})\sqrt{\frac{7\zeta(3)n_{\mathrm{e}}}{8\pi^2T_{\mathrm{c}}^2}} \tag{2.224}$$

and collecting all integrals in equation (2.216), we obtain the phenomenological Ginzburg–Landau equation (section 1.6.1):

$$-\frac{1}{4m}[\nabla+\mathrm{i}2e\boldsymbol{A}(\boldsymbol{r})]^2\phi(\boldsymbol{r})+\beta|\phi(\mathbf{r})|^2\phi(\boldsymbol{r})=-\alpha\phi(\boldsymbol{r}).$$

Now the coefficients α and β are determined microscopically as

$$\alpha=\frac{12\pi^2T_{\mathrm{c}}}{7\zeta(3)mv_{\mathrm{F}}^2}(T-T_{\mathrm{c}}) \tag{2.225}$$

$$\beta=\frac{12\pi^2T_{\mathrm{c}}^2}{7\zeta(3)mv_{\mathrm{F}}^2n_{\mathrm{e}}}.$$

Here $n_{\mathrm{e}}=k_{\mathrm{F}}^3/(3\pi^2)$ is the electron density. Finally let us show that the second GL equation for supercurrent follows from the expansion of GFs (2.215) in powers of $\Delta(\boldsymbol{r})$. The current–density operator can be expressed in terms of the field operators:

$$\hat{\jmath}(\boldsymbol{r})=\sum_s\left[\frac{ie}{2m}(\nabla_{\boldsymbol{r}}-\nabla_{\boldsymbol{r}'})|_{\boldsymbol{r}'\to\boldsymbol{r}}\Psi_s^\dagger(\boldsymbol{r}')\Psi_s(\boldsymbol{r})-\frac{e^2}{m}\boldsymbol{A}(\boldsymbol{r})\Psi_s^\dagger(\boldsymbol{r})\Psi_s(\boldsymbol{r})\right]. \tag{2.226}$$

Its expectation value is

$$j(r) = \frac{ie}{m}(\nabla_r - \nabla_{r'})|_{r' \to r} T \sum_{\omega_n} \mathcal{G}_{\omega_n}(r, r') - \frac{e^2 n_e}{m} A(r). \tag{2.227}$$

There is no current in the normal state in the stationary magnetic field, and the first (normal) term in the expansion of GFs (equation (2.215)) cancels the second term on the right-hand side of equation (2.227). The remaining supercurrent is quadratic with respect to the order parameter:

$$\begin{aligned} j(r) = {} & \frac{ie}{m}(\nabla_{r'} - \nabla_r)|_{r' \to r} T \\ & \times \sum_{\omega_n} \int dx \int dy\, \mathcal{G}^{(n)}_{\omega_n}(r, x)\Delta(x)\mathcal{G}^{(n)}_{-\omega_n}(y, x)\Delta^*(y)\mathcal{G}^{(n)}_{\omega_n}(y, r'). \end{aligned} \tag{2.228}$$

The normal state GF in the magnetic field $\mathcal{G}^{(n)}_{\omega_n}(r, r') = \exp(-ieA \cdot \rho)\mathcal{G}^{(0)}_{\omega_n}(\rho)$ is the product of a slowly varying phase exponent and the zero-field GF, which oscillates in real space with the electron wavelength ($\approx 1/k_F$). Differentiating the oscillating part yields

$$\begin{aligned} j(r) = {} & \frac{ie}{m} T \sum_{\omega_n} \int dx \int dy\, \Delta(x)\Delta^*(y) e^{2iA\cdot(x-y)} \mathcal{G}^{(0)}_{-\omega_n}(|y - x|) \\ & \times [\mathcal{G}^{(0)}_{\omega_n}(|r - x|)\nabla_r \mathcal{G}^{(0)}_{\omega_n}(|y - r|) - \mathcal{G}^{(0)}_{\omega_n}(|y - r|)\nabla_r \mathcal{G}^{(0)}_{\omega_n}(|r - x|)]. \end{aligned} \tag{2.229}$$

Expanding the slowly varying order parameter and the magnetic phase exponent in powers of $|y - r|$ and $|x - r|$, which are of the order of ξ_0,

$$\begin{aligned} \Delta(x) &\approx \Delta(r) + (x - r) \cdot \nabla\Delta(r) \\ \Delta^*(y) &\approx \Delta(r) + (y - r) \cdot \nabla\Delta^*(r) \\ \exp[2ieA \cdot (x - y)] &\approx 1 + 2ieA \cdot (x - y) \end{aligned}$$

we obtain

$$j(r) = \frac{ie}{m} C[\Delta^*(r)\nabla\Delta(r) - \Delta(r)\nabla\Delta^*(r)] - \frac{4e^2 C}{m}|\Delta(r)|^2 A(r). \tag{2.230}$$

Here the constant C is given by

$$C = \frac{T}{3} \sum_{\omega_n} \int dx \int dy\, \mathcal{G}^{(0)}_{-\omega_n}(|y - x|)\mathcal{G}^{(0)}_{\omega_n}(|r - x|)[(x - y) \cdot \nabla_r]\mathcal{G}^{(0)}_{\omega_n}(|y - r|). \tag{2.231}$$

Using the Fourier transform of the ideal gas GF in calculating the integrals over $\boldsymbol{x}$ and $\boldsymbol{y}$ in C leads to

$$C = \frac{T}{3}\sum_{\omega_n}\sum_{\boldsymbol{k}} \frac{1}{\mathrm{i}\omega_n + \xi_{\boldsymbol{k}}} \frac{\partial}{\partial \boldsymbol{k}} \cdot \boldsymbol{k} \frac{1}{(\mathrm{i}\omega_n - \xi_{\boldsymbol{k}})^2}$$

$$= \frac{T}{3m}\sum_{\omega_n}\sum_{\boldsymbol{k}} \frac{k^2}{(\mathrm{i}\omega_n + \xi_{\boldsymbol{k}})(\mathrm{i}\omega_n - \xi_{\boldsymbol{k}})^3}. \tag{2.232}$$

Integrating with respect to $\boldsymbol{k}$ yields

$$C = \frac{2k_{\mathrm{F}}^2}{3m} N(E_{\mathrm{F}}) T \sum_{\omega_n} \int_{-\infty}^{\infty} \mathrm{d}\xi \, \frac{\omega_n^2 - \xi^2}{(\omega_n^2 + \xi^2)^3} \tag{2.233}$$

and we obtain

$$C = \frac{k_{\mathrm{F}}^2 \pi}{6m} N(E_{\mathrm{F}}) T \sum_{\omega_n} \frac{1}{\omega_n^3} = \frac{7\zeta(3) n_{\mathrm{e}}}{16\pi^2 T_{\mathrm{c}}^2}.$$

Replacing the order parameter in equation (2.230) by the 'condensate wavefunction' leads to the second GL equation (see equation (1.40)):

$$\boldsymbol{j}(\boldsymbol{r}) = \frac{\mathrm{i}e^*}{2m^{**}}[\phi^*(\boldsymbol{r})\nabla\phi(\boldsymbol{r}) - \phi(\boldsymbol{r})\nabla\phi^*(\boldsymbol{r})] - \frac{e^{*2}}{m^{**}}\boldsymbol{A}(\boldsymbol{r})|\phi(\mathbf{r})|^2 \tag{2.234}$$

where $e^* = 2e$ and $m^{**} = 2m$ are the charge and mass of the Cooper pair, respectively.

The BCS theory is a mean-field approximation, which is valid if the volume occupied by the correlated electron pair is large compared with the volume per electron. As an example, in *aluminium* the size of the pair (i.e. the coherence length at $T = 0$) is about ten thousand times larger than the distance between electrons. The Cooper pairs in the BCS superconductor disappear above T_{c} and the one-particle excitations are fermions. When the coupling constant λ increases, the critical temperature increases and the coherence length becomes smaller. Hence, one can expect some deviations from the BCS behaviour in the intermediate coupling regime. At the first stage, deviations from BCS theory arise in two ways: (1) the BCS approximation for the interaction between electrons does not provide an adequate representation of the retarded nature of the phonon induced attraction; and (2) the damping rate becomes comparable with the quasi-particle energy. Both the retardation effect and the damping are taken into account in the Eliashberg extension of the BCS theory to the *intermediate* coupling regime [36], which is discussed in the next chapter. However, if λ is larger than 1, the Fermi liquid becomes unstable even in the normal state above T_{c} because of the *polaron collapse* of the electron band (chapter 4). Here we encounter qualitatively different physics [11].

Chapter 3

Intermediate-coupling theory

3.1 Electron–phonon interaction

The attraction between electrons in BCS theory is the result of an 'overscreening' of their Coulomb repulsion by vibrating ions. When the interaction between ion vibrations and electrons (i.e. the electron–phonon interaction) is strong, the electron Bloch states are affected even in the normal phase. Phonons are also affected by conduction electrons. In doped insulators (like high-temperature superconductors), 'bare' phonons are well defined in insulating parent compounds but a separation of electron and phonon degrees of freedom might be a problem in a metal. Here we have to start with the first-principle Hamiltonian describing conduction electrons and ions coupled by the Coulomb forces:

$$H = -\sum_i \frac{\nabla_i^2}{2m_e} + \frac{e^2}{2}\sum_{i\neq i'} \frac{1}{|\boldsymbol{r}_i - \boldsymbol{r}_{i'}|} - Ze^2 \sum_{ij} \frac{1}{|\boldsymbol{r}_i - \boldsymbol{R}_j|} + \frac{Z^2e^2}{2}\sum_{j\neq j'} \frac{1}{|\boldsymbol{R}_j - \boldsymbol{R}_{j'}|} - \sum_j \frac{\nabla_j^2}{2M} \tag{3.1}$$

where $\boldsymbol{r}_i$, $\boldsymbol{R}_j$ are the electron and ion coordinates, respectively, $i = 1, \ldots, N_e$; $j = 1, \ldots, N$; $\nabla_i = \partial/\partial\boldsymbol{r}_i$, $\nabla_j = \partial/\partial\boldsymbol{R}_j$, Ze is the ion charge and M is the ion mass. The system is neutral and $N_e = ZN$. The inner electrons are strongly coupled to the nuclei and follow their motion. Hence, the ions can be considered as rigid charges. To account for their high-energy electron degrees of freedom we can replace the elementary charge in equation (3.1) by $e/\sqrt{\epsilon_\infty}$, where ϵ_∞ is the phenomenological high-frequency dielectric constant. One cannot solve the corresponding Schrödinger equation perturbatively because the Coulomb interaction is strong. The ratio of the characteristic Coulomb energy to the kinetic energy is $r_s = m_e e^2/(4\pi n_e/3)^{1/3} \approx 1$ for the electron density $n_e = ZN/V = 10^{23}$ cm^{-3} (further we take the volume of the system as $V = 1$, unless specified otherwise). However, we can take advantage of the small value of the adiabatic ratio $m_e/M < 10^{-3}$. The ions are heavy and the amplitudes $\langle|\boldsymbol{u}|\rangle \simeq \sqrt{1/M\omega_D}$ of their vibrations near the equilibrium $\boldsymbol{R}_0 \equiv \boldsymbol{l}$ are much smaller than the lattice constant $a = N^{-1/3}$:

$$\frac{\langle|\boldsymbol{u}|\rangle}{a} \approx \left(\frac{m_e}{Mr_s}\right)^{1/4} \ll 1. \tag{3.2}$$

In this estimate we take the characteristic vibration frequency ω_D of the order of the ion plasma frequency $\omega_i = \sqrt{4\pi NZ^2e^2/M}$. Because the vibration amplitudes are small we can expand the Hamiltonian in powers of $|\boldsymbol{u}|$ up to quadratic terms inclusive. Any further progress requires a simplifying physical idea, which is to approach the ground state of the many-electron system via a one-electron picture. This is called the local density approximation (LDA), which replaces the Coulomb electron–electron interaction by an effective one-body potential $\mathcal{V}(\boldsymbol{r})$:

$$\mathcal{V}(\boldsymbol{r}) = -Ze^2\sum_j \frac{1}{|\boldsymbol{r}-\boldsymbol{R}_j|} + e^2\int \mathrm{d}\boldsymbol{r}' \frac{n(\boldsymbol{r}')}{|\boldsymbol{r}-\boldsymbol{r}'|} + \mu_{ex}[n(\boldsymbol{r})] \tag{3.3}$$

where $\mu_{ex}[n(\boldsymbol{r})]$ is the exchange interaction, usually calculated numerically or expressed as $\mu_{ex}[n(\boldsymbol{r})] = -\beta n^{1/3}(\boldsymbol{r})$ with the constant β in a simple approximation. $\mathcal{V}(\boldsymbol{r})$ is the functional of the electron density $n(\boldsymbol{r}) = \sum_s\langle\Psi_s^\dagger(\boldsymbol{r})\Psi_s(\boldsymbol{r})\rangle$. As a result, the Hamiltonian takes the form in the second quantization:

$$H = H_e + H_{ph} + H_{e-ph} + H_{e-e} \tag{3.4}$$

where

$$H_e = \sum_s \int \mathrm{d}\boldsymbol{r}\, \Psi_s^\dagger(\boldsymbol{r})\left[-\frac{\nabla^2}{2m_e} + V(\boldsymbol{r})\right]\Psi_s(\boldsymbol{r}) \tag{3.5}$$

is the electron energy in a periodic crystal field $V(\boldsymbol{r}) = \sum_l v(\boldsymbol{r}-\boldsymbol{l})$ which is $\mathcal{V}(\boldsymbol{r})$ calculated at $\boldsymbol{R}_j = \boldsymbol{l}$ and with the periodic electron density $n^{(0)}(\boldsymbol{r}+\boldsymbol{l}) = n^{(0)}(\boldsymbol{r})$,

$$H_{ph} = \sum_l \left[-\frac{\nabla_u^2}{2M} + \boldsymbol{u}_l \cdot \frac{\partial}{\partial \boldsymbol{l}}\int \mathrm{d}\boldsymbol{r}\, n^{(0)}(\boldsymbol{r})V(\boldsymbol{r})\right] + \frac{1}{2}\sum_{\boldsymbol{l},\boldsymbol{m},\alpha,\beta} u_{\boldsymbol{l}\alpha}u_{\boldsymbol{m}\beta}D_{\alpha\beta}(\boldsymbol{l}-\boldsymbol{m}) \tag{3.6}$$

is the vibration energy. Here $\alpha, \beta = x, y, z$ and

$$D_{\alpha\beta}(\boldsymbol{l}-\boldsymbol{m}) = \frac{\partial^2}{\partial l_\alpha \partial m_\beta}\left[\frac{Z^2e^2}{2}\sum_{\boldsymbol{l}'\neq\boldsymbol{m}'}\frac{1}{|\boldsymbol{l}'-\boldsymbol{m}'|} + \int \mathrm{d}\boldsymbol{r}\, n^{(0)}(\boldsymbol{r})V(\boldsymbol{r})\right] \tag{3.7}$$

is a dynamic matrix. The electron–phonon interaction is given by

$$H_{e-ph} = \sum_l \boldsymbol{u}_l \cdot \frac{\partial}{\partial \boldsymbol{l}}\int \mathrm{d}\boldsymbol{r}\left[\sum_s \Psi_s^\dagger(\boldsymbol{r})\Psi_s(\boldsymbol{r}) - n^{(0)}(\boldsymbol{r})\right]V(\boldsymbol{r})$$

$$+\frac{1}{2}\sum_{l,m,\alpha,\beta} u_{l\alpha}u_{m\beta}\frac{\partial^2}{\partial l_\alpha \partial m_\beta}\int \mathrm{d}\boldsymbol{r}\left[\sum_s \Psi_s^\dagger(\boldsymbol{r})\Psi_s(\boldsymbol{r}) - n^{(0)}(\boldsymbol{r})\right]V(\boldsymbol{r}) \tag{3.8}$$

and the electron–electron correlations are described by

$$H_{\text{e-e}} = \frac{1}{2}\int \mathrm{d}\boldsymbol{r}\int \mathrm{d}\boldsymbol{r}'\frac{e^2}{|\boldsymbol{r}-\boldsymbol{r}'|}\left[\sum_{ss'}\Psi_s^\dagger(\boldsymbol{r})\Psi_{s'}^\dagger(\boldsymbol{r}')\Psi_{s'}'(\boldsymbol{r}')\Psi_s(\boldsymbol{r})\right]$$
$$-\int \mathrm{d}\boldsymbol{r}\left[\int \mathrm{d}\boldsymbol{r}'\frac{e^2 n^{(0)}(\boldsymbol{r}')}{|\boldsymbol{r}-\boldsymbol{r}'|}+\mu_{\text{ex}}[n^{(0)}(\boldsymbol{r})]\right]\sum_s \Psi_s^\dagger(\boldsymbol{r})\Psi_s(\boldsymbol{r})$$
$$+\frac{Z^2e^2}{2}\sum_{l\neq m}\frac{1}{|\boldsymbol{l}-\boldsymbol{m}|}. \tag{3.9}$$

We include the electrostatic repulsive energy of nuclei in $H_{\text{e-e}}$, so that the average of $H_{\text{e-e}}$ is zero in the Hartree approximation.

The vibration Hamiltonian H_{ph} is a quadratic form and, therefore, can be diagonalized with a linear canonical transformation for the displacement operators

$$\boldsymbol{u}_l = \sum_{\boldsymbol{q},\nu}\frac{\boldsymbol{e}_{\boldsymbol{q}\nu}}{\sqrt{2NM\omega_{\boldsymbol{q}\nu}}}d_{\boldsymbol{q}\nu}\exp(\mathrm{i}\boldsymbol{q}\cdot\boldsymbol{l}) + H.c. \tag{3.10}$$
$$\frac{\partial}{\partial \boldsymbol{u}_l} = \sum_{\boldsymbol{q},\nu}\boldsymbol{e}_{\boldsymbol{q}\nu}\sqrt{\frac{M\omega_{\boldsymbol{q}\nu}}{2N}}d_{\boldsymbol{q}\nu}\exp(\mathrm{i}\boldsymbol{q}\cdot\boldsymbol{l}) - H.c.$$

where $\boldsymbol{q}$ is the phonon momentum, $d_{\boldsymbol{q}\nu}$ is the phonon (Bose) annihilation operator, $\boldsymbol{e}_{\boldsymbol{q}\nu}$ and $\omega_{\boldsymbol{q}\nu}$ are the unit polarization vector and the phonon frequency, respectively, of the phonon mode ν. Then H_{ph} takes the following form

$$H_{\text{ph}} = \sum_{\boldsymbol{q},\nu}\omega_{\boldsymbol{q}\nu}(d_{\boldsymbol{q}\nu}^\dagger d_{\boldsymbol{q}\nu} + 1/2) \tag{3.11}$$

if the eigenfrequencies $\omega_{\boldsymbol{q}\nu}$ and the eigenstates $\boldsymbol{e}_{\boldsymbol{q}\nu}$ satisfy

$$M\omega_{\boldsymbol{q}\nu}^2 e_{\boldsymbol{q}\nu}^\alpha = \sum_\beta D_{\boldsymbol{q}}^{\alpha\beta}e_{\boldsymbol{q}\nu}^\beta \tag{3.12}$$

and

$$\sum_{\boldsymbol{q}} e_{\boldsymbol{q}\nu}^{*\alpha}e_{\boldsymbol{q}\nu}^\beta = N\delta_{\alpha\beta}. \tag{3.13}$$

The last equation and the bosonic commutation rules $[d_{\boldsymbol{q}\nu}d_{\boldsymbol{q}'\nu'}^\dagger] = \delta_{\nu\nu'}\delta_{\boldsymbol{q}\boldsymbol{q}'}$ follow from $(\partial/\partial u_l^\alpha)u_l^\beta - u_l^\beta(\partial/\partial u_l^\alpha) = \delta_{\alpha\beta}$. Here

$$D_{\boldsymbol{q}}^{\alpha\beta} = \sum_{\boldsymbol{m}}\exp(\mathrm{i}\boldsymbol{q}\cdot\boldsymbol{m})D_{\alpha\beta}(\boldsymbol{m}) \tag{3.14}$$

is the Fourier transform of the second derivative of the ion potential energy. The first derivative in equation (3.6) is zero in crystals with a centre of symmetry. Different solutions of equation (3.12) are classified with the phonon branch (mode) quantum number ν, which is $1, 2, 3$ for a simple lattice and $1, \ldots, 3k$ for a lattice with k ions per unit cell.

The periodic part of the Hamiltonian H_e is diagonal in the Bloch representation (appendix A):

$$\Psi_s(\boldsymbol{r}) = \sum_{\boldsymbol{k},n,s} \psi_{n\boldsymbol{k}s}(\boldsymbol{r}) c_{n\boldsymbol{k}s} \tag{3.15}$$

where $c_{n\boldsymbol{k}s}$ are the fermion annihilation operators. The Bloch function obeys the Schrödinger equation

$$\left(-\frac{\nabla^2}{2m_e} + V(\boldsymbol{r})\right) \psi_{n\boldsymbol{k}s}(\boldsymbol{r}) = E_{n\boldsymbol{k}s}\psi_{n\boldsymbol{k}s}(\boldsymbol{r}). \tag{3.16}$$

One-particle states are sorted with the momentum $\boldsymbol{k}$ in the Brillouin zone, band index n and spin s. The solutions of this equation allow us to calculate the periodic electron density $n^{(0)}(\boldsymbol{r})$, which determines the crystal field potential $V(\boldsymbol{r})$. The LDA can explain the shape of the Fermi surface of wide-band metals and gaps in narrow-gap semiconductors. The spin-polarized version of LDA can explain a variety of properties of many magnetic materials. This is not the case in narrow d- and f-band metals and oxides (and other ionic lattices), where the electron–phonon interaction and Coulomb correlations are strong. These materials display much less band dispersion and wider gaps compared with the first-principle band structure calculations. Using the phonon and electron annihilation and creation operators, the Hamiltonian is written as

$$H = H_0 + H_{e-ph} + H_{e-e} \tag{3.17}$$

where

$$H_0 = \sum_{\boldsymbol{k},n,s} \xi_{n\boldsymbol{k}s} c^{\dagger}_{n\boldsymbol{k}s} c_{n\boldsymbol{k}s} + \sum_{\boldsymbol{q},\nu} \omega_{\boldsymbol{q}\nu}(d^{\dagger}_{\boldsymbol{q}\nu} d_{\boldsymbol{q}\nu} + 1/2) \tag{3.18}$$

describes independent Bloch electrons and phonons, $\xi_{n\boldsymbol{k}s} = E_{n\boldsymbol{k}s} - \mu$ is the band energy spectrum with respect to the chemical potential. The part of the electron–phonon interaction, which is linear in phonon operators, can be written as

$$H_{e-ph} = \frac{1}{\sqrt{2N}} \sum_{\boldsymbol{k},\boldsymbol{q},n,n',\nu,s} \gamma_{nn'}(\boldsymbol{q},\boldsymbol{k},\nu)\omega_{\boldsymbol{q}\nu} c^{\dagger}_{n\boldsymbol{k}s} c_{n\boldsymbol{k}-\boldsymbol{q}s} d_{\boldsymbol{q}\nu} + H.c. \tag{3.19}$$

where

$$\gamma_{nn'}(\boldsymbol{q},\boldsymbol{k},\nu) = -\frac{N}{M^{1/2}\omega^{3/2}_{\boldsymbol{q}\nu}} \int \mathrm{d}\boldsymbol{r}\, (\boldsymbol{e}_{\boldsymbol{q}\nu} \cdot \nabla v(\boldsymbol{r}))\psi^*_{n\boldsymbol{k}s}(\boldsymbol{r})\psi_{n'\boldsymbol{k}-\boldsymbol{q}s}(\boldsymbol{r}) \tag{3.20}$$

is the dimensionless matrix element. Low-energy physics is often described by a single-band approximation with the matrix element $\gamma_{nn}(\boldsymbol{q}, \boldsymbol{k}, \nu)$ depending only on the momentum transfer $\boldsymbol{q}$ (the Fröhlich interaction):

$$\gamma_{nn'}(\boldsymbol{q}, \boldsymbol{k}, \nu) = \gamma(\boldsymbol{q}, \nu). \tag{3.21}$$

The terms of $H_{\text{e-ph}}$ which are quadratic and higher orders in the phonon operators are small. They have a role to play only for those phonons which are not coupled with electrons by the linear interaction (3.20). The electron–electron correlation energy of a homogeneous electron system is often written as

$$H_{\text{e-e}} = \frac{1}{2} \sum_{\boldsymbol{q}} V_{\text{c}}(\boldsymbol{q}) \rho_{\boldsymbol{q}}^{\dagger} \rho_{\boldsymbol{q}} \tag{3.22}$$

where $V_{\text{c}}(\boldsymbol{q})$ is a matrix element, which is zero for $\boldsymbol{q} = 0$ because of electroneutrality and

$$\rho_{\boldsymbol{q}}^{\dagger} = \sum_{\boldsymbol{k},s} c_{\boldsymbol{k}s}^{\dagger} c_{\boldsymbol{k}+\boldsymbol{q}s} \tag{3.23}$$

is the density fluctuation operator. H_0 should also include a random potential in doped semiconductors and amorphous metals.

3.2 Phonons in metal

In wide-band metals such as Na or K, the correlation energy is relatively small ($r_s \leq 1$) and carriers are almost free. Core electrons together with nuclei form compact ions with an effective Z. The carrier wavefunction outside the core can be approximated by a plane wave

$$\psi_{n\boldsymbol{k}s}(\boldsymbol{r}) \simeq \mathrm{e}^{\mathrm{i}\boldsymbol{k}\cdot\boldsymbol{r}} \tag{3.24}$$

and the carrier density $n^{(0)}(\boldsymbol{r})$ is a constant. Therefore, the only relevant interaction in the dynamic matrix (equation (3.7)) is the Coulomb repulsion between ions, which yields

$$D_{\alpha\beta}(\boldsymbol{l} - \boldsymbol{m}) = \frac{Z^2 e^2}{2} \frac{\partial^2}{\partial l_\alpha \partial m_\beta} \sum_{\boldsymbol{l}' \neq \boldsymbol{m}'} \frac{1}{|\boldsymbol{l}' - \boldsymbol{m}'|}. \tag{3.25}$$

The electron–ion interaction is a pure Coulomb attraction $v(\boldsymbol{r}) = -Ze^2/r$, which is expanded in the Fourier series as

$$\frac{1}{r} = 4\pi \lim_{\kappa \to 0} \sum_{\boldsymbol{q}} \frac{1}{q^2 + \kappa^2} \mathrm{e}^{\mathrm{i}\boldsymbol{q}\cdot\boldsymbol{r}}. \tag{3.26}$$

Substituting this expansion into equations (3.25) and (3.20), we obtain

$$D_{\alpha\beta}(\boldsymbol{m}) = 4\pi \lim_{\kappa \to 0} \sum_{\boldsymbol{q}} \frac{q_\alpha q_\beta}{q^2 + \kappa^2} \cos(\boldsymbol{q} \cdot \boldsymbol{m}) \tag{3.27}$$

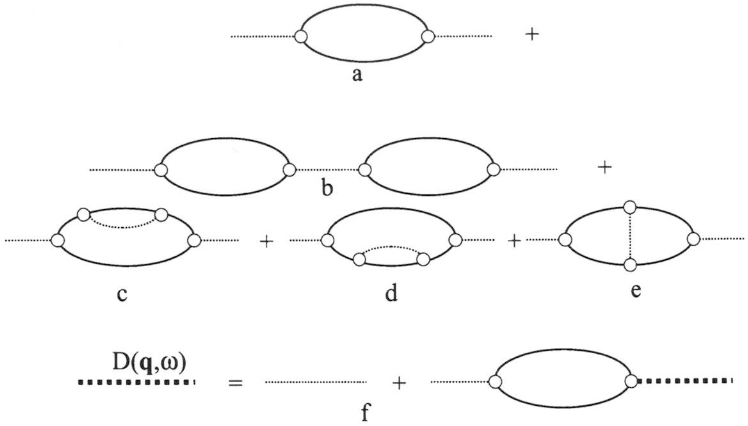

Figure 3.1. Second (*a*) and fourth order (*b*), (*c*), (*d*), (*e*) corrections to the phonon GF. The phonon GF in the Migdal approximation (*f*).

and

$$\gamma(\boldsymbol{q},\nu) = \mathrm{i}\frac{4\pi N Z e^2}{\sqrt{M\omega_{\boldsymbol{q}}^3}} \lim_{\kappa\to 0} \frac{\boldsymbol{e}_{\boldsymbol{q}\nu}\cdot\boldsymbol{q}}{q^2+\kappa^2}. \tag{3.28}$$

Calculating the Fourier transform of equation (3.27), one obtains the following equation for phonon frequencies and polarization vectors:

$$\omega_{\boldsymbol{q}}^2 \boldsymbol{e}_{\boldsymbol{q}\nu} = \omega_i^2 \boldsymbol{q}\frac{\boldsymbol{e}_{\boldsymbol{q}\nu}\cdot\boldsymbol{q}}{q^2}. \tag{3.29}$$

A longitudinal mode with $\boldsymbol{e} \parallel \boldsymbol{q}$ is the ion plasmon

$$\omega_{\boldsymbol{q}} = \omega_i \tag{3.30}$$

and two shear (transverse) modes with $\boldsymbol{e} \perp \boldsymbol{q}$ have zero frequencies, which is the result of our approximation considering ions as rigid charges. In fact, core electrons undergo a polarization, when ions are displaced from their equilibrium positions, which yields a finite shear mode frequency.

According to equation (3.28), the carriers interact only with longitudinal phonons. The interaction gives rise to a significant renormalization of the bare phonon frequency (equation (3.30)). We apply the Green's function (GF) formalism (appendix D) to calculate the renormalized phonon frequency. The Fourier component of the free-electron GF is given by equation (2.163) with $\xi_{\boldsymbol{k}} = k^2/2m_{\mathrm{e}} - \mu$. For interacting electrons, the electron self-energy is introduced as

$$\Sigma(\boldsymbol{k},\omega) = [G^{(0)}(\boldsymbol{k},\omega)]^{-1} - [G(\boldsymbol{k},\omega)]^{-1} \tag{3.31}$$

and

$$G(\boldsymbol{k},\omega) = \frac{1}{\omega - \xi_{\boldsymbol{k}} - \Sigma(\boldsymbol{k},\omega)}. \tag{3.32}$$

The phonon GF is defined as

$$D(\boldsymbol{q}, t) = -\mathrm{i}\frac{\omega_{\boldsymbol{q}}}{2}\langle T_t(d_{\boldsymbol{q}}(t)d^{\dagger}_{\boldsymbol{q}} + d^{\dagger}_{\boldsymbol{q}}(t)d_{\boldsymbol{q}})\rangle, \tag{3.33}$$

and its Fourier transform for free phonons is a dimensionless even function of frequency,

$$D^{(0)}(\boldsymbol{q}, \omega) = \frac{\omega^2_{\boldsymbol{q}}}{\omega^2 - \omega^2_{\boldsymbol{q}} + \mathrm{i}\delta}. \tag{3.34}$$

The phonon self-energy is

$$\Pi(\boldsymbol{q}, \omega) = [D^{(0)}(\boldsymbol{q}, \omega)]^{-1} - [D(\boldsymbol{q}, \omega)]^{-1}. \tag{3.35}$$

The Feynman diagram technique is convenient, see figure 3.1. Thin straight and dotted lines correspond to $G^{(0)}$ and $D^{(0)}$, respectively, a vertex (circle) corresponds to the interaction matrix element $\gamma(\boldsymbol{q})\sqrt{\omega_{\boldsymbol{q}}/N}$ and bold lines represent G and D. The Fröhlich interaction is the sum of two operators describing the emission and absorption of a phonon. Both events are taken into account in the definition of D. Therefore wavy lines have no direction. There are no first- or higher-*odd* orders corrections to D because the Fröhlich interaction is off-diagonal with respect to phonon occupation numbers. The second-order term in D (figure 3.1(*a*)) includes the so-called polarization bubble, $\Pi^{(0)}_{\mathrm{e}}$, which is a convolution of two $G^{(0)}$. Among different fourth-order diagrams the diagram in figure 3.1(*b*) with two polarization loops is the most 'dangerous' one. Differing from others, it is proportional to $1/q^2$, which is large for small q. However, the singularity of internal vertices is 'integrated out' in the diagrams in figures 3.1(*c*)–(*d*). The sum of all dangerous diagrams is given in figure 3.1(*f*) which is

$$\Pi(\boldsymbol{q}, \omega) = \frac{|\gamma(\boldsymbol{q})|^2\omega_{\boldsymbol{q}}}{N}\Pi_{\mathrm{e}}(\boldsymbol{q}, \omega) \tag{3.36}$$

where

$$\Pi_{\mathrm{e}}(\boldsymbol{q}, \omega) = \Pi^{(0)}_{\mathrm{e}}(\boldsymbol{q}, \omega) \tag{3.37}$$

and

$$\Pi^{(0)}_{\mathrm{e}}(\boldsymbol{q}, \omega) = -\frac{2\mathrm{i}}{(2\pi)^4}\int \mathrm{d}\boldsymbol{k}\,\mathrm{d}\epsilon\, G^{(0)}(\boldsymbol{k}+\boldsymbol{q}, \epsilon + \omega)G^{(0)}(\boldsymbol{k}, \epsilon). \tag{3.38}$$

The additional factor 2 in the phonon self-energy is due to a contribution of two electron spin states. It is convenient to integrate over frequency in equation (3.38) first with the following result:

$$\Pi^{(0)}_{\mathrm{e}}(\boldsymbol{q}, \omega) = \frac{1}{4\pi^3}\int \mathrm{d}\boldsymbol{k}\,\frac{\Theta(\xi_{\boldsymbol{k}}) - \Theta(\xi_{\boldsymbol{k}+\boldsymbol{q}})}{\omega + \xi_{\boldsymbol{k}} - \xi_{\boldsymbol{k}+\boldsymbol{q}} + \mathrm{i}\delta\,\mathrm{sign}(\xi_{\boldsymbol{k}+\boldsymbol{q}})}. \tag{3.39}$$

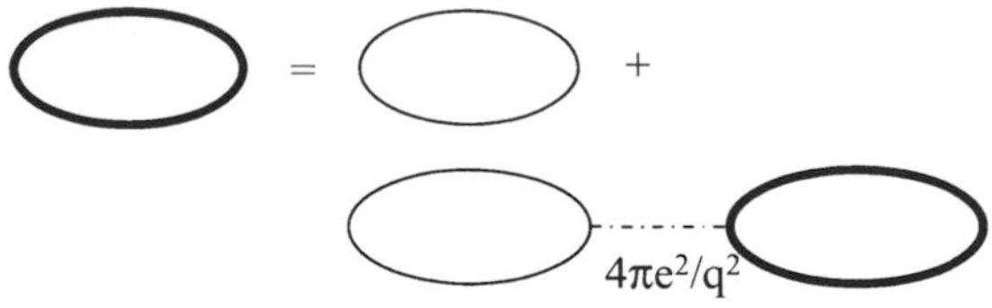

Figure 3.2. Screened polarization bubble $\Pi_e(\boldsymbol{q}, \omega)$.

The phonon frequency ω is small, $\omega \ll \mu$. Thus, we can take the limit $\omega \to 0$ in $\Pi_e^{(0)}$ and obtain

$$\mathrm{Re}\,\Pi_e^0(\boldsymbol{q}, \omega) = -\frac{m_e k_F}{2\pi^2} h\left(\frac{q}{2k_F}\right) \tag{3.40}$$

$$\mathrm{Im}\,\Pi_e^0(\boldsymbol{q}, \omega) = -\frac{m_e^2}{2\pi q}|\omega|\Theta(2k_F - q) \tag{3.41}$$

where

$$h(x) = 1 + \frac{1 - x^2}{2x} \ln\left|\frac{1 + x}{1 - x}\right|.$$

We should also take the Coulomb electron–electron interaction into account because the corresponding vertex is singular in the long-wavelength limit, $V_c(\boldsymbol{q}) = 4\pi e^2/q^2$. This leads to a drastic renormalization of the long wavelength behaviour of Π_e. In the 'bubble' or random phase approximation (RPA), we obtain figure 3.2 as in the case of the electron–ion plasmon interaction, figure 3.1(*f*), but with the Coulomb (dashed-dotted) line instead of the dotted phonon line. In the analytical form, we have

$$\Pi_e(\boldsymbol{q}, \omega) = \frac{\Pi_e^{(0)}(\boldsymbol{q}, \omega)}{1 - (4\pi e^2/q^2)\Pi_e^0(\boldsymbol{q}, \omega)}. \tag{3.42}$$

As a result, in the long-wavelength limit $q \ll q_s$ one obtains

$$\Pi_e(\boldsymbol{q}, \omega) = -\frac{m_e k_F}{\pi^2}\frac{q^2}{q_s^2} \tag{3.43}$$

where $q_s = \sqrt{4 m_e k_F e^2/\pi}$ is the inverse (Debye) screening radius.

While $\Pi_e^{(0)}$ is finite at $q \to 0$, the screened $\Pi_e(\boldsymbol{q}, \omega)$ is zero in this limit. Using the RPA expressions (equations (3.40)–(3.42)) and $\gamma_{\boldsymbol{q}}$ determined in equation (3.28) for $\omega_{\boldsymbol{q}} = \omega_i$, we obtain the phonon GF as

$$D(\boldsymbol{q}, \omega) = \frac{\omega_i^2}{\omega^2 - \tilde{\omega}_{\boldsymbol{q}}^2}. \tag{3.44}$$

The poles of D determine a new phonon dispersion and a damping Γ due to interaction with electrons:

$$\tilde{\omega}_{\boldsymbol{q}} = \frac{\omega_i}{\epsilon(\boldsymbol{q}, \tilde{\omega}_{\boldsymbol{q}})^{1/2}} \tag{3.45}$$

where

$$\epsilon(\boldsymbol{q}, \omega) = 1 - \frac{4\pi e^2}{q^2} \Pi_{\mathrm{e}}^{(0)}(\boldsymbol{q}, \omega) \tag{3.46}$$

is the electron dielectric function. In the long-wavelength limit,

$$\epsilon(\boldsymbol{q}, 0) = 1 + \frac{q_s^2}{q^2} \tag{3.47}$$

and we obtain the sound wave as the real part of $\tilde{\omega}$:

$$\tilde{\omega}_{\boldsymbol{q}} = sq \tag{3.48}$$

where $s = Zk_{\mathrm{F}}/\sqrt{3Mm_{\mathrm{e}}}$ is the sound velocity. The imaginary part of $\tilde{\omega}$ determines the damping of the sound,

$$\Gamma \propto \frac{s}{v_{\mathrm{F}}} \tilde{\omega}. \tag{3.49}$$

Because the ratio of the sound velocity to the Fermi velocity (v_{F}) is adiabatically small ($s/v_{\mathrm{F}} \simeq \sqrt{m_{\mathrm{e}}/M}$), the damping is small ($\Gamma \ll \tilde{\omega}$). Electrons screen the bare ion–ion Coulomb repulsion and the residual short-range dynamic matrix has the sound-wave linear dispersion of the eigenfrequencies in the long-wavelength limit.

3.3 Electrons in metal

The lowest contribution to the electron self-energy is given by two second-order diagrams, see figures 3.3(*a*) and (*b*). The diagram in figure 3.3(*b*) is proportional to $|\gamma(\boldsymbol{q})|^2$ with $\boldsymbol{q} \equiv 0$, which is zero according to equation (3.28).

Higher-order diagrams are taken into account by replacing the bare ionic plasmon GF by a renormalized one, equation (3.44), and the bare electron–phonon interaction $\gamma(\boldsymbol{q})$ by a screened one, $\gamma_{\mathrm{sc}}(\boldsymbol{q}, \omega)$, as shown in figure 3.4. Presented analytically, the diagram in figure 3.4 corresponds to

$$\gamma_{\mathrm{sc}}(\boldsymbol{q}, \omega) = \gamma(\boldsymbol{q}) + \frac{4\pi e^2}{q^2} \Pi_{\mathrm{e}}^{(0)}(\boldsymbol{q}, \omega) \gamma_{\mathrm{sc}}(\boldsymbol{q}, \omega) \tag{3.50}$$

so that

$$\gamma_{\mathrm{sc}}(\boldsymbol{q}, \omega) = \frac{\gamma(\boldsymbol{q})}{\epsilon(\boldsymbol{q}, \omega)}. \tag{3.51}$$

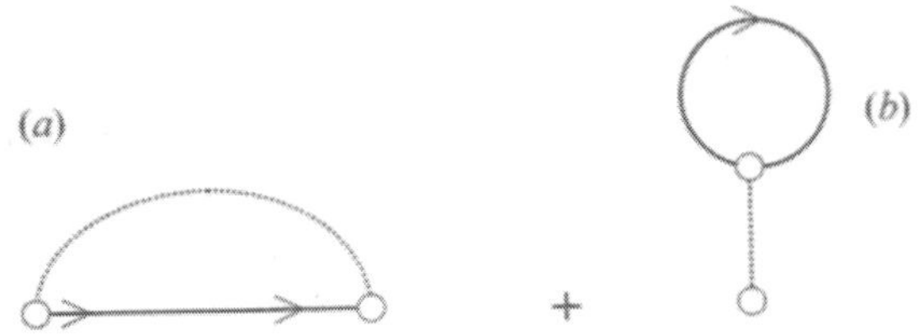

Figure 3.3. Second-order electron self-energy.

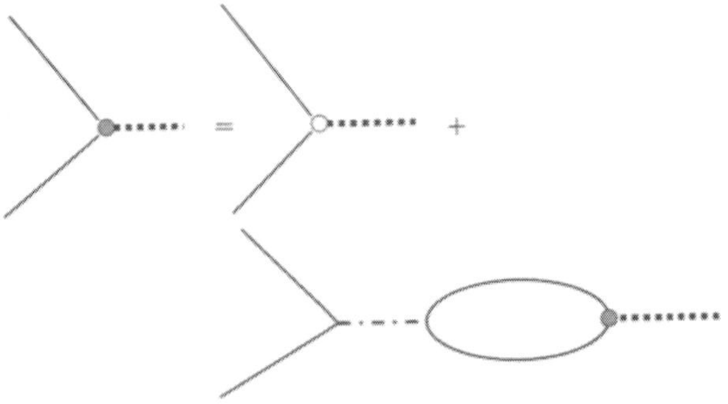

Figure 3.4. Screened electron–phonon interaction (dark circle).

For low-energy excitations ($\omega \ll \mu$) a static approximation of the dielectric function (3.47) is appropriate. Instead of the D and γ_{sc} given by equations (3.44) and (3.51), respectively, we introduce the acoustic phonon GF

$$\tilde{D}(\boldsymbol{q}, \omega) = \frac{\tilde{\omega}_{\boldsymbol{q}}^2}{\omega^2 - \tilde{\omega}_{\boldsymbol{q}}^2} \tag{3.52}$$

and the electron–acoustic phonon vertex $\tilde{\gamma}(\boldsymbol{q})\sqrt{\tilde{\omega}_{\boldsymbol{q}}/N}$, where

$$\tilde{\gamma}(\boldsymbol{q}) = \frac{\gamma(\boldsymbol{q})}{\epsilon(\boldsymbol{q}, 0)} \left(\frac{\omega_i}{\tilde{\omega}_{\boldsymbol{q}}}\right)^{3/2}. \tag{3.53}$$

Finally we obtain the diagram in figure 3.5 for the electron self-energy as a result of the summation of the most divergent diagrams, which is

$$\Sigma(\boldsymbol{k}, \epsilon) = \frac{2\mathrm{i}}{(2\pi)^4 N} \int \mathrm{d}\boldsymbol{q}\, \mathrm{d}\omega\, E_p G(\boldsymbol{k} - \boldsymbol{q}, \epsilon - \omega) \tilde{D}(\boldsymbol{q}, \omega) \tag{3.54}$$

where

$$2E_p = |\tilde{\gamma}(\boldsymbol{q})|^2 \tilde{\omega}_{\boldsymbol{q}} = \frac{2\mu}{Z(1 + q^2/q_s^2)}. \tag{3.55}$$

Because Z is of the order of one, the electron–acoustic phonon interaction E_p is generally of the order of the Fermi energy. Therefore, one has to consider

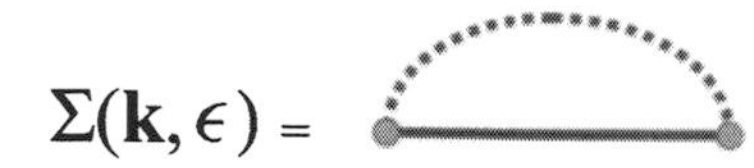

Figure 3.5. Electron self-energy in the Migdal approximation.

Figure 3.6. Adiabatically small corrections to the electron self-energy.

fourth- and higher-order diagrams with the crossing phonon lines as in figure 3.6, which are absent in figure 3.5. These diagrams are known as vertex corrections. Fortunately, as shown by Migdal [48], their contribution is adiabatically small ($\sim s/v_F$) compared with equation (3.54). This result is known as Migdal's 'theorem'.

The electron energy spectrum, renormalized by the electron–phonon interaction, is determined as the pole of the electron GF,

$$\tilde{\xi}_{\boldsymbol{k}} \equiv \tilde{E}_{\boldsymbol{k}} - \tilde{\mu} = v_F(k - k_F) + \delta E_{\boldsymbol{k}} \tag{3.56}$$

where

$$\delta E_{\boldsymbol{k}} = \Sigma(\boldsymbol{k}, \tilde{\xi}_{\boldsymbol{k}}) - \Sigma(k_F, 0) \tag{3.57}$$

and $\tilde{\mu} = \mu + \Sigma(k_F, 0)$ is the renormalized Fermi energy. The region of large $q \simeq k_F \gg q_s$ contributes mostly to the integral in equation (3.54). The dimensionless coupling constant $\lambda = 2E_p N(E_F)$ is small in this region,

$$\lambda \approx r_s \ll 1 \tag{3.58}$$

and the second order is sufficient in our calculations of Σ. Hence, we can replace the exact GF in the integral equation (3.54) by the free GF. This is appropriate for almost all quasi-particle energies. The difference between G and $G^{(0)}$ appears to be important only for the damping calculations in a very narrow region $|\tilde{\xi}_{\boldsymbol{k}}| \ll \omega_D s/v_F$ near the Fermi surface. As a result, we obtain

$$\delta E_{\boldsymbol{k}} = \frac{2\mathrm{i}E_p}{(2\pi)^4 N} \int \mathrm{d}\boldsymbol{q}\,\mathrm{d}\omega\, \tilde{D}(\boldsymbol{q}, \omega)[G^{(0)}(\boldsymbol{k} - \boldsymbol{q}, \tilde{\xi}_{\boldsymbol{k}} - \omega) - G^{(0)}(\boldsymbol{k}_F - \boldsymbol{q}, -\omega)]. \tag{3.59}$$

To simplify the calculations, we take E_p as a constant and apply the Debye approximation $\tilde{\omega}_{\boldsymbol{q}} = sq$ for $q < q_D$ where $q_D \simeq \pi/a$ is the Debye momentum. We also consider a half-filled band, $k_F \approx \pi/2a$ with the energy-independent DOS near the Fermi level $N(E_F) = m_e a^2/4\pi$.

The main contribution to the integral in equation (3.59) comes from the momentum region close to the Fermi surface,

$$|\boldsymbol{k} - \boldsymbol{q}| \simeq k_F. \tag{3.60}$$

It is convenient to introduce a new variable $k' = |\boldsymbol{k} - \boldsymbol{q}|$ instead of the angle Θ between $\boldsymbol{k}$ and $\boldsymbol{q}$, and extend the integration to $\pm\infty$ for $\xi = v_F(k' - k_F)$. Then the angular integration in equation (3.59) yields

$$\int \mathrm{d}\Theta \sin\Theta(\ldots) \sim \int_{-\infty}^{\infty} \mathrm{d}\xi \frac{\tilde{\xi}}{[\tilde{\xi} - \omega - \xi + \mathrm{i}\delta\,\mathrm{sign}(\xi)][\omega + \xi - \mathrm{i}\delta\,\mathrm{sign}(\xi)]}. \tag{3.61}$$

This integral is non-zero only if $\tilde{\xi} > \omega > 0$ or $\tilde{\xi} < \omega < 0$. It is $-2\pi\mathrm{i}$ in the first region and $2\pi\mathrm{i}$ in the second region. Taking into account that $\tilde{D}$ is an even function of ω, we obtain

$$\delta E_{\boldsymbol{k}} = \frac{2E_p}{(2\pi)^2 v_F N} \int_0^{q_D} \mathrm{d}q\, q \int_0^{|\tilde{\xi}|} \mathrm{d}\omega\, \mathrm{sign}(\tilde{\xi}) \frac{\tilde{\omega}_q^2}{\omega^2 - \tilde{\omega}_q^2 + \mathrm{i}\delta}. \tag{3.62}$$

The real and imaginary parts of equation (3.62) determine the renormalized spectrum and the lifetime of quasi-particles, respectively:

$$\mathrm{Re}(\delta E_{\boldsymbol{k}}) = \frac{E_p}{4\pi^2 v_F N} \int_0^{q_D} \mathrm{d}q\, q\tilde{\omega}_q \ln\left|\frac{\tilde{\omega}_q - \tilde{\xi}}{\tilde{\omega}_q + \tilde{\xi}}\right| \tag{3.63}$$

$$\mathrm{Im}(\delta E_{\boldsymbol{k}}) = \frac{E_p}{4\pi v_F N} \int_0^{q_m} \mathrm{d}q\, q\tilde{\omega}_q\, \mathrm{sign}(\tilde{\xi}) \tag{3.64}$$

with $q_m = |\tilde{\xi}|/s$ if $|\tilde{\xi}| < \omega_D$ and $q_m = q_D$ if $|\tilde{\xi}| > \omega_D$. For excitations far away from the Fermi surface ($|\tilde{\xi}| \gg \omega_D$), we find

$$\mathrm{Re}(\delta E_{\boldsymbol{k}}) = -\lambda \frac{\omega_D^2}{2\tilde{\xi}} \tag{3.65}$$

and for low-energy excitations with $|\tilde{\xi}| \ll \omega_D$,

$$\mathrm{Re}(\delta E_{\boldsymbol{k}}) = -\lambda\tilde{\xi}. \tag{3.66}$$

This means an increase in the effective mass of the electron due to the electron–phonon interaction,

$$\tilde{\xi} = \frac{k_F}{m^*}(k - k_F) \tag{3.67}$$

where the renormalized mass is

$$m^* = (1 + \lambda)m_e. \tag{3.68}$$

Hence, the excitation spectrum of metals has two different regions with two different values of the effective mass. The thermodynamic properties of a metal at low temperatures $T \ll \omega_D$ involve m^* but the optical properties in the frequency range $\nu \gg \omega_D$ are determined by high-energy excitations, where, according to equation (3.65), corrections are small and the mass is equal to the band mass m_e.

Damping shows just the opposite behaviour. The integral in equation (3.64), yields

$$\operatorname{Im}(\delta E_{\boldsymbol{k}}) = \operatorname{sign}(\tilde{\xi})\frac{\pi\lambda\omega_{\mathrm{D}}}{3} \tag{3.69}$$

if $|\tilde{\xi}| > \omega_{\mathrm{D}}$ and

$$\operatorname{Im}(\delta E_{\boldsymbol{k}}) = \operatorname{sign}(\tilde{\xi})\frac{\pi\lambda|\tilde{\xi}|^3}{3\omega_{\mathrm{D}}^2} \tag{3.70}$$

if $|\tilde{\xi}| \ll \omega_{\mathrm{D}}$.

These expressions describe the rate of decay of quasi-particles due to the emission of phonons. In the immediate neighbourhood of the Fermi surface, $|\tilde{\xi}| \ll \omega_{\mathrm{D}}$, the decay is small compared with the quasi-particle energy $|\tilde{\xi}|$ even for a relatively strong coupling $\lambda \sim 1$ and the concept of well-defined quasi-particles has a definite meaning. Within the Migdal approximation the electron–phonon interaction does not destroy the Fermi-liquid behaviour of electrons. The Pauli exclusion principle is responsible for the stability of the Fermi liquid. In the intermediate-energy region $|\tilde{\xi}| \sim \omega_{\mathrm{D}}$, the decay is comparable with the energy and the quasi-particle spectrum loses its meaning. In the high-energy region $|\tilde{\xi}| \gg \omega_{\mathrm{D}}$, the decay becomes small again in comparison with $|\tilde{\xi}|$ and the quasi-particle concept recovers its meaning.

Going beyond the Migdal approximation, we have to consider adiabatically small higher-order diagrams, that is to solve the Hamiltonian of free electrons and acoustic phonons coupled by an interaction:

$$H_{\mathrm{e-ph}} = \frac{1}{\sqrt{2N}}\sum_{\boldsymbol{k},\boldsymbol{q},s} \tilde{\gamma}(\boldsymbol{q})\tilde{\omega}_{\boldsymbol{q}} c^{\dagger}_{\boldsymbol{k}s} c_{\boldsymbol{k}-\boldsymbol{q}s}\tilde{d}_{\boldsymbol{q}} + H.c. \tag{3.71}$$

where $\tilde{d}_{\boldsymbol{q}}$ is the acoustic-phonon annihilation operator. From our consideration, it follows that, in applying this Hamiltonian to electrons, one should not consider the acoustic phonon self-energy. Acoustic phonons in a metal appear as a result of the electron-plasmon coupling and the Coulomb screening, so their frequency already includes the self-energy effect (section 3.2).

3.4 Eliashberg equations

Based on Migdal's theorem, Eliashberg [36] extended BCS theory towards the intermediate-coupling regime ($\lambda \lesssim 1$), applying the Gor'kov formalism. The condensed state is described by a classical field, which is the average of the product of two annihilation field operators $\mathcal{F} \propto \langle\psi\psi\rangle$ or two creation operators $\mathcal{F}^{+} \propto \langle\psi^{\dagger}\psi^{\dagger}\rangle$. These averages are macroscopically large below T_{c}. The appearance of anomalous averages cannot be seen perturbatively but they should be included in the self-energy diagram, figure 3.5, from the very beginning. This can be done in a compact form by introducing the matrix GF [49]:

$$\mathcal{G}_s(\boldsymbol{k},\tau) = -\begin{pmatrix} \langle\langle T_\tau c_{\boldsymbol{k}\uparrow}(\tau)c^\dagger_{\boldsymbol{k}\uparrow}\rangle\rangle & \langle\langle T_\tau c_{\boldsymbol{k}\uparrow}(\tau)c_{-\boldsymbol{k}\downarrow}\rangle\rangle \\ \langle\langle T_\tau c^\dagger_{-\boldsymbol{k}\downarrow}(\tau)c^\dagger_{\boldsymbol{k}\uparrow}\rangle\rangle & \langle\langle T_\tau c^\dagger_{-\boldsymbol{k}\downarrow}(\tau)c_{-\boldsymbol{k}\downarrow}\rangle\rangle \end{pmatrix} \tag{3.72}$$

and the matrix self-energy

$$\hat{\Sigma}(\boldsymbol{k},\omega_n) = (\mathcal{G}^{(0)}(\boldsymbol{k},\omega_n))^{-1} - \mathcal{G}_s^{-1}(\boldsymbol{k},\omega_n) \tag{3.73}$$

where $\mathcal{G}^{(0)}(\boldsymbol{k},\omega_n) = (\mathrm{i}\omega_n\tau_0 - \xi_{\boldsymbol{k}}\tau_3)^{-1}$ is the normal-state matrix GF. Here $\tau_{0,1,2,3}$ are the Pauli matrices,

$$\tau_0 = \begin{pmatrix} 1 & 0 \\ 0 & 1 \end{pmatrix} \qquad \tau_1 = \begin{pmatrix} 0 & 1 \\ 1 & 0 \end{pmatrix}$$

$$\tau_2 = \begin{pmatrix} 0 & -i \\ i & 0 \end{pmatrix} \qquad \tau_3 = \begin{pmatrix} 1 & 0 \\ 0 & -1 \end{pmatrix}.$$

The generalized equation for the matrix $\hat{\Sigma}$ is given by the same diagram as in the normal state (figure 3.5) but replacing $\tilde{\gamma}(\mathbf{q})$ by $\tilde{\gamma}(\boldsymbol{q})\tau_3$ and the integral over ω by the sum over the Matsubara frequencies:

$$\hat{\Sigma}(\boldsymbol{k},\omega_n) = -\frac{T}{(2\pi)^3 N}\sum_{\omega_{n'}}\int \mathrm{d}\boldsymbol{q}\,\tilde{\gamma}^2(\boldsymbol{q})\tilde{\omega}_{\boldsymbol{q}}\tau_3\mathcal{G}_s(\boldsymbol{k}-\boldsymbol{q},\omega_{n'})\tau_3\mathcal{D}(\boldsymbol{q},\omega_n-\omega_{n'}). \tag{3.74}$$

Here the temperature GF of phonons is given by

$$\mathcal{D}(\boldsymbol{q},\Omega_n) = -\frac{\tilde{\omega}_{\boldsymbol{q}}^2}{\Omega_n^2 + \tilde{\omega}_{\boldsymbol{q}}^2} \tag{3.75}$$

where $\Omega_n = 2\pi T n$ are the Matsubara frequencies for bosons. The most important difference between equation (3.74), figure 3.7, and the normal state equation (3.54), figure 3.5, is a finite value of the anomalous GF in the self-consistent solution. If we replace $\mathcal{G}_s$ by $\mathcal{G}^{(0)}$ on the right-hand side of equation (3.74), we do not find any anomalous $\mathcal{F}$. Hence, there are no anomalous averages and no phase transition in the second or in any finite order of the perturbation theory. However, if we solve equation (3.74) self-consistently, we find the finite anomalous averages.

The diagrams in figure 3.7 are expressed in analytical form as a set of Eliashberg equations, similar to Gor'kov's equations (2.186) but with a microscopically determined pairing potential. They include all non-crossing ('ladder') diagrams in every order of the perturbation theory. For simplicity, let us consider a momentum-independent $\tilde{\gamma}^2(\boldsymbol{q})\tilde{\omega}_{\boldsymbol{q}}$ as in the previous section and approximate the phonon GF in equation (3.74) as

$$\mathcal{D}(\boldsymbol{q},\Omega_n) \approx -1 \tag{3.76}$$

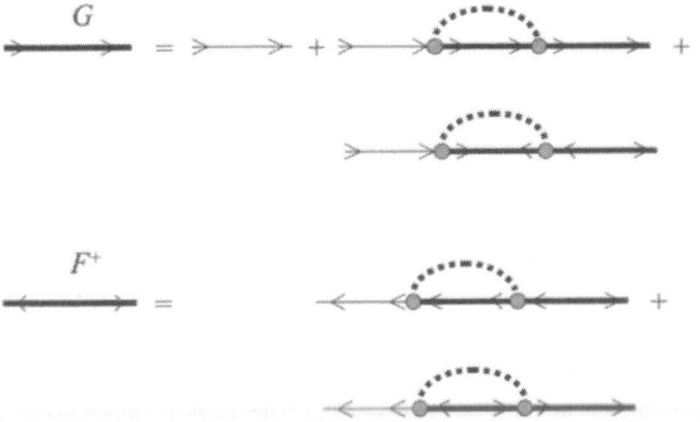

Figure 3.7. Normal $\mathcal{G}$ and anomalous $\mathcal{F}^+$ GFs of the BCS superconductor in the Eliashberg theory.

if $|\Omega_n| < \omega_D$, and zero otherwise. $\hat{\Sigma}$ is the sum of three Pauli matrices $\tau_{0,1,3}$ with the coefficients $(1 - Z)\mathrm{i}\omega_n$, Δ and χ, respectively, which in this case are functions of frequency alone,

$$\hat{\Sigma}(\boldsymbol{k}, \omega_n) = \mathrm{i}(1 - Z)\omega_n \tau_0 + \Delta \tau_1 + \chi \tau_3 \tag{3.77}$$

and

$$\mathcal{G}_s^{-1}(\boldsymbol{k}, \omega_n) = \mathrm{i}Z\omega_n \tau_0 - \Delta \tau_1 - \tilde{\xi}_{\boldsymbol{k}} \tau_3. \tag{3.78}$$

Transforming the inverse matrix (equation (3.78)) back into the original one yields

$$\mathcal{G}_s(\boldsymbol{k}, \omega_n) = -\frac{\mathrm{i}Z\omega_n + \Delta\tau_1 + \tilde{\xi}_{\boldsymbol{k}}\tau_3}{Z^2\omega_n^2 + \tilde{\xi}_{\boldsymbol{k}}^2 + |\Delta|^2} \tag{3.79}$$

where $\tilde{\xi}_{\boldsymbol{k}} = \xi_{\boldsymbol{k}} + \chi$. Substituting equations (3.76) and (3.79) into the master equation (3.74) leads to the following simplified Eliashberg equations:

$$[1 - Z(\omega_n)]\mathrm{i}\omega_n = -\lambda T \int_{-\infty}^{\infty} \mathrm{d}\tilde{\xi} \sum_{\omega_{n'}} \frac{\Theta(\omega_D - |\omega_n - \omega_{n'}|)\mathrm{i}\omega_{n'} Z}{Z^2\omega_{n'}^2 + \tilde{\xi}^2 + |\Delta|^2} \tag{3.80}$$

$$\chi(\omega_n) = -\lambda T \sum_{\omega_{n'}} \int_{-\infty}^{\infty} \mathrm{d}\tilde{\xi} \frac{\Theta(\omega_D - |\omega_n - \omega_{n'}|)\tilde{\xi}}{Z^2\omega_{n'}^2 + \tilde{\xi}^2 + |\Delta|^2} \tag{3.81}$$

and

$$\Delta(\omega_n) = \lambda T \sum_{\omega_{n'}} \int_{-\infty}^{\infty} \mathrm{d}\tilde{\xi} \frac{\Theta(\omega_D - |\omega_n - \omega_{n'}|)\Delta(\omega_{n'})}{Z^2\omega_{n'}^2 + \tilde{\xi}^2 + |\Delta|^2}. \tag{3.82}$$

We can satisfy equations (3.80) and (3.81) with $Z = 1$ and $\chi = 0$. In the last equation, we extend the summation over frequencies to infinity but cut the integral over $|\tilde{\xi}|$ at ω_D. Then applying equation (2.54) for tanh yields the familiar BCS equation for the order parameter,

$$1 = \lambda \int_0^{\omega_D} \frac{d\tilde{\xi}}{\sqrt{\tilde{\xi}^2 + |\Delta|^2}} \tanh \frac{\sqrt{\tilde{\xi}^2 + |\Delta|^2}}{2T}. \tag{3.83}$$

Hence the Migdal–Eliashberg theory reproduces the BCS results, if a similar approximation for the attraction between electrons is adopted. The critical temperature and the BCS gap are adiabatically small ($\lesssim \omega_D$) compared with the Fermi energy and we could worry about the adiabatically small crossing diagrams in figure 3.6, which are neglected in the master equation, (3.74). However, the BCS state is essentially the same as the normal state outside the narrow momentum region around the Fermi surface. The outside regions mainly contribute to the integrals of the crossing diagrams, which makes them small as in the normal state. As a result, the vertex corrections are small and Migdal's theorem holds in the BCS state as well.

Within a more general consideration the master equation (3.74) takes properly into account the phonon spectrum, retardation and realistic matrix elements of the electron–phonon interaction in metals [50–52]. The important feature of the intermediate-coupling theory is the explicit frequency dependence of the order parameter $\Delta(\omega_n)$, which is transformed into the energy-dependent gap in the quasi-particle DOS as

$$\rho(\epsilon) = \frac{\epsilon}{\sqrt{\epsilon^2 - |\Delta(\epsilon)|^2}}. \tag{3.84}$$

The DOS (equation (3.84)) can be measured in the tunnelling experiments (section 2.4). As the gap depends on the phonon spectrum, phonons affect the tunnelling I–V characteristics. Let us define the phonon density of states

$$F(\omega) = \frac{1}{N} \sum_{\mathbf{q}\nu} \delta(\omega - \tilde{\omega}_{\mathbf{q}\nu}) \tag{3.85}$$

and the so-called Eliashberg function $\alpha^2 F$,

$$\alpha^2(\omega) F(\omega) = \frac{1}{N(E_F) N} \sum_{\mathbf{k},\mathbf{k}'} \tilde{\gamma}^2(\mathbf{k} - \mathbf{k}', \nu) \tilde{\omega}^2_{\mathbf{k}-\mathbf{k}',\nu} \delta(\omega - \tilde{\omega}_{\mathbf{k}-\mathbf{k}',\nu}) \delta(\xi_{\mathbf{k}}) \delta(\xi_{\mathbf{k}'}) \tag{3.86}$$

which is an average over the Fermi surface of the interaction matrix element squared multiplied by the phonon spectral density. The quasi-particle DOS (equation (3.84)) can be obtained in a direct fashion using the tunnelling conductivity of the superconducting tunnel junctions. Then the function $\alpha^2 F$ can be calculated to fit the experimental energy dependence of the gap $\Delta(\epsilon)$ [54] and compared with the phonon density of states measured independently, for example, in neutron scattering experiments. The comparison shows that frequency dependence of the Eliashberg function and the phonon density of states are similar in many low-temperature superconductors with a weak or intermediate electron–phonon coupling, for example in lead. The observation of the characteristic phonon frequencies in tunnelling (usually in the second derivative of the current *versus* voltage) and the isotope effect are used to verify the phonon-mediated pairing mechanism.

3.5 Coulomb pseudopotential

The theory of superconductivity has to take account of the Coulomb repulsive correlations between electrons, which might be much stronger than the attraction induced by phonons. There is no adiabatic parameter for this interaction because the electron plasma frequency $\omega_{\mathrm{e}} = \sqrt{4\pi n_{\mathrm{e}} e^2/m_{\mathrm{e}}}$ has about the same order of magnitude as the Fermi energy in metals. Nevertheless, we can account for the Coulomb repulsion in the same fashion as for the electron–phonon interaction but replacing $\tilde{\gamma}^2(\boldsymbol{q})\tilde{\omega}_{\boldsymbol{q}} D(\boldsymbol{q}, \omega_n - \omega_{n'})$ in equation (3.74) by the Fourier component of the Coulomb potential V_{c}. The Coulomb interaction is non-retarded for frequencies less than ω_{e}. Then the kernel $K(\omega_n - \omega_{n'})$ in the BCS equation,

$$\Delta(\omega_n) = T \int \mathrm{d}\xi \sum_{\omega_{n'}} K(\omega_n - \omega_{n'}) \frac{\Delta(\omega_{n'})}{\omega_{n'}^2 + \xi^2 + |\Delta(\omega_{n'})|^2} \tag{3.87}$$

can be parametrized as

$$K(\omega_n - \omega_{n'}) = \lambda\Theta(\omega_{\mathrm{D}} - |\omega_n - \omega_{n'}|) - \mu_{\mathrm{c}}\Theta(\omega_{\mathrm{e}} - |\omega_n - \omega_{n'}|) \tag{3.88}$$

where $\mu_{\mathrm{c}} = V_{\mathrm{c}} N(E_{\mathrm{F}})$. At $T = T_{\mathrm{c}}$, we neglect second and higher powers of the order parameter and integrating over ξ obtain

$$\Delta(\omega_n) = \pi T_{\mathrm{c}} \sum_{\omega_{n'}} K(\omega_n - \omega_{n'}) \frac{\Delta(\omega_{n'})}{|\omega_{n'}|}. \tag{3.89}$$

Let us adopt the BCS-like approximation of the kernel:

$$\begin{aligned} K(\omega_n - \omega_{n'}) \simeq\ & \lambda\Theta(2\omega_{\mathrm{D}} - |\omega_n|)\Theta(2\omega_{\mathrm{D}} - |\omega_{n'}|) \\ & - \mu_{\mathrm{c}}\Theta(2\omega_{\mathrm{e}} - |\omega_n|)\Theta(2\omega_{\mathrm{e}} - |\omega_{n'}|) \end{aligned}$$

and replace the summation in equation (3.89) by the integral:

$$\pi T_{\mathrm{c}} \sum \approx \int_{\pi T_{\mathrm{c}}}^{\infty} \mathrm{d}\omega \tag{3.90}$$

because $T_{\mathrm{c}} \ll \omega_{\mathrm{D}} \ll \omega_{\mathrm{e}}$. Then the solution is found in the form

$$\Delta(\omega) = \Delta_1\Theta(2\omega_{\mathrm{D}} - |\omega|) + \Delta_2\Theta(2\omega_{\mathrm{e}} - |\omega|)\Theta(|\omega| - 2\omega_{\mathrm{D}}) \tag{3.91}$$

with constant but different values of the order parameter Δ_1 and Δ_2 below and above the cut-off energy $2\omega_D$, respectively. Substituting equation (3.91) into equation (3.89) yields the following equations for $\Delta_{1,2}$:

$$\Delta_1 \left[1 - (\lambda - \mu_c) \ln \frac{2\omega_D}{\pi T_c} \right] + \Delta_2 \mu_c \ln \frac{\omega_e}{\omega_D} = 0 \tag{3.92}$$

$$\Delta_1 \mu_c \ln \frac{2\omega_D}{\pi T_c} + \Delta_2 \left[1 + \mu_c \ln \frac{\omega_e}{\omega_D} \right] = 0. \tag{3.93}$$

A non-trivial solution of these coupled equations is found, if

$$T_c = \frac{2\omega_D}{\pi} \exp\left(-\frac{1}{\lambda - \mu_c^*} \right) \tag{3.94}$$

where

$$\mu_c^* = \frac{\mu_c}{1 + \mu_c \ln(\omega_e/\omega_D)} \tag{3.95}$$

is the so-called Coulomb pseudopotential [53]. This is a remarkable result. It shows that even a large Coulomb repulsion $\mu_c > \lambda$ does not destroy the Cooper pairs because its contribution is suppressed down to the value about $\ln^{-1}(\omega_e/\omega_D) \ll 1$. The retarded attraction mediated by phonons acts well after two electrons meet each other. This time delay is sufficient for two electrons to be separated by a relative distance, at which the Coulomb repulsion is small. The Coulomb correlations also lead to a damping of excitations of the order of ξ^2/μ, which is relevant only in a narrow region around the Fermi surface $|\xi| \lesssim \omega_D \sqrt{m_e/M}$. The damping due to the Fröhlich interaction dominates outside this region.

Computational analysis of the Eliashberg equations led McMillan [54] to suggest an empirical formula for T_c, which works well for simple metals and their alloys,

$$T_c = \frac{\omega_D}{1.45} \exp\left(-\frac{1.04(1+\lambda)}{\lambda - \mu_c^*(1 + 0.62\lambda)} \right). \tag{3.96}$$

However, in materials with a moderate $T_c \propto 20$ K (like Nb_3Sn, V_3Si and in other A-15 compounds), the discrepancy between the value of λ, estimated from this equation and from the first-principle band-structure calculations, exceeds the limit allowed by the experimental and computation accuracy by several times [55].

In the original papers, Migdal [48] and Eliashberg [36] restricted the applicability of their approach to the intermediate region of coupling $\lambda < 1$. With the typical values of $\lambda = 0.5$ and $\mu_c^* = 0.14$ and with the Debye temperature as high as $\omega_D = 400$ K, McMillan's formula predicts $T_c \approx 2$ K, clearly too low to explain high T_c values in novel superconductors. One can formally compute T_c and the gap using the Eliashberg equations (3.74) also in the strong-coupling regime $\lambda > 1$. In particular, Allen and Dynes [56] found that in the extreme strong-coupling limit ($\lambda \gg 1$), the critical temperature may rise as

$$T_c \approx \frac{\omega_D \lambda^{1/2}}{2\pi}. \tag{3.97}$$

However, the Migdal–Eliashberg theory is based on the assumption that the Fermi liquid is stable and the adiabatic condition $\mu \gg \omega_D$ is satisfied. As we shall later discuss in chapter 4, this assumption cannot be applied in the strong-coupling regime and the proper extension of the BCS theory to $\lambda > 1$ inevitably involves small polarons and bipolarons.

3.6 Cooper pairing of repulsive fermions

The phenomenon of superconductivity is due to the interaction of electrons with vibrating ions, which mediates an effective attractive potential $V(\boldsymbol{k}, \boldsymbol{k}') < 0$ in equation (2.98). In recent years, great attention has been paid to a possibility of superconductivity mediated by strong electron–electron correlations without any involvement of phonons. Some time ago, Kohn and Luttinger [57] pointed out that such possibility exists at least theoretically. They found that a dilute Fermi gas cannot remain normal down to absolute zero of temperature even in the case of purely repulsive short-range interaction between the particles. A system of fermions with purely repulsive short-range forces will inevitably be superfluid at zero temperature.

To understand what is involved, one should consider the screening of a charge placed in a metal. It has long been known [58] that if fermions are degenerate (i.e. their Fermi surface is well defined), the screening produces an oscillatory potential of the form $\cos(2k_F r + \varphi)/r^3$ at the distance r from the charge (here φ is a constant) which has attractive regions. Using these regions of screened interaction, unconventional Cooper pairs can form with non-zero orbital momentum (section 2.10). Following Kohn and Luttinger, let us consider a simplified model of spin-$\frac{1}{2}$ fermions with a weak short-range repulsive interaction between them. The critical temperature of unconventional Cooper pairing with the orbital momentum l is found as (see equation (2.104))

$$T_c \approx \mu \exp\left(-\frac{1}{\lambda_l}\right). \tag{3.98}$$

Here we replaced the Debye temperature by the Fermi energy μ because the fermion–fermion interaction is non-retarded. The coupling constant λ_l is expressed in terms of spherical harmonics $V(l)$ of the Fourier transform of the interaction potential,

$$\lambda_l \equiv -\frac{V(l) m k_F}{2\pi(2l+1)} \tag{3.99}$$

on condition that $V(l)$ is negative. For simplicity, we choose the repulsive bare potential as $V(\boldsymbol{r} - \boldsymbol{r}') = U$ with a positive U, if $|\boldsymbol{r} - \boldsymbol{r}'| \leqslant r_0$, and $V(\boldsymbol{r} - \boldsymbol{r}') = 0$, if $|\boldsymbol{r} - \boldsymbol{r}'| > r_0$. We also assume that the gas is diluted, that is the radius of the potential is small compared with the characteristic wave-length of fermions,

$$k_F r_0 \ll 1. \tag{3.100}$$

It is easy to estimate $V(l)$ for such a potential,

$$V(l) = \frac{2l+1}{2}\int_0^{\pi} \mathrm{d}\Theta \, \sin\Theta \, P_l(\cos\Theta) V(q) \tag{3.101}$$

where $q^2 = 2k_{\mathrm{F}}(1-\cos\Theta)$ is about $2k_{\mathrm{F}}$ or less and

$$V(q) = U\int_{r\leqslant a} \mathrm{d}\boldsymbol{r}\, \exp(\mathrm{i}\boldsymbol{q}\cdot\boldsymbol{r}) \approx \frac{4\pi}{3}Ur_0^3\left\{1 - \frac{(qr_0)^2}{10} + \frac{(qr_0)^4}{40} + \mathcal{O}[(k_{\mathrm{F}}r_0)^6]\right\}. \tag{3.102}$$

Substituting the Fourier transform (3.102) into equation (3.101), we obtain a *positive* $V(l)$ of the order of

$$V(l) \propto Ur_0^3(k_{\mathrm{F}}r_0)^{2l}. \tag{3.103}$$

We conclude that the repulsive bare potential does not produce any pairing. Let us now take screening into account. It is sufficient to consider it perturbatively, if the potential is weak, $U \ll \mu$. The diagrams contributing to the effective interaction up to the second order in U are shown in figure 3.8. In the analytical form, the Fourier transform of the effective interaction of two electrons on the Fermi surface is given by

$$\begin{aligned}\tilde{V}(\boldsymbol{k},\boldsymbol{k}') = V(q) &- \frac{T}{(2\pi)^3}\sum_{\omega_n}\int \mathrm{d}\boldsymbol{p}[2V^2(q) - 2V(q)V(\boldsymbol{k}'-\boldsymbol{p})]\mathcal{G}^{(0)}_{\omega_n}(\boldsymbol{p},\omega_n)\\ &\times \mathcal{G}^{(0)}_{\omega_n}(\boldsymbol{p}+\boldsymbol{q},\omega_n)\\ &+ \frac{T}{(2\pi)^3}\sum_{\omega_n}\int \mathrm{d}\boldsymbol{p}\, V(\boldsymbol{k}-\boldsymbol{p})V(\boldsymbol{k}'-\boldsymbol{p})\mathcal{G}^{(0)}_{\omega_n}(\boldsymbol{p},\omega_n)\\ &\times \mathcal{G}^{(0)}_{\omega_n}(\boldsymbol{p}-\boldsymbol{k}'-\boldsymbol{k},\omega_n).\end{aligned} \tag{3.104}$$

Because the density is low (equation (3.100)) we can neglect the q-dependence of $V(q)$ and take $V(q) = 4\pi Ur_0^3/3 \equiv v$ in the second-order terms of equation (3.104). Then the diagrams (*b*) and (*c*) + (*d*) cancel each other and the second term in the right-hand side of equation (3.104) vanishes. The remaining last term yields the familiar polarization bubble (equation (3.40)) which finally leads to

$$\begin{aligned}\tilde{\lambda}_l &\equiv -\frac{\tilde{V}(l)mk_{\mathrm{F}}}{2\pi(2l+1)}\\ &= \lambda_l - \lambda(-1)^l\int_0^{\pi} \mathrm{d}\Theta\, \sin\Theta\, P_l(\cos\Theta)\left[1 + \frac{4k_{\mathrm{F}}^2 - q^2}{4k_{\mathrm{F}}q}\ln\left|\frac{2k_{\mathrm{F}}+q}{2k_{\mathrm{F}}-q}\right|\right]\end{aligned} \tag{3.105}$$

where $\lambda = (ak_{\mathrm{F}}/\pi)^2 \ll 1$. Here the first repulsive term is about $(ak_{\mathrm{F}})(k_{\mathrm{F}}r_0)^{2l}$, where $a = vm_{\mathrm{e}}/(2\pi)$ is the s-wave scattering amplitude in

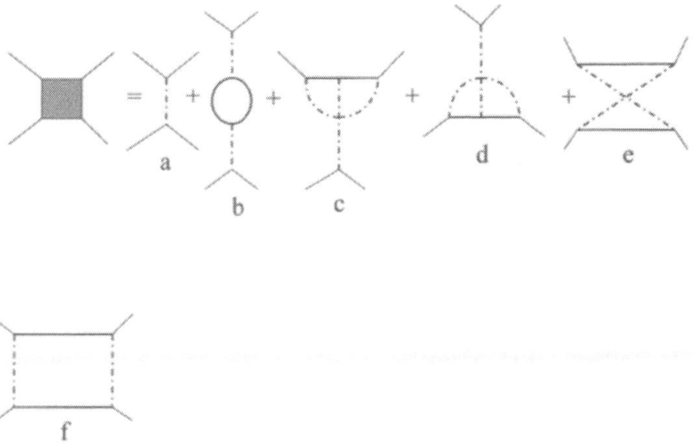

Figure 3.8. The first (a) and second-order diagrams (b–e) which contribute to the screened electron–electron interaction. Note that the non-crossing second-order diagram (f) as well as all higher-order non-crossing diagrams are fully taken into account in the BCS equation (section 3.4) and should not be included in the irreducible scattering vertex.

the Born approximation. The second 'polarization' contribution to effective interaction is of the order of $(ak_F)^2$. Hence, if the gas parameter is sufficiently small,

$$(k_F r_0)^{2l-3} \ll \frac{U}{\mu} \tag{3.106}$$

the polarization contribution overcomes direct repulsion for all $l \geqslant 2$ and it is attractive. For example, for $l = 2$ (d-wave pairing), the integral in equation (3.105) yields

$$\tilde{\lambda}_2 = \lambda \frac{4}{105}(8 - 11 \ln 2) \approx 0.015\lambda. \tag{3.107}$$

The corresponding critical temperature is

$$T_c \approx \mu \exp\left(-\frac{1}{0.015\lambda}\right) \tag{3.108}$$

which is practically zero at any $\lambda < 1$. The situation is slightly better when the repulsive potential U is strong ($U \gg \mu$) (i.e. for hard-core spheres) [59]. Partial scattering amplitudes f_l in a vacuum are of the order of $r_0(k_F r_0)^{2l}$ in this limit. They are small compared with the attractive polarization contribution to the amplitudes ($\propto r_0(k_F r_0)$) starting from $l = 1$. Hence, at $T = 0$, hard-core fermions with repulsive scattering in a vacuum are necessarily in a superfluid p-wave state. Calculating the integral in equation (3.105) for $l = 1$, we obtain

$$\tilde{\lambda}_1 = \lambda \frac{2}{5}(2 \ln 2 - 1) \approx 0.15\lambda \tag{3.109}$$

and

$$T_c \approx \mu \exp\left(-\frac{1}{0.15\lambda}\right). \tag{3.110}$$

The Fermi energy could be as large as $\mu = 10^4$ K but T_c still remains very low because $\lambda \ll 1$ in the dilute limit.

Chapter 4

Strong-coupling theory

The electron–phonon coupling constant λ is about the ratio of the electron–phonon interaction energy E_p to the half-bandwidth $D \propto N(E_F)^{-1}$ (appendix A). We expect [11] that when the coupling is strong ($\lambda > 1$), all electrons in the Bloch band are 'dressed' by phonons because their kinetic energy ($<D$) is small compared with the potential energy due to a local lattice deformation, E_p, caused by an electron. If phonon frequencies are very low, the local lattice deformation traps the electron. This *self-trapping* phenomenon was predicted by Landau [60]. It has been studied in greater detail by Pekar [61], Fröhlich [62], Feynman [63], Devreese [64] and other authors in the effective mass approximation, which leads to the so-called *large polaron.* The large polaron propagates through the lattice like a free electron but with the enhanced effective mass. In the strong-coupling regime ($\lambda > 1$), the finite bandwidth becomes important, so that the effective mass approximation cannot be applied. The electron is called a *small polaron* in this regime. The self-trapping is never 'complete', that is any polaron can tunnel through the lattice. Only in the extreme *adiabatic* limit, when the phonon frequencies tend to zero, is the self-trapping complete and the polaron motion no longer translationally continuous (section 4.2). The main features of the small polaron were understood by Tjablikov [65], Yamashita and Kurosava [66], Sewell [67], Holstein [68] and his school [69, 70], Lang and Firsov [71], Eagles [72] and others and described in several review papers and textbooks [13, 64, 73–76]. The exponential reduction of the bandwidth at large values of λ is one of those features (section 4.3). The small polaron bandwidth decreases with increasing temperature up to a crossover region from the coherent small polaron tunnelling to a thermally activated hopping. The crossover from the polaron Bloch states to the incoherent hopping takes place at temperatures $T \approx \omega_0/2$ or higher, where ω_0 is the characteristic phonon frequency. In this chapter, we extend BCS theory to the strong-coupling regime ($\lambda > 1$) with the itinerant (Bloch) states of polarons and bipolarons.

4.1 Electron–phonon and Coulomb interactions in the Wannier representation

For doped semiconductors and metals with a strong electron–phonon (e–ph) interaction, it is convenient to transform the Bloch states to the site (Wannier) states using the canonical linear transformation of the electron operators (appendix C):

$$c_i = \frac{1}{\sqrt{N}} \sum_{\boldsymbol{k}} \mathrm{e}^{\mathrm{i}\boldsymbol{k}\cdot\boldsymbol{m}} c_{\boldsymbol{k}s} \tag{4.1}$$

where $i \equiv (\boldsymbol{m}, s)$ includes both site $\boldsymbol{m}$ and spin s quantum numbers. In the new representation, the periodic part of the Hamiltonian (3.17) takes the following form:

$$H_{\mathrm{e}} = \sum_{i,j} [T(\boldsymbol{m} - \boldsymbol{m}')\delta_{ss'} - \mu\delta_{ij}] c_i^{\dagger} c_j, \tag{4.2}$$

where

$$T(\boldsymbol{m}) = \frac{1}{N} \sum_{\boldsymbol{k}} E_{n\boldsymbol{k}} \mathrm{e}^{\mathrm{i}\boldsymbol{k}\cdot\boldsymbol{m}}$$

is the 'bare' hopping integral (appendix A). Here $i = (\boldsymbol{m}, s)$ and $j = (\boldsymbol{n}, s')$.

The electron–phonon interaction and the Coulomb correlations acquire simple forms in the Wannier representation, if their matrix elements in the momentum representation depend only on the momentum transfer $\boldsymbol{q}$, equation (3.21):

$$H_{\mathrm{e-ph}} = \sum_{\boldsymbol{q},\nu,i} \omega_{\boldsymbol{q}\nu} \hat{n}_i [u_i(\boldsymbol{q}, \nu) d_{\boldsymbol{q}\nu} + H.c.] \tag{4.3}$$

$$H_{\mathrm{e-e}} = \tfrac{1}{2} \sum_{i \neq j} V_{\mathrm{c}}(\boldsymbol{m} - \boldsymbol{n}) \hat{n}_i \hat{n}_j. \tag{4.4}$$

Here

$$u_i(\boldsymbol{q}, \nu) = \frac{1}{\sqrt{2N}} \gamma(\boldsymbol{q}, \nu) \mathrm{e}^{\mathrm{i}\boldsymbol{q}\cdot\boldsymbol{m}} \tag{4.5}$$

and

$$V_{\mathrm{c}}(\boldsymbol{m}) = \frac{1}{N} \sum_{\boldsymbol{q}} V_{\mathrm{c}}(\boldsymbol{q}) \mathrm{e}^{\mathrm{i}\boldsymbol{q}\cdot\boldsymbol{m}} \tag{4.6}$$

are the matrix elements of the electron–phonon and Coulomb interactions, respectively, in the Wannier representation for electrons, and $\hat{n}_i = c_i^{\dagger} c_i$ is the density operator. Assuming the interaction matrix elements depend only on the momentum transfer, we neglect the terms in the electron–phonon and Coulomb interactions which are proportional to the overlap integrals of the Wannier orbitals on different sites. This approximation is justified for narrow-band materials, whose bandwidth $2D$ is less than the characteristic value of the crystal field. As a result, in the Wannier representation, the Hamiltonian is

$$H = \sum_{i,j}[T(\boldsymbol{m}-\boldsymbol{m}')\delta_{ss'} - \mu\delta_{ij}]c_i^\dagger c_j + \sum_{\boldsymbol{q},\nu,i}\omega_{\boldsymbol{q}\nu}\hat{n}_i[u_i(\boldsymbol{q},\nu)d_{\boldsymbol{q}\nu} + H.c.]$$
$$+\tfrac{1}{2}\sum_{i\neq j}V_{\mathrm{c}}(\boldsymbol{m}-\boldsymbol{n})\hat{n}_i\hat{n}_j + \sum_{\boldsymbol{q}}\omega_{\boldsymbol{q}\nu}(d_{\boldsymbol{q}\nu}^\dagger d_{\boldsymbol{q}\nu} + 1/2). \tag{4.7}$$

This Hamiltonian should be treated as a 'bare' one for metals, where the matrix elements and phonon frequencies are ill defined. In contrast, the bare phonons $\omega_{\boldsymbol{q}\nu}$ and the electron band structure $E_{n\boldsymbol{k}}$ are well defined in doped semiconductors, which have their 'parent' dielectric compounds. Here, the effect of carriers on the crystal field and on the dynamic matrix is small while the carrier density is much less than the atomic one. Therefore, we can use the band structure and the crystal field of the parent insulators to calculate the parameters of the Hamiltonian (4.7). Depending on the particular phonon branch, the interaction constant $\gamma(\boldsymbol{q},\nu)$ has a different q-dependence. For example, in the long-wavelength limit ($q \ll \pi/a$),

$$\gamma(\boldsymbol{q},\nu) \propto \frac{1}{q}$$
$$\propto \text{constant}$$
$$\propto \frac{1}{\sqrt{q}}$$

for optical, molecular ($\omega_{\boldsymbol{q}} \propto$ constant) and acoustic ($\omega_{\boldsymbol{q}} \propto q$) phonons, respectively. We can transform the e–ph interaction further using the site-representation also for phonons. Replacing the Bloch functions in the definition of $\gamma(\boldsymbol{q},\nu)$ (equation (3.20)) by their Wannier series yields

$$\gamma(\boldsymbol{q},\nu) = -\frac{1}{M^{1/2}\omega_{\boldsymbol{q}\nu}^{3/2}}\sum_{\boldsymbol{n}} \mathrm{e}^{-\mathrm{i}\boldsymbol{q}\cdot\boldsymbol{n}}\boldsymbol{e}_{\boldsymbol{q}\nu}\cdot\nabla_{\boldsymbol{n}}v(\boldsymbol{n}). \tag{4.8}$$

This result is obtained by neglecting the overlap integrals of the Wannier orbitals on different sites and by assuming that the single-ion potential $v(\boldsymbol{r})$ varies over the distance, which is much larger than the radius of the orbital. After substituting equation (4.8) into equation (4.5) and using the displacement operators (equation (3.10)), one arrives at the following expression:

$$H_{\mathrm{e-ph}} = \sum_{\boldsymbol{m},\boldsymbol{n},s}\hat{n}_{\boldsymbol{m}s}\boldsymbol{u}_{\boldsymbol{n}}\cdot\nabla_{\boldsymbol{n}}v(\boldsymbol{m}-\boldsymbol{n}) \tag{4.9}$$

which can also be derived by replacing the field operators by the Wannier series in equation (3.8). The site representation of $H_{\mathrm{e-ph}}$ (equation (4.9)) is particularly convenient for the interaction with dispersionless local modes whose $\omega_{\boldsymbol{q}\nu} = \omega_\nu$ and $\boldsymbol{e}_{\boldsymbol{q}\nu} = \boldsymbol{e}_\nu$ are q independent. Introducing the phonon site operators

$$d_{\boldsymbol{n}\nu} = \frac{1}{\sqrt{N}}\sum_{\boldsymbol{k}}\mathrm{e}^{\mathrm{i}\boldsymbol{q}\cdot\boldsymbol{n}}d_{\boldsymbol{q}\nu} \tag{4.10}$$

we obtain in this case

$$u_n = \sum_{\nu} \frac{e_\nu}{\sqrt{2M\omega_\nu}}(d_{n\nu} + d^\dagger_{n\nu})$$

$$H_{\text{ph}} = \sum_{n,\nu} \omega_\nu (d^\dagger_{n\nu} d_{n\nu} + 1/2)$$

and

$$H_{\text{e-ph}} = \sum_{n,m,\nu} \omega_\nu g_\nu(m-n)(e_\nu \cdot e_{m-n})\hat{n}_{ms}(d^\dagger_{n\nu} + d_{n\nu}) \tag{4.11}$$

where

$$g_\nu(m) = \frac{1}{\omega_\nu\sqrt{2M\omega_\nu}} \frac{\mathrm{d}v(m)}{\mathrm{d}m}$$

is a dimensionless *force* acting between the electron on site m and the displacement of ion n, and $e_{m-n} \equiv (m-n)/|m-n|$ is the unit vector in the direction from the electron m to the ion n. The 'real space' representation (4.11) is convenient in modelling the electron–phonon interaction in complex lattices. Atomic orbitals of an ion adiabatically follow its motion. Therefore, the electron does not interact with the displacement of the ion, whose orbital it occupies, that is $g_\nu(0) = 0$.

4.2 Breakdown of Migdal–Eliashberg theory in the strong-coupling regime

The perturbative approach to the e–ph interaction fails when $\lambda > 1$. But one might expect that the self-consistent Migdal–Eliashberg (ME) theory is still valid in the strong-coupling regime because it sums the infinite set of particular (non-crossing) diagrams in the electron self-energy (chapter 3). One of the problems with such an extension of ME theory is lattice instability. The same theory applied to phonons yields the renormalized phonon frequency $\tilde{\omega} = \omega(1-2\lambda)^{1/2}$ [48]. The frequency turns out to be zero at $\lambda = 0.5$. Because of this lattice instability, Migdal [48] and Eliashberg [36] restricted the applicability of their approach to $\lambda < 1$. However, it was then shown that there was no lattice instability but only a small renormalization of the phonon frequencies of the order of the adiabatic ratio, $\omega/\mu \ll 1$, for *any* value of λ, if the adiabatic Born–Oppenheimer approach was properly applied [77]. The conclusion was that the Fröhlich Hamiltonian (4.3) correctly describes the electron self-energy for any value of λ but it should not be applied to further renormalize phonons (see also chapter 3). As a result, many authors used ME theory with λ much larger than 1 [50].

In fact, the Migdal–Eliashberg theory cannot be applied at $\lambda > 1$ for a reason, which has nothing to do with lattice instability. The inverse $(1/\lambda)$ expansion technique [11] showed that the many-electron system collapses into a small polaron regime at $\lambda \approx 1$ for any adiabatic ratio. This regime is beyond

ME theory. It cannot be described by a summation of the standard Feynman–Dyson perturbation diagrams even including the vertex corrections (section 3.3), because of the broken translation symmetry in the strong-coupling limit. The major problem with the extension of ME theory to strong coupling originates from its basic assumption that the electron and phonon GFs are translation invariants. One assumes that $G(\boldsymbol{r}, \boldsymbol{r}', \tau) = G(\boldsymbol{r} - \boldsymbol{r}', \tau)$ prior to solving the self-energy equation. This assumption excludes the possibility of the breakdown of the local translation symmetry due to lattice deformation, similar to the absence of the anomalous Bogoliubov averages in any order of the perturbation theory (section 3.4). To enable the electron relaxation into the lowest polaron states, one has to introduce an infinitesimal translation-non-invariant potential in the Hamiltonian which should be set equal to zero only in the final solution for the GFs [78]. As in the case of the off-diagonal superconducting order parameter, a small translation symmetry-breaking potential drives the system into a new ground state, if the e–ph coupling is sufficiently strong ($\lambda \gtrsim 1$). Setting the potential equal to zero in the solution of the equations of motion restores the translation symmetry but in a new polaron band (section 4.3) rather than in the bare electron band.

To illustrate the point let us compare the Migdal solution of the molecular-chain (Holstein) model of the e–ph interaction [68] with the exact solution in the adiabatic limit, $\omega/\mu \to 0$. The Hamiltonian of the model is

$$H = -t\sum_{\langle ij\rangle} c_i^\dagger c_j + H.c. + 2(\lambda k t)^{1/2}\sum_i x_i c_i^\dagger c_i + \sum_i \left(-\frac{1}{2M}\frac{\partial^2}{\partial x_i^2} + \frac{k x_i^2}{2}\right) \tag{4.12}$$

where t is the nearest-neighbour hopping integral, x_i is the normal coordinate of the molecule (site) i and $k = M\omega^2$. The Migdal theorem is exact in this limit. Hence, in the framework of ME theory, one would expect Fermi-liquid behaviour above T_c and the BCS ground state below T_c at any value of λ. In fact, the exact ground state is a self-trapped insulator at any filling of the band if $\lambda \gtrsim 1$.

First, we consider a two-site case (zero-dimensional limit), $i, j = 1, 2$ with one electron and then generalize the result for an infinite lattice with many electrons. The transformation $X = (x_1 + x_2)$, $\xi = x_1 - x_2$ allows us to eliminate the coordinate X, which is coupled only with the total density ($n_1 + n_2 = 1$). That leaves the following Hamiltonian to be solved in the extreme adiabatic limit $M \to \infty$:

$$H = -t(c_1^\dagger c_2 + c_2^\dagger c_1) + (\lambda k t)^{1/2}\xi(c_1^\dagger c_1 - c_2^\dagger c_2) + \frac{k\xi^2}{4}. \tag{4.13}$$

The solution is

$$\psi = (\alpha c_1^\dagger + \beta c_2^\dagger)|0\rangle \tag{4.14}$$

where

$$\alpha = \frac{t}{[t^2 + ((\lambda kt)^{1/2}\xi + (t^2 + \lambda kt\xi^2)^{1/2})^2]^{1/2}} \tag{4.15}$$

$$\beta = -\frac{(\lambda kt)^{1/2}\xi + (t^2 + \lambda kt\xi^2)^{1/2}}{[t^2 + ((\lambda kt)^{1/2}\xi + (t^2 + \lambda kt\xi^2)^{1/2})^2]^{1/2}} \tag{4.16}$$

and the energy is

$$E = \frac{k\xi^2}{4} - (t^2 + \lambda kt\xi^2)^{1/2}. \tag{4.17}$$

In the extreme adiabatic limit the displacement ξ is classical, so the ground-state energy E_0 and the ground-state displacement ξ_0 are obtained by minimizing equation (4.17) with respect to ξ. If $\lambda \geq 0.5$, we obtain

$$E_0 = -t\left(\lambda + \frac{1}{4\lambda}\right) \tag{4.18}$$

and

$$\xi_0 = \left[\frac{t(4\lambda^2 - 1)}{\lambda k}\right]^{1/2}. \tag{4.19}$$

The symmetry-breaking 'order' parameter is

$$\Delta \equiv \beta^2 - \alpha^2 = \frac{[2\lambda + (4\lambda^2 - 1)^{1/2}]^2 - 1}{[2\lambda + (4\lambda^2 - 1)^{1/2}]^2 + 1}. \tag{4.20}$$

If $\lambda < 0.5$, the ground state is translationally invariant, $\Delta = 0$ and $E_0 = -t$, $\xi = 0$, $\beta = -\alpha$. Precisely this state is the 'Migdal' solution of the Holstein model. Indeed, the Fourier transform of the GF should be diagonal in the Migdal approximation, $G(\boldsymbol{k}, \boldsymbol{k}', \tau) = G(\boldsymbol{k}, \tau)\delta_{\boldsymbol{k},\boldsymbol{k}'}$. The site operators are transformed into momentum space as

$$c_k = N^{-1/2} \sum_j c_j \exp(ikaj) \tag{4.21}$$

where $k = 2\pi n/Na$, $-N/2 < n \leq N/2$. Then the off-diagonal GF with $k = 0$ and $k' = \pi/a$ of the two-site chain ($N = 2$) is given by

$$G(k, k', -0) = \frac{\mathrm{i}}{2}\langle (c_1^\dagger - c_2^\dagger)(c_1 + c_2)\rangle \tag{4.22}$$

at $\tau = -0$. Calculating this average, we obtain

$$G(k, k', -0) = \frac{\mathrm{i}}{2}(\alpha^2 - \beta^2) \tag{4.23}$$

which should vanish in the Migdal theory. Hence, the theory provides only a symmetric (translation invariant) solution with $|\alpha| = |\beta|$. When $\lambda > 0.5$, this

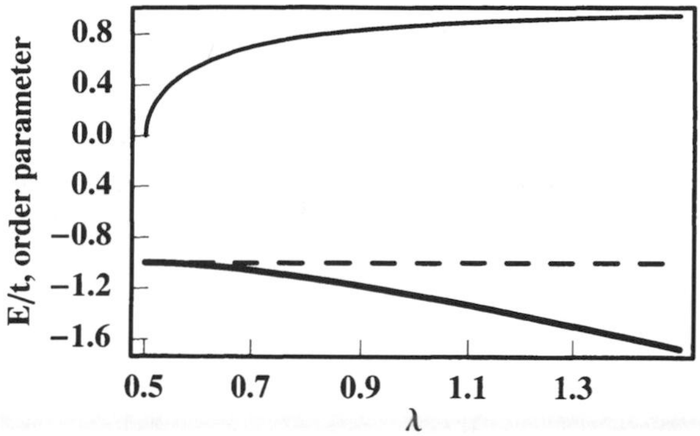

Figure 4.1. The ground-state energy (in units of t, full line) and the order parameter (thin full line) of the adiabatic Holstein model. The Migdal solution is shown as the broken line.

solution is *not* the ground state of the system (figure 4.1). The system collapses into a localized adiabatic polaron trapped on the 'right-hand' or 'left-hand' site due to the finite lattice deformation ξ_0. Alternatively, when $\lambda < 0.5$, the Migdal solution is the *only* solution.

The generalization to a multi-polaron system on an infinite lattice of any dimension is straightforward in the extreme adiabatic regime. The adiabatic solution of the infinite one-dimensional chain with one electron was obtained by Rashba [79] in a continuous (i.e. effective mass) approximation and by Holstein [68] and Kabanov and Mashtakov [80] for a discrete lattice. The latter authors also studied the Holstein two-dimensional and three-dimensional lattices in the adiabatic limit. According to [80], the self-trapping of a single electron occurs at $\lambda \geq 0.875$ and at $\lambda \geq 0.92$ in two and three dimensions, respectively. The radius of the self-trapped adiabatic polaron, r_p, is readily derived from its continuous wavefunction [79]

$$\psi(x) \simeq 1/\cosh(\lambda x/a). \tag{4.24}$$

It becomes smaller than the lattice constant ($r_p = a/\lambda$) for $\lambda \geq 1$. That is why a multi-polaron system remains in a self-trapped insulating state in the strong-coupling adiabatic regime, no matter how many polarons it has. The only instability which might occur in this regime is the formation of on-site self-trapped bipolarons (section 4.6), if the on-site attractive interaction, $2\lambda zt$, is larger than the repulsive Hubbard U [14]. On-site self-trapped bipolarons form a charge-ordered state due to a weak repulsion between them [10]. The asymptotically exact many-particle ground state of the half-filled Holstein model in the strong-coupling limit ($\lambda \to \infty$) is

$$\psi = \prod_{j \in B} c_{j\uparrow}^{\dagger} c_{j\downarrow}^{\dagger} |0\rangle \tag{4.25}$$

for any value of the adiabatic ratio, ω/zt [10, 81]. Here the js are B-sites of a bipartite lattice $A + B$. It is an insulating state rather than a Fermi liquid, which is expected in the Migdal approximation at any value of λ in the adiabatic limit, $\omega \to 0$.

The non-adiabatic corrections (phonons) allow polarons and bipolarons to propagate as Bloch states in a new narrow band (sections 4.3 and 4.6). Thus, under certain conditions, the multi-polaron system is metallic with polaronic or bipolaronic carriers rather than with bare electrons. There is a qualitative difference between the ordinary Fermi liquid and the polaronic one. In particular, the renormalized (effective) mass of electrons is independent of the ion mass M in ordinary metals (equation (3.68)) because λ does not depend on the isotope mass. In contrast, the polaron effective mass m^* will depend on M (section 4.3). Hence, there is a large isotope effect on the carrier mass in polaronic metals [82] (section 4.7.5) while there is no carrier mass isotope effect in ordinary metals. Likewise, the bipolaron superconducting state is essentially different from the BCS superconductor (section 4.5).

In the last years, quite a few numerical and analytical studies have confirmed these conclusions (see, for example, [80–96]). In particular, Takada [86, 88] applied the gauge-invariant self-consistent method neglecting the momentum dependence of the vertex. Benedetti and Zeyher [91] applied the dynamical mean-field theory in infinite dimensions. As in the $1/\lambda$ expansion technique, both approaches avoided the problem of broken translation symmetry by using the non-dispersive vertex and GFs as the starting point. As a result, they arrived at the same conclusion about the applicability of the Migdal approach (in [91] the critical value of λ was found to be 1.3 in the adiabatic limit).

The transition into the self-trapped state due to the broken translational symmetry is expected at $0.5 < \lambda < 1.3$ (depending on the lattice dimensionality) for any electron–phonon interaction conserving the on-site electron occupation numbers. For example, Hiramoto and Toyozawa [97] calculated the strength of the deformation potential, which transforms electrons into small polarons and bipolarons. They found that the transition of two electrons into a self-trapped small bipolaron occurs at the electron–acoustic phonon coupling $\lambda \simeq 0.5$, that is half of the critical value of λ at which the transition of the electron into the small acoustic polaron takes place in the extreme adiabatic limit, $sq_{\mathrm{D}} \ll zt$. The effect of the adiabatic ratio sq_{D}/zt on the critical value of λ was found to be negligible. The radius of the acoustic polaron and bipolaron is about the lattice constant, so that the critical value of λ does not very much depend on the number of electrons in this case either.

4.3 Polaron dynamics

4.3.1 Polaron band

The kinetic energy is smaller than the interaction energy as long as $\lambda > 1$. Hence, a self-consistent approach to the *many*-polaron problem is possible with the '$1/\lambda$'

expansion technique [98], which treats the kinetic energy as a perturbation. The technique is based on the fact, known for a long time, that there is an analytical exact solution of a *single* polaron problem in the strong-coupling limit $\lambda \to \infty$ [71]. Following Lang and Firsov, we apply the canonical transformation e^S to diagonalize the Hamiltonian. The diagonalization is exact if $T(\boldsymbol{m}) = 0$ (or $\lambda = \infty$):

$$\tilde{H} = e^S H e^{-S} \tag{4.26}$$

where

$$S = -\sum_{\boldsymbol{q},\nu,i} \hat{n}_i [u_i(\boldsymbol{q},\nu) d_{\boldsymbol{q}\nu} - H.c.] \tag{4.27}$$

is such that $S^\dagger = -S$. The electron and phonon operators are transformed as

$$\tilde{c}_i = c_i \exp\left[\sum_{\boldsymbol{q},\nu} u_i(\boldsymbol{q},\nu) d_{\boldsymbol{q}\nu} - H.c.\right] \tag{4.28}$$

and

$$\tilde{d}_{\boldsymbol{q}\nu} = d_{\boldsymbol{q}\nu} - \sum_i \hat{n}_i u_i^*(\boldsymbol{q},\nu) \tag{4.29}$$

respectively (appendix E). It follows from equation (4.29) that the Lang–Firsov canonical transformation shifts the ions to new equilibrium positions. In a more general sense, it changes the boson vacuum. As a result, the transformed Hamiltonian takes the following form:

$$\tilde{H} = \sum_{i,j} [\hat{\sigma}_{ij} - \mu\delta_{ij}] c_i^\dagger c_j - E_p \sum_i \hat{n}_i + \sum_{\boldsymbol{q},\nu} \omega_{\boldsymbol{q}\nu} (d_{\boldsymbol{q}\nu}^\dagger d_{\boldsymbol{q}\nu} + 1/2) + \tfrac{1}{2} \sum_{i \neq j} v_{ij} \hat{n}_i \hat{n}_j \tag{4.30}$$

where

$$\hat{\sigma}_{ij} = T(\boldsymbol{m}-\boldsymbol{n}) \delta_{ss'} \exp\left(\sum_{\boldsymbol{q},\nu} [u_j(\boldsymbol{q},\nu) - u_i(\boldsymbol{q},\nu)] d_{\boldsymbol{q}\nu} - H.c.\right) \tag{4.31}$$

is the renormalized hopping integral depending on the phonon operators and

$$\begin{aligned} v_{ij} &\equiv v(\boldsymbol{m}-\boldsymbol{n}) \\ &= V_c(\boldsymbol{m}-\boldsymbol{n}) - \frac{1}{N} \sum_{\boldsymbol{q},\nu} |\gamma(\boldsymbol{q},\nu)|^2 \omega_{\boldsymbol{q}\nu} \cos[\boldsymbol{q}\cdot(\boldsymbol{m}-\boldsymbol{n})] \end{aligned} \tag{4.32}$$

is the interaction of polarons comprising their Coulomb repulsion and the interaction via a local lattice deformation. In the extreme infinite-coupling limit ($\lambda \to \infty$) we can neglect the hopping term of the transformed Hamiltonian. The rest has analytically determined eigenstates and eigenvalues. The eigenstates $|\tilde{N}\rangle = |n_i, n_{\boldsymbol{q}\nu}\rangle$ are sorted by the polaron $n_{\boldsymbol{m}s}$ and phonon $n_{\boldsymbol{q}\nu}$ occupation numbers. The energy levels are

$$E = -(\mu + E_p) \sum_i n_i + \tfrac{1}{2} \sum_{i \neq j} v_{ij} n_i n_j + \sum_{\boldsymbol{q}} \omega_{\boldsymbol{q}\nu} (n_{\boldsymbol{q}\nu} + 1/2) \tag{4.33}$$

where $n_i = 0, 1$ and $n_{q\nu} = 0, 1, 2, 3, \ldots, \infty$.

The Hamiltonian (4.30), in zero order with respect to the hopping describes localized polarons and independent phonons, which are vibrations of ions relative to new equilibrium positions, which depend on the polaron occupation numbers. The phonon frequencies remain unchanged in this limit. The middle of the electron band falls by the polaron level-shift E_p due to a potential well created by lattice deformation,

$$E_p = \frac{1}{2N}\sum_{q,\nu}|\gamma(q,\nu)|^2\omega_{q\nu}. \tag{4.34}$$

Now let us discuss the $1/\lambda$ expansion. First we restrict the discussion to a single-polaron problem with no polaron–polaron interaction and $\mu = 0$. The finite hopping term leads to the polaron tunnelling because of degeneracy of the zero-order Hamiltonian with respect to the site position of the polaron. To see how the tunnelling occurs, we apply the perturbation theory using $1/\lambda$ as a small parameter, where

$$\lambda \equiv \frac{E_p}{D}. \tag{4.35}$$

Here $D = zT(a)$, z is the coordination lattice number and $T(a)$ is the nearest-neighbour hopping integral. The proper Bloch set of N-degenerate zero-order eigenstates with the lowest energy $(-E_p)$ of the unperturbed Hamiltonian is

$$|\boldsymbol{k}, 0\rangle = \frac{1}{\sqrt{N}}\sum_{\boldsymbol{m}} c_{\boldsymbol{m}s}^{\dagger}\exp(\mathrm{i}\boldsymbol{k}\cdot\boldsymbol{m})|0\rangle \tag{4.36}$$

where $|0\rangle$ is the vacuum. By applying the textbook perturbation theory, one readily calculates the perturbed energy levels. Up to the second order in the hopping integral, they are given by

$$E(\boldsymbol{k}) = -E_p + \epsilon_{\boldsymbol{k}} - \sum_{\boldsymbol{k}',n_{q\nu}}\frac{|\langle\boldsymbol{k},0|\sum_{i,j}\hat{\sigma}_{ij}c_i^{\dagger}c_j|\boldsymbol{k}',n_{q\nu}\rangle|^2}{\sum_{q,\nu}\omega_{q\nu}n_{q\nu}} \tag{4.37}$$

where $|\boldsymbol{k}', n_{q\nu}\rangle$ are the excited states of the unperturbed Hamiltonian with one electron and at least one real phonon. The second term in equation (4.37), which is linear with respect to the bare hopping $T(\boldsymbol{m})$, describes a polaron-band dispersion,

$$\epsilon_{\boldsymbol{k}} = \sum_{\boldsymbol{m}} T(\boldsymbol{m})\mathrm{e}^{-g^2(\boldsymbol{m})}\exp(-\mathrm{i}\boldsymbol{k}\cdot\boldsymbol{m}) \tag{4.38}$$

where

$$g^2(\boldsymbol{m}) = \frac{1}{2N}\sum_{q,\nu}|\gamma(q,\nu)|^2[1-\cos(q\cdot\boldsymbol{m})] \tag{4.39}$$

is the *band-narrowing factor* at zero temperature. The third term in equation (4.37), quadratic in $T(\boldsymbol{m})$, yields a negative $\boldsymbol{k}$*-independent* correction

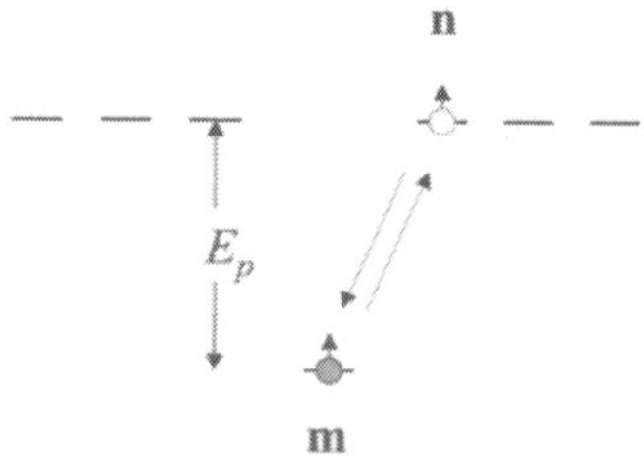

Figure 4.2. 'Back and forth' virtual transitions of the polaron without any transfer of the lattice deformation from one site to another. These transitions shift the middle of the band down without any real charge delocalization.

to the polaron level-shift of the order of $1/\lambda^2$. The origin of this correction, which could be much larger than the first-order contribution (equation (4.38)) (containing a small exponent), is understood in figure 4.2. The polaron localized in the potential well of depth E_p on site $\boldsymbol{m}$ hops onto a neighbouring site $\boldsymbol{n}$ with no deformation around and comes back. As in any second-order correction, this transition shifts the energy down by an amount of about $-T^2(\boldsymbol{m})/E_p$. It has little to do with the polaron effective mass and the polaron tunnelling mobility because the lattice deformation around $\boldsymbol{m}$ does not follow the electron. The electron hops back and forth many times (about e^{g^2}) waiting for a sufficient lattice deformation to appear around the site $\boldsymbol{n}$. Only after the deformation around $\boldsymbol{n}$ is created does the polaron tunnel onto the next site together with the deformation.

4.3.2 Damping of the polaron band

The polaron band is exponentially narrow, see equation (4.38). Hence, one can raise a concern about its existence in real solids. At zero temperature the perturbation term of the transformed Hamiltonian conserves the momentum because all off-diagonal matrix elements vanish:

$$\langle \boldsymbol{k}, 0| \sum_{i,j} \hat{\sigma}_{i,j} c_i^\dagger c_j |\boldsymbol{k}', 0\rangle = 0 \tag{4.40}$$

if $\boldsymbol{k} \neq \boldsymbol{k}'$. The emission of a single high-frequency phonon is impossible for any $\boldsymbol{k}$ because of the energy conservation. The *polaron* half-bandwidth is exponentially reduced,

$$w \approx D e^{-g^2} \tag{4.41}$$

and it is usually less than the optical phonon energy ω_0 (g^2 is of the order of $D\lambda/\omega_0$). Hence, there is no damping of the polaron band at $T = 0$ caused by optical phonons, no matter how strong the interaction is. These phonons 'dress' the electron and coherently follow its motion. However, at finite temperatures, the simultaneous emission and absorption of phonons (figure 4.3) become possible.

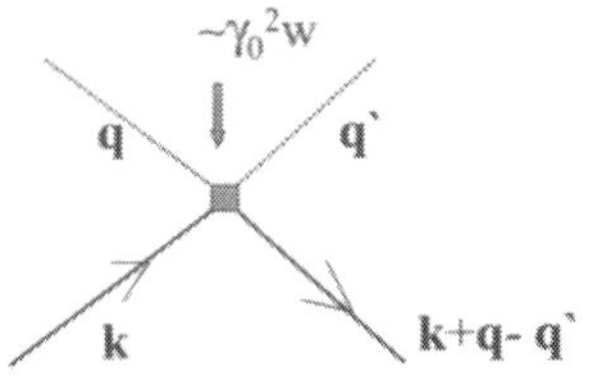

Figure 4.3. Two-phonon scattering responsible for the damping of the polaron band.

Moreover, the polaron bandwidth shrinks with increasing temperature because the phonon-averaged hopping integrals depend on temperature (appendix E):

$$\langle \hat{\sigma}_{ij} \rangle_{\rm ph} = T(\boldsymbol{m} - \boldsymbol{n})\delta_{ss'} \exp\left(- \frac{1}{2N} \sum_{\boldsymbol{q},\nu} |\gamma(\boldsymbol{q},\nu)|^2 [1 - \cos(\boldsymbol{q} \cdot \boldsymbol{m})] \coth \frac{\omega_{\boldsymbol{q}\nu}}{2T} \right).$$

For high temperatures ($T \gg \omega_0/2$) the band narrows exponentially:

$$w \approx D e^{-T/T_0} \tag{4.42}$$

where

$$T_0^{-1} = \frac{1}{N} \sum_{\boldsymbol{q},\nu} |\gamma(\boldsymbol{q},\nu)|^2 \omega_{\boldsymbol{q}\nu}^{-1} [1 - \cos(\boldsymbol{q} \cdot \boldsymbol{m})].$$

However, the two-phonon scattering of polarons (figure 4.3) becomes more important with increasing temperature.

We can estimate the scattering rate by applying the Fermi–Dirac golden rule:

$$\frac{1}{\tau} = 2\pi \left\langle \sum_{\boldsymbol{q},\boldsymbol{q}'} |M_{\boldsymbol{q}\boldsymbol{q}'}|^2 \delta(\epsilon_{\boldsymbol{k}} - \epsilon_{\boldsymbol{k}+\boldsymbol{q}-\boldsymbol{q}'}) \right\rangle_{\rm ph} \tag{4.43}$$

where the corresponding matrix element is

$$M_{\boldsymbol{q}\boldsymbol{q}'} = \sum_{i,j} \langle \boldsymbol{k} + \boldsymbol{q} - \boldsymbol{q}', n_{\boldsymbol{q}} - 1, n_{\boldsymbol{q}'} + 1 | \hat{\sigma}_{i,j} c_i^\dagger c_j | \boldsymbol{k}, n_{\boldsymbol{q}}, n_{\boldsymbol{q}'} \rangle.$$

For simplicity we drop the phonon branch index ν and consider the momentum independent $\gamma(\boldsymbol{q}) = \gamma_0$ and $\omega_{\boldsymbol{q}} = \omega_0$. Expanding $\hat{\sigma}_{ij}$-operators in powers of the phonon creation and annihilation operators, we estimate the matrix element of the two-phonon scattering as

$$M_{\boldsymbol{q}\boldsymbol{q}'} \approx \frac{1}{N} w \gamma_0^2 \sqrt{n_{\boldsymbol{q}}(n_{\boldsymbol{q}'} + 1)}. \tag{4.44}$$

Using this estimate and the polaron density of states,

$$N_p(\xi) \equiv \frac{1}{N} \sum_{\boldsymbol{k}} \delta(\xi - \epsilon_{\boldsymbol{k}}) \approx \frac{1}{2w} \tag{4.45}$$

we obtain

$$\frac{1}{\tau} \approx w\gamma_0^4 n_\omega(1+n_\omega) \tag{4.46}$$

where $n_\omega = [\exp(\omega_0/T) - 1]^{-1}$ is the phonon distribution function. The polaron band is well defined if

$$\frac{1}{\tau} < w \tag{4.47}$$

which is satisfied for a wide temperature range

$$T \leq T_{\min} \approx \frac{\omega_0}{\ln \gamma_0^4} \tag{4.48}$$

about half of the characteristic phonon frequency for relevant values of γ_0^2. Phonons dominate in the scattering at finite temperatures if the number of impurities is sufficiently low. Therefore, the polaron mobility decreases when the temperature increases from zero to $T_{\min}$ due to an increasing number of phonons. At higher temperatures, the incoherent thermal activated hopping dominates in the polaron dynamics [66–68, 71] and the polaron states are no longer the Bloch states. Hence, the mobility increases above $T_{\min}$, where it is at minimum, due to thermal activated hopping. There is a temperature range around $T_{\min}$ where the thermal activated hopping still makes a small contribution to the conductivity but the uncertainty in the polaron band is already significant [71]. The polaron transport theory requires a special diagrammatic technique in this region [74, 75]. The optical phonon frequencies are exceptionally high, about 1000 K or even higher and polarons are in Bloch states in the whole relevant range of temperatures in novel superconductors.

4.3.3 Small Holstein polaron and small Fröhlich polaron

The narrowing of the band and the polaron effective mass strongly depend on the radius of the electron–phonon interaction [100]. Let us compare the small Holstein polaron (SHP) formed by the short-range e–ph interaction and a small polaron formed by the long-range (Fröhlich) interaction, which we refer as the small Fröhlich polaron (SFP). Introducing a normal coordinate at site $\boldsymbol{n}$ as

$$\xi_{\boldsymbol{n}} = \sum_{\boldsymbol{q}} (2NM\omega_{\boldsymbol{q}})^{-1/2} \mathrm{e}^{\mathrm{i}\boldsymbol{q}\cdot\boldsymbol{n}} d_{\boldsymbol{q}} + H.c. \tag{4.49}$$

and a 'force' between the electron at site $\boldsymbol{m}$ and the normal coordinate $\xi_{\boldsymbol{n}}$,

$$f(\boldsymbol{m}) = N^{-1} \sum_{\boldsymbol{q}} \gamma(\boldsymbol{q})(M\omega_{\boldsymbol{q}}^3)^{1/2} \mathrm{e}^{\mathrm{i}\boldsymbol{q}\cdot\boldsymbol{m}} \tag{4.50}$$

we rewrite the e–ph interaction, equation (4.3), as

$$H_{\text{e-ph}} = \sum_{\boldsymbol{n},i} f(\boldsymbol{m}-\boldsymbol{n}) \xi_{\boldsymbol{n}} \hat{n}_i. \tag{4.51}$$

For simplicity we consider the interaction with a single phonon branch and $\gamma(-\boldsymbol{q}) = \gamma(\boldsymbol{q})$. In general, there is no simple relation between the polaron level shift E_p and the exponent g^2 of the mass enhancement. This relation depends on the form of the electron–phonon interaction. Indeed, for dispersionless phonons, $\omega_{\boldsymbol{q}} = \omega_0$, using equations (4.34) and (4.39) we obtain

$$E_p = \frac{1}{2M\omega_0^2} \sum_{\boldsymbol{m}} f^2(\boldsymbol{m}) \tag{4.52}$$

and

$$g^2 = \frac{1}{2M\omega_0^3} \sum_{\boldsymbol{m}} [f^2(\boldsymbol{m}) - f(\boldsymbol{m}) f(\boldsymbol{m} + \boldsymbol{a})] \tag{4.53}$$

where $\boldsymbol{a}$ is the primitive lattice vector. In the nearest-neighbour approximation, the effective mass renormalization is given by

$$m^*/m = \mathrm{e}^{g^2}$$

where m is the bare band mass and $1/m^* = \partial^2 \epsilon_{\boldsymbol{k}}/\partial k^2$ at $k \to 0$ is the inverse polaron mass.

If the interaction is short range, $f(\boldsymbol{m}) = \kappa \delta_{\boldsymbol{m},0}$ (the Holstein model), then $g^2 = E_p/\omega$. Here κ is a constant. In general, we have $g^2 = \gamma E_p/\omega$ with the numerical coefficient

$$\gamma = \frac{1 - \sum_{\boldsymbol{m}} f(\boldsymbol{m}) f(\boldsymbol{m} + \boldsymbol{a})}{\sum_{\boldsymbol{n}} f^2(\boldsymbol{n})} \tag{4.54}$$

which might be less than 1. To estimate γ, let us consider a one-dimensional chain model with a long-range Coulomb interaction between the electron on the chain ($\times$) and ion vibrations of the chain ($\circ$), polarized in a direction perpendicular to the chains [94] (figure 4.4). The corresponding force is given by

$$f(\boldsymbol{m} - \boldsymbol{n}) = \frac{\kappa}{(|\boldsymbol{m} - \boldsymbol{n}|^2 + 1)^{3/2}}. \tag{4.55}$$

Here the distance along the chains, $|\boldsymbol{m} - \boldsymbol{n}|$, is measured in units of the lattice constant, a, the inter-chain distance is also a and we take $a = 1$. For this long-range interaction, we obtain $E_p = 1.27\kappa^2/(2M\omega^2)$, $g^2 = 0.49\kappa^2/(2M\omega^3)$ and

$$g^2 = 0.39 E_p/\omega. \tag{4.56}$$

Thus the effective mass renormalization is much smaller than in the Holstein model, roughly $m^*_{\mathrm{SFP}} \propto (m^*_{\mathrm{SHP}})^{1/2}$.

Not only does the small polaron mass strongly depend on the radius of the electron–phonon interaction but the range of the applicability of the analytical $1/\lambda$ expansion theory also does. The theory appears almost exact in a wide region of parameters for the Fröhlich interaction. The exact polaron mass in a

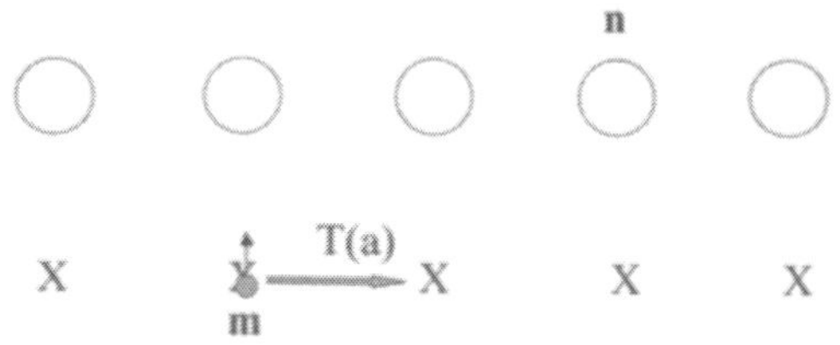

Figure 4.4. A one-dimensional model of a small polaron on the chain interacting with the ion displacements of another chain.

wide region of the adiabatic parameter $\omega/T(a)$ and coupling was calculated with the continuous-time path-integral quantum Monte Carlo (QMC) algorithm [94]. This method is free from any systematic finite-size, finite-time-step and finite-temperature errors and allows for an *exact* (in the QMC sense) calculation of the ground-state energy and the effective mass of the lattice polaron for any electron–phonon interaction described by the Hamiltonian (4.51).

At large λ (>1.5), the SFP was found to be much lighter than the SHP, while the large Fröhlich polaron (i.e. at $\lambda < 1$) was *heavier* than the large Holstein polaron with the same binding energy (figure 4.5). The mass ratio $m^*_{\text{FP}}/m^*_{\text{HP}}$ is a non-monotonic function of λ. The effective mass of *Fröhlich* polarons, $m^*_{\text{FP}}(\lambda)$ is well fitted by a single exponent, which is $\mathrm{e}^{0.73\lambda}$ for $\omega_0 = T(a)$ and $\mathrm{e}^{1.4\lambda}$ for $\omega = 0.5T(a)$. The exponents are remarkably close to those obtained with the Lang–Firsov transformation, $\mathrm{e}^{0.78\lambda}$ and $\mathrm{e}^{1.56\lambda}$, respectively. Hence, in the case of the Fröhlich interaction the transformation is perfectly accurate even in the moderate adiabatic regime, $\omega/T(a) \leq 1$ for *any* coupling strength. This is not the case for the Holstein polaron. If the interaction is short range, the same analytical technique is applied only in the non-adiabatic regime $\omega/T(a) > 1$.

Another interesting point is that the size of the SFP and the length over which the distortion spreads are *different*. In the strong-coupling limit, the polaron is almost localized on one site $\boldsymbol{m}$. Hence, the size of its wavefunction is the atomic size. In contrast, the ion displacements, proportional to the displacement force $f(\boldsymbol{m}-\boldsymbol{n})$, spread over a large distance. Their amplitude at a site $\boldsymbol{n}$ falls with distance as $|\boldsymbol{m}-\boldsymbol{n}|^{-3}$ in our one-dimensional model. The polaron cloud (i.e. lattice distortion) is more extended than the polaron itself [66, 72, 100]. Such a polaron tunnels with a larger probability than the Holstein polaron due to a smaller *relative* lattice distortion around two neighbouring sites. For a short-range e–ph interaction, the *entire* lattice deformation disappears at one site and then forms at its neighbour, when the polaron tunnels from site to site. Therefore, $\gamma = 1$ and the polaron is very heavy already at $\lambda \approx 1$. In contrast, if the interaction is long-ranged, only a fraction of the total deformation changes every time the polaron tunnels from one site to its neighbour and γ is smaller than 1. The fact that the Lang–Firsov transformation is rather accurate for the long-range interaction in a wide region of parameters, allows us to generalize this result. Including all

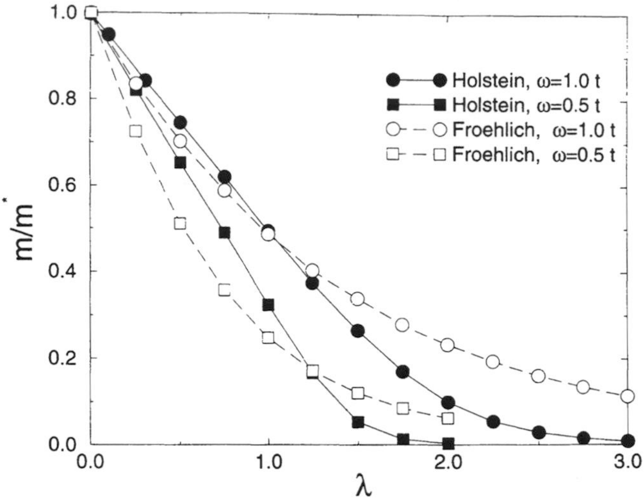

Figure 4.5. Inverse effective polaron mass in units of $1/m = 2T(a)a^2$.

phonon branches in the three-dimensional lattice, we obtain $m^*_{\mathrm{SFP}} = (m^*_{\mathrm{SHP}})^{\gamma}$, where the masses are measured in units of the band mass, and

$$\gamma = \frac{\sum_{q,\nu} \gamma^2(\boldsymbol{q}, \nu)[1 - \cos(\boldsymbol{q} \cdot \boldsymbol{m})]}{\sum_{q,\nu} \gamma^2(\boldsymbol{q}, \nu)}. \tag{4.57}$$

For the Fröhlich interaction (i.e. $\gamma(\boldsymbol{q}) \propto 1/q$), we find $\gamma = 0.57$ in a cubic lattice and $\gamma = 0.255$ for the in-plane oxygen hole in cuprates (Part II). A lighter mass of SFP compared with the non-dispersive SHP is a generic feature of any dispersive electron–phonon interaction. As an example, a short-range interaction with dispersive acoustic phonons ($\gamma(\boldsymbol{q}) \sim 1/q^{1/2}$, $\omega_{\boldsymbol{q}} \sim q$) also leads to a lighter polaron in the strong-coupling regime compared with the SHP. Actually, Holstein [68] pointed out in his original paper that the dispersion is a vital ingredient of the polaron theory.

4.3.4 Polaron spectral and Green's functions

The multi-polaron problem has an exact solution in the extreme infinite-coupling limit ($\lambda = \infty$) for any type of e–ph interaction conserving the on-site occupation numbers of electrons (equation (4.3)). For finite coupling, the $1/\lambda$ perturbation expansion is applied. The expansion parameter is actually [72, 74, 98, 99]

$$\frac{1}{2z\lambda^2} \ll 1$$

so that the analytical perturbation theory has a wider region of applicability than one can expect using a semiclassical estimate $E_p > D$. However, the expansion convergency is different for different e–ph interactions. Exact numerical diagonalizations of vibrating clusters, variational calculations [83, 85, 87, 89, 92, 93], dynamical mean-field approach in infinite dimensions [91], and quantum Monte Carlo simulations [94] revealed that the ground state energy ($\approx -E_p$) is not very sensitive to the parameters. In contrast, the effective mass, the bandwidth and the polaron density of states strongly depend on the adiabatic ratio $\omega/T(a)$ and on the radius of the interaction. The first order in $1/\lambda$ perturbation theory is practically exact in the non-adiabatic regime ($\omega/T(a) > 1$) *for any value* of the coupling constant and any type of e–ph interaction. However, it *overestimates* the polaron mass by a few orders of magnitude in the adiabatic case ($\omega/T(a) < 1$), if the interaction is short-ranged [83]. A much lower effective mass of the adiabatic Holstein polaron compared with that estimated using first-order perturbation theory is the result of the poor convergency of the perturbation expansion owing to the double-well potential [68] in the adiabatic limit. The tunnelling probability is extremely sensitive to the shape of this potential. However, the analytical theory is practically exact in a wider range of the adiabatic parameter and of the coupling constant for the long-range Fröhlich interaction.

Keeping this in mind, let us calculate the one-particle GF in the first order in $1/\lambda$. Applying the canonical transformation we write the transformed Hamiltonian as

$$\tilde{H} = H_p + H_{\text{ph}} + H_{\text{int}} \tag{4.58}$$

where

$$H_p = \sum_{\boldsymbol{k}} \xi(\boldsymbol{k}) c_{\boldsymbol{k}}^\dagger c_{\boldsymbol{k}} \tag{4.59}$$

is the 'free' *polaron* contribution and

$$H_{\text{ph}} = \sum_{\boldsymbol{q}} \omega_{\boldsymbol{q}} (d_{\boldsymbol{q}}^\dagger d_{\boldsymbol{q}} + 1/2) \tag{4.60}$$

is the free phonon part. For simplicity, we drop the spin and phonon branch indexes. Here $\xi(\boldsymbol{k}) = Z' E(\boldsymbol{k}) - \mu$ is the renormalized polaron band dispersion. The chemical potential μ includes the polaron level shift $-E_p$. It might also include all higher orders in $1/\lambda$ corrections to the spectrum independent of $\boldsymbol{k}$. $E(\boldsymbol{k}) = \sum_{\boldsymbol{m}} T(\boldsymbol{m}) \exp(-\mathrm{i}\boldsymbol{k} \cdot \boldsymbol{m})$ is the bare dispersion in the rigid lattice and

$$Z' = \frac{\sum_{\boldsymbol{m}} T(\boldsymbol{m}) \mathrm{e}^{-g^2(\boldsymbol{m})} \exp(-\mathrm{i}\boldsymbol{k} \cdot \boldsymbol{m})}{\sum_{\boldsymbol{m}} T(\boldsymbol{m}) \exp(-\mathrm{i}\boldsymbol{k} \cdot \boldsymbol{m})} \tag{4.61}$$

is the band-narrowing factor ($Z' = \exp(-\gamma E_p/\omega)$) as discussed earlier). The interaction term H_{int} comprises the polaron–polaron interaction, equation (4.32), and the residual polaron–phonon interaction

$$H_{\text{p-ph}} \equiv \sum_{i \neq j} [\hat{\sigma}_{ij} - \langle \hat{\sigma}_{ij} \rangle_{\text{ph}}] c_i^\dagger c_j. \tag{4.62}$$

We can neglect $H_{\mathrm{p-ph}}$ in the first order in $1/\lambda \ll 1$. To better understand the spectral properties of a single polaron, let us also neglect the polaron–polaron interaction. If $H_{\mathrm{int}} = 0$, the energy levels are

$$E_{\{\tilde{m}\}} = \sum_{k} \xi_{k} n_{k} + \sum_{q} \omega_{q}[n_{q} + 1/2] \tag{4.63}$$

and the transformed eigenstates $|\tilde{m}\rangle$ are sorted by the polaron Bloch-state occupation numbers, $n_{k} = 0, 1$, and the phonon occupation numbers, $n_{q} = 0, 1, 2, \ldots, \infty$. The spectral function (equation (D.10)) is defined by the matrix element $\langle n|c_{k}|m\rangle$. It can be written as

$$\langle n|c_{k}|m\rangle = \frac{1}{\sqrt{N}} \sum_{m} \mathrm{e}^{-\mathrm{i}k\cdot m} \langle \tilde{n}|c_{i}\hat{X}_{i}|\tilde{m}\rangle \tag{4.64}$$

by the use of the Wannier representation and the Lang–Firsov transformation. Here

$$\hat{X}_{i} = \exp\left[\sum_{q} u_{i}(q) d_{q} - H.c.\right].$$

Now, applying the Fourier transform of the δ-function in equation (D.10),

$$\delta(\omega_{nm} + \omega) = \frac{1}{2\pi} \int_{-\infty}^{\infty} \mathrm{d}t\, \mathrm{e}^{\mathrm{i}(\omega_{nm}+\omega)t}$$

the spectral function is expressed as

$$A(k,\omega) = \tfrac{1}{2} \int_{-\infty}^{\infty} \mathrm{d}t\, \mathrm{e}^{\mathrm{i}\omega t} \frac{1}{N} \sum_{m,n} \mathrm{e}^{\mathrm{i}k\cdot(n-m)}$$
$$\times \{\langle\langle c_{i}(t)\hat{X}_{i}(t)c_{j}^{\dagger}\hat{X}_{j}^{\dagger}\rangle\rangle + \langle\langle c_{j}^{\dagger}\hat{X}_{j}^{\dagger}c_{i}(t)\hat{X}_{i}(t)\rangle\rangle\}. \tag{4.65}$$

Here the quantum and statistical averages are performed for independent polarons and phonons, therefore

$$\langle\langle c_{i}(t)\hat{X}_{i}(t)\hat{X}_{j}^{\dagger}c_{i}^{\dagger}\rangle\rangle = \langle\langle c_{i}(t)c_{j}^{\dagger}\rangle\rangle\langle\langle\hat{X}_{i}(t)\hat{X}_{j}^{\dagger}\rangle\rangle. \tag{4.66}$$

The Heisenberg free-polaron operator evolves with time as

$$c_{k}(t) = c_{k}\mathrm{e}^{-\mathrm{i}\xi_{k}t} \tag{4.67}$$

and

$$\langle\langle c_{i}(t)c_{i}^{\dagger}\rangle\rangle = \frac{1}{N} \sum_{k',k''} \mathrm{e}^{\mathrm{i}(k'\cdot m - k''\cdot n)} \langle\langle c_{k'}(t)c_{k''}^{\dagger}\rangle\rangle$$
$$= \frac{1}{N} \sum_{k'} [1 - \bar{n}(k')]\mathrm{e}^{\mathrm{i}k'\cdot(m-n)-\mathrm{i}\xi_{k'}t} \tag{4.68}$$

$$\langle\langle c_{i}^{\dagger}c_{i}(t)\rangle\rangle = \frac{1}{N} \sum_{k'} \bar{n}(k')\mathrm{e}^{\mathrm{i}k'\cdot(m-n)-\mathrm{i}\xi_{k'}t} \tag{4.69}$$

where $\bar{n}(\boldsymbol{k}) = [1 + \exp \xi_{\boldsymbol{k}}/T]^{-1}$ is the Fermi–Dirac distribution function of polarons. The Heisenberg free-phonon operator evolves in a similar way,

$$d_{\boldsymbol{q}}(t) = d_{\boldsymbol{q}} \mathrm{e}^{-\mathrm{i}\omega_{\boldsymbol{q}} t}$$

and

$$\langle\langle \hat{X}_i(t) \hat{X}_j^\dagger \rangle\rangle = \prod_{\boldsymbol{q}} \langle\langle \exp[u_i(\boldsymbol{q}, t) d_{\boldsymbol{q}} - H.c.] \exp[-u_j(\boldsymbol{q}) d_{\boldsymbol{q}} - H.c.] \rangle\rangle \tag{4.70}$$

where $u_{i,j}(\boldsymbol{q}, t) = u_{i,j}(\boldsymbol{q}) \mathrm{e}^{-\mathrm{i}\omega_{\boldsymbol{q}} t}$. This average is calculated using the operator identity (appendix E)

$$\mathrm{e}^{\hat{A}+\hat{B}} = \mathrm{e}^{\hat{A}} \mathrm{e}^{\hat{B}} \mathrm{e}^{-[\hat{A}, \hat{B}]/2} \tag{4.71}$$

which is applied for any two operators $\hat{A}$ and $\hat{B}$, whose commutator $[\hat{A}, \hat{B}]$ is a number. Because $[d_{\boldsymbol{q}}, d_{\boldsymbol{q}}^\dagger] = 1$, we can apply this identity in equation (4.70) to obtain

$$\mathrm{e}^{[u_i(\boldsymbol{q},t) d_{\boldsymbol{q}} - H.c.]} \mathrm{e}^{[-u_j(\boldsymbol{q}) d_{\boldsymbol{q}} - H.c.]} = \mathrm{e}^{(\alpha^* d_{\boldsymbol{q}}^\dagger - \alpha d_{\boldsymbol{q}})} \times \mathrm{e}^{[u_i(\boldsymbol{q},t) u_j^*(\boldsymbol{q}) - u_i^*(\boldsymbol{q},t) u_j(\boldsymbol{q})]/2}$$

where $\alpha \equiv u_j(\boldsymbol{q}, t) - u_i(\boldsymbol{q})$. Applying once again the same identity yields

$$\mathrm{e}^{[u_i(\boldsymbol{q},t) d_{\boldsymbol{q}} - H.c.]} \mathrm{e}^{[-u_j(\boldsymbol{q}) d_{\boldsymbol{q}} - H.c.]} = \mathrm{e}^{\alpha^* d_{\boldsymbol{q}}^\dagger} \mathrm{e}^{-\alpha d_{\boldsymbol{q}}} \mathrm{e}^{-|\alpha|^2/2} \times \mathrm{e}^{[u_i(\boldsymbol{q},t) u_j^*(\boldsymbol{q}) - u_i^*(\boldsymbol{q},t) u_j(\boldsymbol{q})]/2}. \tag{4.72}$$

Now the quantum and statistical averages are calculated by the use of (appendix E)

$$\langle\langle \mathrm{e}^{\alpha^* d_{\boldsymbol{q}}^\dagger} \mathrm{e}^{-\alpha d_{\boldsymbol{q}}} \rangle\rangle = \mathrm{e}^{-|\alpha|^2 n_\omega} \tag{4.73}$$

where $n_\omega = [\exp(\omega_{\boldsymbol{q}}/T) - 1]^{-1}$ is the Bose–Einstein distribution function of phonons. Collecting all multiplies in equation (4.72), we arrive at

$$\langle\langle \hat{X}_i(t) \hat{X}_j^\dagger \rangle\rangle = \exp\left\{ -\frac{1}{2N} \sum_{\boldsymbol{q}} |\gamma(\boldsymbol{q})|^2 f_{\boldsymbol{q}}(\boldsymbol{m} - \boldsymbol{n}, t) \right\} \tag{4.74}$$

where

$$f_{\boldsymbol{q}}(\boldsymbol{m}, t) = [1 - \cos(\boldsymbol{q} \cdot \boldsymbol{m}) \cos(\omega_{\boldsymbol{q}} t)] \coth \frac{\omega_{\boldsymbol{q}}}{2T} + i \cos(\boldsymbol{q} \cdot \boldsymbol{m}) \sin(\omega_{\boldsymbol{q}} t). \tag{4.75}$$

Here we have used the symmetry of $\gamma(-\boldsymbol{q}) = \gamma(\boldsymbol{q})$, and, hence, the terms containing $\sin(\boldsymbol{q} \cdot \boldsymbol{m})$ have disappeared. The average $\langle\langle \hat{X}_j^\dagger \hat{X}_i(t) \rangle\rangle$, which is a multiplier in the second term in the brackets of equation (4.65), is obtained by replacing $u_i(\boldsymbol{q}, t) \rightleftarrows u_j(\boldsymbol{q})$ in the previous expressions. The result is

$$\langle\langle \hat{X}_j^\dagger \hat{X}_i(t) \rangle\rangle = \langle\langle \hat{X}_i(t) \hat{X}_j^\dagger \rangle\rangle^*. \tag{4.76}$$

To proceed with the analytical results, we consider low temperatures, $T \ll \omega_{\boldsymbol{q}}$, when $\coth(\omega_{\boldsymbol{q}}/2T) \approx 1$. Then expanding the exponent in equation (4.74) yields

$$\langle\langle \hat{X}_i(t)\hat{X}_j^{\dagger}\rangle\rangle = Z\sum_{l=0}^{\infty}\frac{\{\sum_{\boldsymbol{q}}|\gamma(\boldsymbol{q})|^2 \mathrm{e}^{\mathrm{i}[\boldsymbol{q}\cdot(\boldsymbol{m}-\boldsymbol{n})-\omega_{\boldsymbol{q}}t]}\}^l}{(2N)^l l!} \tag{4.77}$$

where

$$Z = \exp\left[-\frac{1}{2N}\sum_{\boldsymbol{q}}|\gamma(\boldsymbol{q})|^2\right]. \tag{4.78}$$

Substituting equations (4.77) and (4.68) into equation (4.66) and performing summation with respect to $\boldsymbol{m}$, $\boldsymbol{n}$, $\boldsymbol{k}'$ and integration with respect to time in equation (4.65), we arrive at [101]

$$A(\boldsymbol{k},\omega) = \sum_{l=0}^{\infty}[A_l^{(-)}(\boldsymbol{k},\omega) + A_l^{(+)}(\boldsymbol{k},\omega)] \tag{4.79}$$

where

$$A_l^{(-)}(\boldsymbol{k},\omega) = \pi Z\sum_{\boldsymbol{q}_1,\ldots,\boldsymbol{q}_l}\frac{\prod_{r=1}^{l}|\gamma(\boldsymbol{q}_r)|^2}{(2N)^l l!} \times\left[1-\bar{n}\left(\boldsymbol{k}-\sum_{r=1}^{l}\boldsymbol{q}_r\right)\right]\delta\left(\omega-\sum_{r=1}^{l}\omega_{\boldsymbol{q}_r}-\xi_{\boldsymbol{k}-\sum_{r=1}^{l}\boldsymbol{q}_r}\right) \tag{4.80}$$

and

$$A_l^{(+)}(\boldsymbol{k},\omega) = \pi Z\sum_{\boldsymbol{q}_1,\ldots,\boldsymbol{q}_l}\frac{\prod_{r=1}^{l}|\gamma(\boldsymbol{q}_r)|^2}{(2N)^l l!} \times\bar{n}\left(\boldsymbol{k}+\sum_{r=1}^{l}\boldsymbol{q}_r\right)\delta\left(\omega+\sum_{r=1}^{l}\omega_{\boldsymbol{q}_r}-\xi_{\boldsymbol{k}+\sum_{r=1}^{l}\boldsymbol{q}_r}\right). \tag{4.81}$$

Obviously, equation (4.79) is in the form of a perturbative multi-phonon expansion. Each contribution $A_l^{(\pm)}(\boldsymbol{k},\omega)$ to the spectral function describes the transition from the initial state $\boldsymbol{k}$ of the polaron band to the final state $\boldsymbol{k}\pm\sum_{r=1}^{l}\boldsymbol{q}_r$ with the emission (or absorption) of l phonons. The $1/\lambda$ expansion result (equation (4.79)) is applied to *low-energy* polaron excitations in the strong-coupling limit. In the case of the long-range Fröhlich interaction with high-frequency phonons, it is also applied in the weak-coupling and intermediate regimes (section 4.3.3). Differing from the canonical Migdal GF (chapter 3), there is no damping of polaronic excitations in equation (4.79). Instead the e–ph coupling leads to the coherent dressing of electrons by phonons because of the energy conservation (section 4.3.2). The dressing can be seen as the phonon 'side-bands' with $l \geq 1$. While the major sum rule (equation (D.11)) is satisfied,

$$\begin{aligned}\frac{1}{\pi}\int_{-\infty}^{\infty} d\omega\, A(\boldsymbol{k},\omega) &= Z\sum_{l=0}^{\infty}\sum_{\boldsymbol{q}_1,\ldots,\boldsymbol{q}_l}\frac{\prod_{r=1}^{l}|\gamma(\boldsymbol{q}_r)|^2}{(2N)^l l!}\\ &= Z\sum_{l=0}^{\infty}\frac{1}{l!}\left\{\frac{1}{2N}\sum_{\boldsymbol{q}}|\gamma(\boldsymbol{q})|^2\right\}^l\\ &= Z\exp\left[\frac{1}{2N}\sum_{\boldsymbol{q}}|\gamma(\boldsymbol{q})|^2\right] = 1 \qquad (4.82)\end{aligned}$$

the higher-momentum integrals, $\int_{-\infty}^{\infty} d\omega\, \omega^{p} A(\boldsymbol{k},\omega)$ with $p > 0$, calculated using equation (4.79), differ from the exact value by an amount proportional to $1/\lambda$. The difference is due to a partial 'undressing' of high-energy excitations in the side-bands, which is beyond the first-order $1/\lambda$ expansion.

The spectral function of the polaronic carriers comprises two different parts. The first ($l = 0$) $\boldsymbol{k}$-dependent *coherent* term arises from the polaron band tunnelling,

$$A_{\text{coh}}(\boldsymbol{k},\omega) = [A_0^{(-)}(\boldsymbol{k},\omega) + A_0^{(+)}(\boldsymbol{k},\omega)] = \pi Z\delta(\omega - \xi_{\boldsymbol{k}}). \qquad (4.83)$$

The spectral weight of the coherent part is suppressed as $Z \ll 1$. However, in the case of the Fröhlich interaction, the effective mass is less enhanced, $\xi_{\boldsymbol{k}} = Z'E_{\boldsymbol{k}} - \mu$, because $Z \ll Z' < 1$ (section 4.3.3). The second *incoherent* part $A_{\text{incoh}}(\boldsymbol{k},\omega)$ comprises all the terms with $l \geq 1$. It describes the excitations accompanied by emission and absorption of phonons. We note that its spectral density spreads over a wide energy range of about twice the polaron level shift E_p, which might be larger than the unrenormalized bandwidth $2D$ in the rigid lattice without phonons. In contrast, the coherent part shows a dispersion only in the energy window of the order of the polaron bandwidth, $2w = 2Z'D$. It is interesting that there is some $\boldsymbol{k}$ dependence of the *incoherent* background as well, if the matrix element of the e–ph interaction and/or phonon frequencies depend on q. Only in the Holstein model with the short-range dispersionless e–ph interaction ($\gamma(\boldsymbol{q}) = \gamma_0$ and $\omega_{\boldsymbol{q}} = \omega_0$) is the incoherent part momentum independent. Replacing $\boldsymbol{k} \pm \sum_{r=1}^{l}\boldsymbol{q}_r$ by $\boldsymbol{k}'$ in equations (4.80) and (4.81), we obtain the following expression in this case:

$$\begin{aligned}A_{\text{incoh}}(\boldsymbol{k},\omega) &= \pi\frac{Z}{N}\sum_{l=1}^{\infty}\frac{\gamma_0^{2l}}{2^l l!}\\ &\times\sum_{\boldsymbol{k}'}\{[1-\bar{n}(\boldsymbol{k}')]\delta(\omega - l\omega_0 - \xi_{\boldsymbol{k}'}) + \bar{n}(\boldsymbol{k}')\delta(\omega + l\omega_0 - \xi_{\boldsymbol{k}'})\}\end{aligned} \qquad (4.84)$$

which has no $\boldsymbol{k}$-dependence.

As soon as we know the spectral function, polaron GFs are easily obtained using their analytical properties (appendix D). For example, the temperature polaron GF is given by the integral (D.26). Calculating the integral with the spectral density equation (4.79), we find, in the Holstein model [102], that

$$\mathcal{G}(\boldsymbol{k}, \omega_n) = \frac{Z}{\mathrm{i}\omega_n - \xi_{\boldsymbol{k}}} + \frac{Z}{N}\sum_{l=1}^{\infty}\frac{\gamma_0^{2l}}{2^l l!}\sum_{\boldsymbol{k}'}\left\{\frac{1-\bar{n}(\boldsymbol{k}')}{\mathrm{i}\omega_n - l\omega_0 - \xi_{\boldsymbol{k}'}} + \frac{\bar{n}(\boldsymbol{k}')}{\mathrm{i}\omega_n + l\omega_0 - \xi_{\boldsymbol{k}'}}\right\}. \tag{4.85}$$

Here the first term describes the coherent tunnelling in the narrow polaron band while the second $\boldsymbol{k}$-independent sum is due to the phonon cloud 'dressing' the electron.

4.4 Polaron-polaron interaction and bipolaron

Polarons interact with each other (equation (4.32)). The range of the deformation surrounding the Fröhlich polarons is quite large and their deformation fields overlap at finite density. Taking into account both the long-range attraction of polarons owing to the lattice deformations *and* their direct Coulomb repulsion, the residual *long-range* interaction turns out to be rather weak and repulsive in ionic crystals [13]. The Fourier component of the polaron–polaron interaction, $v(\boldsymbol{q})$, comprising the direct Coulomb repulsion and the attraction mediated by phonons, is

$$v(\boldsymbol{q}) = \frac{4\pi e^2}{\epsilon_\infty q^2} - |\gamma(\boldsymbol{q})|^2\omega_{\boldsymbol{q}}. \tag{4.86}$$

In the long-wave limit ($q \ll \pi/a$), the Fröhlich interaction dominates in the attractive part, which is described by [62]

$$|\gamma(\boldsymbol{q})|^2\omega_0 = \frac{4\pi e^2(\epsilon_\infty^{-1} - \epsilon_0^{-1})}{q^2}. \tag{4.87}$$

Here ϵ_∞ and ϵ_0 are the high-frequency and static dielectric constants, respectively, of the host ionic insulator which are usually well known from experiment. Fourier transforming equation (4.86) yields the repulsive interaction in real space,

$$v(\boldsymbol{m} - \boldsymbol{n}) = \frac{e^2}{\epsilon_0|\boldsymbol{m} - \boldsymbol{n}|} > 0. \tag{4.88}$$

We see that optical phonons nearly nullify the bare Coulomb repulsion in ionic solids, where $\epsilon_0 \gg 1$, but cannot overscreen it at large distances.

Considering the polaron–phonon interaction in the multi-polaron system, we have to take into account the dynamic properties of the polaron response function. One can erroneously believe that the long-range Fröhlich interaction becomes a short-range (Holstein) one due to the screening of ions by heavy polaronic carriers. In fact, small polarons cannot screen high-frequency optical vibrations because their renormalized plasma frequency is comparable with or even less than the phonon frequency. In the absence of bipolarons (see later), we can apply the ordinary bubble approximation (chapter 3) to calculate the dielectric response function of polarons at the frequency Ω:

$$\epsilon(\boldsymbol{q}, \Omega) = 1 - 2v(\boldsymbol{q}) \sum_{\boldsymbol{k}} \frac{\bar{n}(\boldsymbol{k}+\boldsymbol{q}) - \bar{n}(\boldsymbol{k})}{\Omega - \epsilon_{\boldsymbol{k}} + \epsilon_{\boldsymbol{k}+\boldsymbol{q}}}. \tag{4.89}$$

This expression describes the response of small polarons to any external field of the frequency $\Omega \lesssim \omega_0$, when phonons in the polaron cloud follow the polaron motion. In the static limit, we obtain the usual Debye screening at large distances ($q \to 0$). For a temperature larger than the polaron half-bandwidth ($T > w$), we can approximate the polaron distribution function as

$$\bar{n}(\boldsymbol{k}) \approx \frac{n}{2a^3}\left(1 - \frac{(2-n)\epsilon_{\boldsymbol{k}}}{2T}\right) \tag{4.90}$$

and obtain

$$\epsilon(q, 0) = 1 + \frac{q_s^2}{q^2} \tag{4.91}$$

where

$$q_s = \left[\frac{2\pi e^2 n(2-n)}{\epsilon_0 T a^3}\right]^{1/2}$$

and n is the number of polarons per a unit cell. For a finite but rather low frequency ($\omega_0 \gtrsim \Omega \gg w$), the polaron response becomes dynamic:

$$\epsilon(\boldsymbol{q}, \Omega) = 1 - \frac{\omega_p^2(\boldsymbol{q})}{\Omega^2} \tag{4.92}$$

where

$$\omega_p^2(\boldsymbol{q}) = 2v(\boldsymbol{q}) \sum_{\boldsymbol{k}} n(\boldsymbol{k})(\epsilon_{\boldsymbol{k}+\boldsymbol{q}} - \epsilon_{\boldsymbol{k}}) \tag{4.93}$$

is the temperature-dependent polaron plasma frequency squared, which is about

$$\omega_p^2(\boldsymbol{q}) \lesssim \omega_{\mathrm{e}}^2 \frac{m_{\mathrm{e}}\epsilon_\infty}{m^*\epsilon_0} \ll \omega_{\mathrm{e}}^2.$$

The polaron plasma frequency is very low due to the large static dielectric constant ($\epsilon_0 \gg 1$) and the enhanced polaron mass m^*.

Now replacing the bare electron–phonon interaction vertex $\gamma(\boldsymbol{q})$ by a screened one, $\gamma_{\mathrm{sc}}(\boldsymbol{q}, \omega_0)$, as shown in figure 3.4, we obtain

$$\gamma_{\mathrm{sc}}(\boldsymbol{q}, \omega_0) = \frac{\gamma(\boldsymbol{q})}{\epsilon(\boldsymbol{q}, \omega_0)} \approx \gamma(\boldsymbol{q}) \tag{4.94}$$

because $\omega_0 > \omega_p$. Therefore, the singular behaviour of $\gamma(\boldsymbol{q}) \sim 1/q$ is unaffected by screening. Polarons are too slow to screen high-frequency crystal field oscillations. As a result, the strong interaction with high-frequency optical phonons in ionic solids remains unscreened at any density of small polarons.

Another important point is the possibility of the Wigner crystallization of the polaronic liquid. Because the net long-range repulsion is relatively weak, the relevant dimensionless parameter $r_s = m^*e^2/\epsilon_0(4\pi n/3)^{1/3}$ is not very large in doped cuprates. The Wigner crystallization appears around $r_s \simeq 100$ or larger, which corresponds to the atomic density of polarons $n \leq 10^{-6}$ with $\epsilon_0 = 30$ and $m^* = 5m_e$. This estimate tells us that polaronic carriers are usually in the liquid state.

At large distance, polarons repel each other (equation (4.88)). Nevertheless two *large* polarons can be bound into a *large* bipolaron by an exchange interaction even with no additional e–ph interaction other than the Fröhlich one [64, 103]. When a short-range deformation potential and molecular-type (i.e. Jahn–Teller [104]) e–ph interactions are taken into account together with the Fröhlich interaction, they overcome the Coulomb repulsion at a short distance of about the lattice constant. Then, owing to a narrow band, two heavy polarons easily form a bound state, i.e. a *small* bipolaron. Let us estimate the coupling constant λ and the adiabatic ratio $\omega_0/T(a)$, at which the small 'bipolaronic' instability occurs. The characteristic attractive potential is $V = D/(\lambda - \mu_c)$, where μ_c is the dimensionless Coulomb repulsion (section 3.5), and λ includes the interaction with all phonon branches. The radius of the potential is about a. In three dimensions, a bound state of two attractive particles appears, if

$$V \geq \frac{\pi^2}{8m^*a^2}. \tag{4.95}$$

Substituting the polaron mass, $m^* = [2a^2T(a)]^{-1}\exp(\gamma\lambda D/\omega_0)$, we find

$$\frac{T(a)}{\omega_0} \leq (\gamma z\lambda)^{-1}\ln\left[\frac{\pi^2}{4z(\lambda - \mu_c)}\right]. \tag{4.96}$$

The corresponding 'phase' diagram is shown in figure 4.6, where $t \equiv T(a)$ and $\omega \equiv \omega_0$. Small bipolarons form at $\lambda \geq \mu_c + \pi^2/4z$ almost independently of the adiabatic ratio. In the Fröhlich interaction, there is no sharp transition between small and large polarons, as one can see in figure 4.5 and the first-order $1/\lambda$ expansion is accurate in the whole region of coupling, if the adiabatic parameter is not very small (down to $\omega/T(a) \approx 0.5$). Hence, we can say that the carriers are small polarons *independent* of the value of λ in this case. It means that they tunnel together with the entire phonon cloud no matter how 'thin' the cloud is.

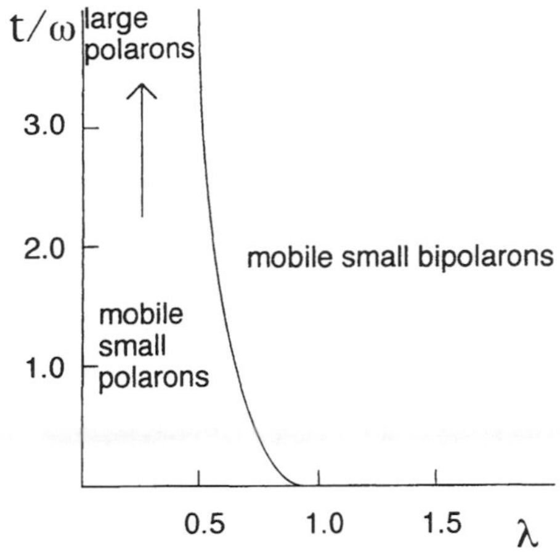

Figure 4.6. The 't/ω–λ' diagram with a small bipolaron domain and a region of unbound small polarons for $z = 6$, $\gamma = 0.4$ and Coulomb potential $\mu_c = 0.5$.

4.5 Polaronic superconductivity

The polaron–polaron interaction is the sum of two large contributions of the opposite sign (equation (4.32)). It is generally larger than the polaron bandwidth and the polaron Fermi energy, $\epsilon_F = Z'E_F$ (equation (4.61)). This condition is opposite to the weak-coupling BCS regime, where the Fermi energy is the largest. However, there is still a narrow window of parameters, where bipolarons are 'extended' enough and pairs of two small polarons overlap similarly to Cooper pairs. Here the BCS approach is applied to non-adiabatic carriers with a *non-retarded* attraction, so that bipolarons are the Cooper pairs formed by two *small polarons* [11]. The size of the bipolaron is estimated as

$$r_b \approx \frac{1}{(m^*\Delta)^{1/2}} \tag{4.97}$$

where Δ is the binding energy of the order of an attraction potential $|V|$. The BCS approach is applied if $r_b \gg n^{-1/3}$, which puts a severe constraint on the value of the attraction

$$|V| \ll \epsilon_F. \tag{4.98}$$

There is no 'Tolmachev' logarithm (section 3.5) in the case of non-adiabatic carriers, because the attraction is non-retarded if $\epsilon_F \lesssim \omega_0$. Hence, a superconducting state of small polarons is possible only if $\lambda > \mu_c$. This consideration leaves a rather narrow *crossover* region from the normal polaron Fermi liquid to a superconductor, where one can still apply the BCS mean-field approach,

$$0 < \lambda - \mu_c \ll Z' < 1. \tag{4.99}$$

In the case of the Fröhlich interaction, Z' is about 0.1–0.3 for typical values of λ (figure 4.5). Hence, the region, equation (4.99), is on the border-line in figure 4.6.

In the crossover region polarons behave like fermions in a narrow band with a weak non-retarded attraction. As long as $\lambda \gg 1/\sqrt{2z}$, we can drop their residual interaction with phonons in the transformed Hamiltonian,

$$\tilde{H} \approx \sum_{i,j} [(\langle \hat{\sigma}_{ij} \rangle_{\text{ph}} - \mu\delta_{ij}) c_i^\dagger c_j + \tfrac{1}{2} v_{ij} c_i^\dagger c_j^\dagger c_j c_i] \tag{4.100}$$

written in the Wannier representation. If the condition (4.99) is satisfied, we can treat the polaron–polaron interaction approximately by the use of BCS theory (chapter 2). For simplicity, let us keep only the on-site v_0 and the nearest-neighbour v_1 interactions. At least one of them should be attractive to ensure that the ground state is superconducting. Introducing two order parameters

$$\Delta_0 = -v_0 \langle c_{\mathbf{m},\uparrow} c_{\mathbf{m},\downarrow} \rangle \tag{4.101}$$

$$\Delta_1 = -v_1 \langle c_{\mathbf{m},\uparrow} c_{\mathbf{m}+\mathbf{a},\downarrow} \rangle \tag{4.102}$$

and transforming to the $\boldsymbol{k}$-space results in the usual BCS Hamiltonian,

$$H_p = \sum_{\mathbf{k},s} \xi_{\mathbf{k}} c_{\mathbf{k}s}^\dagger c_{\mathbf{k}s} + \sum_{\mathbf{k}} [\Delta_{\mathbf{k}} c_{\mathbf{k}\uparrow}^\dagger c_{-\mathbf{k}\downarrow}^\dagger + H.c.] \tag{4.103}$$

where $\xi_{\mathbf{k}} = \epsilon_{\mathbf{k}} - \mu$ is the renormalized kinetic energy and

$$\Delta_{\mathbf{k}} = \Delta_0 - \Delta_1 \frac{\xi_{\mathbf{k}} + \mu}{w} \tag{4.104}$$

is the order parameter.

Applying the Bogoliubov diagonalization procedure (chapter 2), one obtains the following expressions:

$$\langle c_{\mathbf{k},\uparrow} c_{-\mathbf{k},\downarrow} \rangle = \frac{\Delta_{\mathbf{k}}}{2\sqrt{\xi_{\mathbf{k}}^2 + \Delta_{\mathbf{k}}^2}} \tanh \frac{\sqrt{\xi_{\mathbf{k}}^2 + \Delta_{\mathbf{k}}^2}}{2T} \tag{4.105}$$

and

$$\Delta_0 = -\frac{v_0}{N} \sum_{\mathbf{k}} \frac{\Delta_{\mathbf{k}}}{2\sqrt{\xi_{\mathbf{k}}^2 + \Delta_{\mathbf{k}}^2}} \tanh \frac{\sqrt{\xi_{\mathbf{k}}^2 + \Delta_{\mathbf{k}}^2}}{2T} \tag{4.106}$$

$$\Delta_1 = -\frac{v_1}{Nw} \sum_{\mathbf{k}} \frac{\Delta_{\mathbf{k}}(\xi_{\mathbf{k}} + \mu)}{2\sqrt{\xi_{\mathbf{k}}^2 + \Delta_{\mathbf{k}}^2}} \tanh \frac{\sqrt{\xi_{\mathbf{k}}^2 + \Delta_{\mathbf{k}}^2}}{2T}. \tag{4.107}$$

These equations are equivalent to a single BCS equation for $\Delta_{\mathbf{k}} = \Delta(\xi_{\mathbf{k}})$ but with the half polaron bandwidth w cutting the integral, rather than the Debye temperature:

$$\Delta(\xi) = \int_{-w-\mu}^{w-\mu} d\eta\, N_p(\eta) V(\xi, \eta) \frac{\Delta(\eta)}{2\sqrt{\eta^2 + \Delta^2(\eta)}} \tanh \frac{\sqrt{\eta^2 + \Delta^2(\eta)}}{2T}. \quad (4.108)$$

Here $V(\xi, \eta) = -v_0 - zv_1(\xi + \mu)(\eta + \mu)/w^2$.

The critical temperature T_c of the polaronic superconductor is determined by two linearized equations (4.106) and (4.107) in the limit $\Delta_{0,1} \to 0$:

$$\left[1 + A\left(\frac{v_0}{zv_1} + \frac{\mu^2}{w^2}\right)\right]\Delta - \frac{B\mu}{w}\Delta_1 = 0 \quad (4.109)$$

$$-\frac{A\mu}{w}\Delta + (1 + B)\Delta_1 = 0 \quad (4.110)$$

where $\Delta = \Delta_0 - \Delta_1\mu/w$ and

$$A = \frac{zv_1}{2w}\int_{-w-\mu}^{w-\mu} d\eta \frac{\tanh(\eta/2T_c)}{\eta}$$

$$B = \frac{zv_1}{2w}\int_{-w-\mu}^{w-\mu} d\eta \frac{\eta \tanh(\eta/2T_c)}{w^2}.$$

These equations are applied only if the polaron–polaron coupling is small ($|v_{0,1}| < w$). A non-trivial solution is found at

$$T_c \approx 1.14w\sqrt{1 - \frac{\mu^2}{w^2}} \exp\left(\frac{2w}{v_0 + zv_1\mu^2/w^2}\right) \quad (4.111)$$

if $v_0 + zv_1\mu^2/w^2 < 0$, so that superconductivity exists even in the case of on-site repulsion ($v_0 > 0$), if this repulsion is less than the total inter-site attraction, $z|v_1|$. There is a non-trivial dependence of T_c on doping. With a constant density of states in the polaron band, the Fermi energy $\epsilon_F \approx \mu$ is expressed via the number of polarons per atom n as

$$\mu = w(n - 1) \quad (4.112)$$

so that

$$T_c \simeq 1.14w\sqrt{n(2 - n)} \exp\left(\frac{2w}{v_0 + zv_1[n - 1]^2}\right). \quad (4.113)$$

T_c have two maxima as a function of n separated by a deep minimum in a half-filled band ($n = 1$), where the nearest-neighbour contributions to pairing cancel each other.

4.6 Mobile small bipolarons

The attractive energy of two small polarons is generally larger than the polaron bandwidth, $\lambda - \mu_c \gg Z'$. When this condition is fulfilled, small bipolarons are not overlapped. Consideration of particular lattice structures shows that small bipolarons are mobile even when the electron–phonon coupling is strong and the bipolaron binding energy is large [100]. Hence the polaronic Fermi liquid transforms into a Bose liquid of double-charged carriers in the strong-coupling regime, rather than into the BCS-like ground state of the previous section. The Bose liquid is stable because bipolarons repel each other (see later). Here we encounter a novel electronic state of matter, a charged Bose liquid, qualitatively different from the normal Fermi liquid and from the BCS superfluid.

4.6.1 On-site bipolarons and bipolaronic Hamiltonian

The small parameter $Z'/(\lambda - \mu_c) \ll 1$ allows for a consistent treatment of bipolaronic systems [10, 11]. Under this condition the hopping term in the transformed Hamiltonian $\tilde{H}$ is a small perturbation of the ground state of immobile bipolarons and free phonons:

$$\tilde{H} = H_0 + H_{\text{pert}} \tag{4.114}$$

where

$$H_0 = \frac{1}{2}\sum_{i,j} v_{ij} c_i^\dagger c_j^\dagger c_j c_i + \sum_{\mathbf{q},\nu} \omega_{\mathbf{q}\nu}[d_{\mathbf{q}\nu}^\dagger d_{\mathbf{q}\nu} + \tfrac{1}{2}] \tag{4.115}$$

and

$$H_{\text{pert}} = \sum_{i,j} \hat{\sigma}_{ij} c_i^\dagger c_j. \tag{4.116}$$

Let us first discuss the dynamics of *on-site* bipolarons, which are the ground state of the system with the Holstein non-dispersive e–ph interaction. The on-site bipolaron is formed if

$$2E_p > U \tag{4.117}$$

where U is the on-site Coulomb correlation energy (the so-called Hubbard U). The inter-site polaron–polaron interaction (4.32) is purely the Coulomb repulsion because the phonon-mediated attraction between two polarons on different sites is zero in the Holstein model. Two or more on-site bipolarons as well as three or more polarons cannot occupy the same site because of the Pauli exclusion principle. Hence, bipolarons repel single polarons and each other. Their binding energy, $\Delta = 2E_p - U$, is larger than the polaron half-bandwidth, $\Delta \gg w$, so that there are no unbound polarons in the ground state. H_{pert} (equation (4.116)) destroys bipolarons in the first order. Hence, it has no diagonal matrix elements. Then the bipolaron dynamics, including superconductivity, is described by the use of a new canonical transformation, $\exp(S_2)$ [10], which eliminates the first order of H_{pert}:

$$(S_2)_{\text{fp}} = \sum_{i,j} \frac{\langle f|\hat{\sigma}_{ij} c_i^\dagger c_j|p\rangle}{E_{\text{f}} - E_{\text{p}}}. \tag{4.118}$$

Here $E_{\text{f,p}}$ and $|f\rangle$, $|p\rangle$ are the energy levels and the eigenstates of H_0. Neglecting the terms of the orders higher than $(w/\Delta)^2$, we obtain

$$(H_{\text{b}})_{\text{ff}'} \equiv (\text{e}^{S_2}\tilde{H}\text{e}^{-S_2})_{\text{ff}'} \tag{4.119}$$

$$(H_{\text{b}})_{\text{ff}'} \approx (H_0)_{\text{ff}'} - \tfrac{1}{2}\sum_{\nu}\sum_{i\neq i',j\neq j'} \langle f|\hat{\sigma}_{ii'} c_i^\dagger c_{i'}|p\rangle\langle p|\hat{\sigma}_{jj'} c_j^\dagger c_{j'}|f'\rangle \times \left(\frac{1}{E_{\text{p}} - E_{\text{f}'}} + \frac{1}{E_{\text{p}} - E_{\text{f}}}\right).$$

S_2 couples a localized on-site bipolaron and a state of two unbound polarons on different sites. The expression (4.119) determines the matrix elements of the transformed *bipolaronic* Hamiltonian H_{b} in the subspace $|f\rangle$, $|f'\rangle$ with no single (unbound) polarons. However, the intermediate bra $\langle p|$ and ket $|p\rangle$ in equation (4.119) refer to configurations involving two unpaired polarons and any number of phonons. Hence, we have

$$E_{\text{p}} - E_{\text{f}} = \Delta + \sum_{\boldsymbol{q},\nu} \omega_{\boldsymbol{q}\nu}(n_{\boldsymbol{q}\nu}^{\text{p}} - n_{\boldsymbol{q}\nu}^{\text{f}}) \tag{4.120}$$

where $n_{\boldsymbol{q}\nu}^{\text{f,p}}$ are phonon occupation numbers $(0, 1, 2, 3, \ldots, \infty)$. This equation is an explicit definition of the bipolaron binding energy Δ which takes into account the residual inter-site repulsion between bipolarons and between two unpaired polarons. The lowest eigenstates of H_{b} are in the subspace, which has only doubly occupied $c_{\boldsymbol{m}s}^\dagger c_{\boldsymbol{m}s'}^\dagger|0\rangle$ or empty $|0\rangle$ sites. On-site bipolaron tunnelling is a two-step transition. It takes place via a single polaron tunnelling to a neighbouring site. The subsequent tunnelling of its 'partner' to the same site restores the initial energy state of the system. There are no *real* phonons emitted or absorbed because the bipolaron band is narrow (see later). Hence, we can average H_{b} with respect to phonons. Replacing the energy denominators in the second term in equation (4.119) by the integrals with respect to time,

$$\frac{1}{E_{\text{p}} - E_{\text{f}}} = \text{i}\int_0^\infty \text{d}t\, \text{e}^{\text{i}(E_{\text{f}} - E_{\text{p}} + \text{i}\delta)t}$$

we obtain

$$H_{\text{b}} = H_0 - \text{i}\sum_{\boldsymbol{m}\neq\boldsymbol{m}',s}\sum_{\boldsymbol{n}\neq\boldsymbol{n}',s'} T(\boldsymbol{m}-\boldsymbol{m}')T(\boldsymbol{n}-\boldsymbol{n}') \times c_{\boldsymbol{m}s}^\dagger c_{\boldsymbol{m}'s} c_{\boldsymbol{n}s'}^\dagger c_{\boldsymbol{n}'s'} \int_0^\infty \text{d}t\, \text{e}^{-\text{i}\Delta t}\Phi_{\boldsymbol{m}\boldsymbol{m}'}^{\boldsymbol{n}\boldsymbol{n}'}(t). \tag{4.121}$$

Here $\Phi^{nn'}_{mm'}(t)$ is a multi-phonon correlator,

$$\Phi^{nn'}_{mm'}(t) \equiv \langle\langle \hat{X}^{\dagger}_{i}(t)\hat{X}_{i'}(t)\hat{X}^{\dagger}_{j}\hat{X}_{j'}\rangle\rangle \tag{4.122}$$

which is calculated using the commutation relations in section 4.3.4. $\hat{X}^{\dagger}_{i}(t)$ and $\hat{X}_{i'}(t)$ commute for any $\gamma(\boldsymbol{q},\nu) = \gamma(-\boldsymbol{q},\nu)$ (appendix E). $\hat{X}^{\dagger}_{j}$ and $\hat{X}_{j'}$ commute also, so that we can write

$$\hat{X}^{\dagger}_{i}(t)\hat{X}_{i'}(t) = \prod_{\boldsymbol{q}} \mathrm{e}^{[u_{i'}(\boldsymbol{q},t)-u_{i}(\boldsymbol{q},t)]d_{\boldsymbol{q}}-H.c.]} \tag{4.123}$$

$$\hat{X}^{\dagger}_{j}\hat{X}_{j'} = \prod_{\boldsymbol{q}} \mathrm{e}^{[u_{j'}(\boldsymbol{q})-u_{j}(\boldsymbol{q})]d_{\boldsymbol{q}}-H.c.]} \tag{4.124}$$

where the phonon branch index ν is dropped for simplicity. Applying the identity (4.71) twice yields

$$\begin{aligned}\hat{X}^{\dagger}_{i}(t)\hat{X}_{i'}(t)\hat{X}^{\dagger}_{j}\hat{X}_{j'} = &\prod_{\boldsymbol{q}} \mathrm{e}^{\beta^{*}d^{\dagger}_{\boldsymbol{q}}}\mathrm{e}^{-\beta d_{\boldsymbol{q}}}\mathrm{e}^{-|\beta|^2/2} \\ &\times \mathrm{e}^{[u_{i'}(\boldsymbol{q},t)-u_{i}(\boldsymbol{q},t)][u^{*}_{j'}(\boldsymbol{q})-u^{*}_{j}(\boldsymbol{q})]/2-H.c.}\end{aligned} \tag{4.125}$$

where

$$\beta = u_{i}(\boldsymbol{q},t) - u_{i'}(\boldsymbol{q},t) + u_{j}(\boldsymbol{q}) - u'_{j}(\boldsymbol{q}).$$

Finally using the average equation (4.73), we find

$$\begin{aligned}\Phi^{nn'}_{mm'}(t) = &\,\mathrm{e}^{-g^2(\boldsymbol{m}-\boldsymbol{m}')}\mathrm{e}^{-g^2(\boldsymbol{n}-\boldsymbol{n}')} \\ &\times \exp\left\{\frac{1}{2N}\sum_{\boldsymbol{q},\nu}|\gamma(\boldsymbol{q},\nu)|^2 F_{\boldsymbol{q}}(\boldsymbol{m},\boldsymbol{m}',\boldsymbol{n},\boldsymbol{n}')\frac{\cosh[\omega_{\boldsymbol{q}\nu}((1/2T)-\mathrm{i}t)]}{\sinh[\omega_{\boldsymbol{q}\nu}/2T]}\right\}\end{aligned} \tag{4.126}$$

where

$$\begin{aligned}F_{\boldsymbol{q}}(\boldsymbol{m},\boldsymbol{m}',\boldsymbol{n},\boldsymbol{n}') = &\cos[\boldsymbol{q}\cdot(\boldsymbol{n}'-\boldsymbol{m})] + \cos[\boldsymbol{q}\cdot(\boldsymbol{n}-\boldsymbol{m}')] \\ &- \cos[\boldsymbol{q}\cdot(\boldsymbol{n}'-\boldsymbol{m}')] - \cos[\boldsymbol{q}\cdot(\boldsymbol{n}-\boldsymbol{m})].\end{aligned} \tag{4.127}$$

Taking into account that there are only bipolarons in the subspace where H_{b} operates, we finally rewrite the Hamiltonian in terms of the creation $b^{\dagger}_{\boldsymbol{m}} = c^{\dagger}_{\boldsymbol{m}\uparrow}c^{\dagger}_{\boldsymbol{m}\downarrow}$ and annihilation $b_{\boldsymbol{m}} = c_{\boldsymbol{m}\downarrow}c_{\boldsymbol{m}\uparrow}$ operators of singlet pairs as

$$\begin{aligned}H_{\mathrm{b}} = &-\sum_{\boldsymbol{m}}\left[\Delta + \tfrac{1}{2}\sum_{\boldsymbol{m}'} v^{(2)}(\boldsymbol{m}-\boldsymbol{m}')\right]n_{\boldsymbol{m}} \\ &+ \sum_{\boldsymbol{m}\neq\boldsymbol{m}'}[t(\boldsymbol{m}-\boldsymbol{m}')b^{\dagger}_{\boldsymbol{m}}b_{\boldsymbol{m}'} + \tfrac{1}{2}\bar{v}(\boldsymbol{m}-\boldsymbol{m}')n_{\boldsymbol{m}}n_{\boldsymbol{m}'}].\end{aligned} \tag{4.128}$$

There are no triplet pairs in the Holstein model because the Pauli exclusion principle does not allow two electrons with the same spin to occupy the same site. Here $n_{\mathbf{m}} = b^{\dagger}_{\mathbf{m}} b_{\mathbf{m}}$ is the bipolaron site-occupation operator,

$$\bar{v}(\mathbf{m} - \mathbf{m}') = 4v(\mathbf{m} - \mathbf{m}') + v^{(2)}(\mathbf{m} - \mathbf{m}') \tag{4.129}$$

is the bipolaron–bipolaron interaction including a direct polaron–polaron interaction $v(\mathbf{m} - \mathbf{m}')$ and a second order in $T(\mathbf{m})$ repulsive correction:

$$v^{(2)}(\mathbf{m} - \mathbf{m}') = 2\mathrm{i}T^2(\mathbf{m} - \mathbf{m}') \int_0^{\infty} \mathrm{d}t\, \mathrm{e}^{-\mathrm{i}\Delta t} \Phi^{\mathbf{m}'\mathbf{m}}_{\mathbf{m}\mathbf{m}'}(t). \tag{4.130}$$

This additional repulsion appears because a virtual hop of one of two polarons of the pair is forbidden if the neighbouring site is occupied by another pair. The bipolaron transfer integral is of the second order in $T(\mathbf{m})$:

$$t(\mathbf{m} - \mathbf{m}') = -2\mathrm{i}T^2(\mathbf{m} - \mathbf{m}') \int_0^{\infty} \mathrm{d}t\, \mathrm{e}^{-\mathrm{i}\Delta t} \Phi^{\mathbf{m}\mathbf{m}'}_{\mathbf{m}\mathbf{m}'}(t). \tag{4.131}$$

The *bipolaronic* Hamiltonian (4.128) describes the low-energy physics of strongly coupled electrons and phonons. We use the explicit form of the multi-phonon correlator, equation (4.126), to calculate $t(\mathbf{m})$ and $v^{(2)}(\mathbf{m})$. If the phonon frequency is dispersionless, we obtain

$$\Phi^{\mathbf{m}\mathbf{m}'}_{\mathbf{m}\mathbf{m}'}(t) = \mathrm{e}^{-2g^2(\mathbf{m}-\mathbf{m}')} \exp[-2g^2(\mathbf{m} - \mathbf{m}')\mathrm{e}^{-\mathrm{i}\omega_0 t}]$$
$$\Phi^{\mathbf{m}'\mathbf{m}}_{\mathbf{m}\mathbf{m}'}(t) = \mathrm{e}^{-2g^2(\mathbf{m}-\mathbf{m}')} \exp[2g^2(\mathbf{m} - \mathbf{m}')\mathrm{e}^{-\mathrm{i}\omega_0 t}]$$

at $T \ll \omega_0$. Expanding the time-dependent exponents in the Fourier series and calculating the integrals in equations (4.131) and (4.130) yield [105]

$$t(\mathbf{m}) = -\frac{2T^2(\mathbf{m})}{\Delta} \mathrm{e}^{-2g^2(\mathbf{m})} \sum_{l=0}^{\infty} \frac{[-2g^2(\mathbf{m})]^l}{l!(1 + l\omega_0/\Delta)} \tag{4.132}$$

and

$$v^{(2)}(\mathbf{m}) = \frac{2T^2(\mathbf{m})}{\Delta} \mathrm{e}^{-2g^2(\mathbf{m})} \sum_{l=0}^{\infty} \frac{[2g^2(\mathbf{m})]^l}{l!(1 + l\omega_0/\Delta)}. \tag{4.133}$$

When $\Delta \ll \omega_0$, we can keep the first term only with $l = 0$ in the bipolaron hopping integral in equation (4.132). In this case, the bipolaron half-bandwidth $zt(\mathbf{a})$ is of the order of $2w^2/(z\Delta)$. However, if the bipolaron binding energy is large ($\Delta \gg \omega_0$), the bipolaron bandwidth dramatically decreases proportionally to $\mathrm{e}^{-4g^2} \lll 1$ in the limit $\Delta \to \infty$. However, this limit is not realistic because $\Delta = 2E_p - V_c < 2g^2\omega_0$. In a more realistic regime, $\omega_0 < \Delta < 2g^2\omega_0$, equation (4.132) yields

$$t(\mathbf{m}) \approx \frac{2\sqrt{2\pi}T^2(\mathbf{m})}{\sqrt{\omega_0 \Delta}} \exp\left[-2g^2 - \frac{\Delta}{\omega_0}\left(1 + \ln \frac{2g^2(\mathbf{m})\omega_0}{\Delta}\right)\right]. \tag{4.134}$$

In contrast, the bipolaron–bipolaron repulsion (equation (4.133)) has no small exponent in the limit $\Delta \to \infty$, $v^{(2)} \propto D^2/\Delta$. Together with the direct Coulomb repulsion, the second order $v^{(2)}$ ensures the stability of the bipolaronic liquid against clustering.

The high-temperature behaviour of the bipolaron bandwidth is just the opposite to that of the small polaron bandwidth. While the polaron band collapses with increasing temperature (equation (4.42)), the bipolaron band becomes wider [106]:

$$t(\boldsymbol{m}) \propto \frac{1}{\sqrt{T}} \exp\left[-\frac{E_p + \Delta}{2T}\right] \tag{4.135}$$

for $T > \omega_0$.

4.6.2 Inter-site bipolaron in the chain model

On-site bipolarons are very heavy for realistic values of the on-site attractive energy $2E_p$ and phonon frequencies. Indeed, to bind two polarons on a single site $2E_p$ should overcome the on-site Coulomb energy, which is typically of the order of 1 eV or higher. Optical phonon frequencies are about 0.1–0.2 eV in novel superconductors like oxides and doped fullerenes. Therefore, in the framework of the Holstein model, the mass enhancement exponent of on-site bipolarons in equation (4.134) is rather large ($\gtrsim \exp(2E_p/\omega_0) > 150$), so that on-site bipolarons could hardly account for high values of the superconducting critical temperature [100].

But the Holstein model is not a typical model. The Fröhlich interaction with optical phonons, which is unscreened in polaronic systems (section 4.4), is much stronger. This longer-range interaction leads to a lighter polaron in the strong-coupling regime (section 4.3.3). Indeed, the polaron is heavy because it has to carry the lattice deformation with it, the same deformation that forms the polaron itself. Therefore, there exists a generic relation between the polaron stabilization energy, E_p, and the renormalization of its mass, $m \propto \exp(\gamma E_p/\omega_0)$, where the numeric coefficient γ depends on the radius of the interaction. For a short-range e–ph interaction, the *entire* lattice deformation disappears and then forms at another site, when the polaron moves between the nearest lattices sites. Therefore, $\gamma = 1$ and polarons and on-site bipolarons are very heavy for the characteristic values of E_p and ω_0. In contrast, in a long-range interaction, only a fraction of the total deformation changes every time the polaron moves and γ could be as small as 0.25 (part 2). Clearly, this results in a dramatic lightening of the polaron since γ enters the exponent. Thus the small polaron mass could be $\leq 10m_e$ where a Holstein-like estimate would yield a huge mass of $10\,000m_e$. The lower mass has important consequences because lighter polarons are more likely to remain mobile and less likely to trap on impurities.

The bipolaron also becomes much lighter, if the e–ph interaction is long range. There are two reasons for the lowering of its mass with an increasing radius for the e–ph interaction. The first one is the same as in the case of the

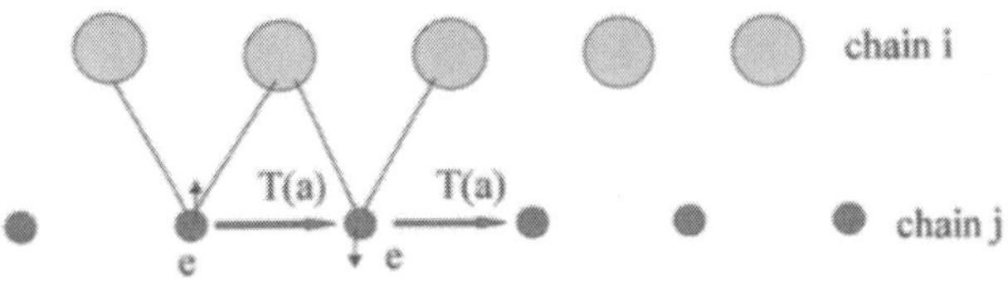

Figure 4.7. Simplified chain model with two electrons on the chain interacting with nearest-neighbour ions of another chain. Second-order inter-site bipolaron tunnelling is shown by arrows.

single polaron discussed earlier. The second reason is the possibility to form *inter-site* bipolarons which, in certain lattice structures, already tunnel coherently in the first-order in $T(\boldsymbol{m})$ [100] (section 4.6.3), in contrast with on-site bipolarons, which tunnel only in the second order, equation (4.134).

To illustrate the essential dynamic properties of bipolarons formed by the longer-range e–ph interaction let us discuss a few simplified models. Following Bonča and Trugman [107], we first consider a single bipolaron in the chain model of section 4.3.3 (figure 4.7). One can further simplify the chain model by placing ions in the interstitial sites located between the Wannier orbitals of one chain and allowing for the e–ph interaction only with the nearest neighbours of another chain, as shown in figure 4.7. The Coulomb interaction is represented by the on-site Hubbard U term.

The model Hamiltonian is

$$H = T(a)\sum_{j,s}[c^{\dagger}_{j+1,s}c_{js} + H.c.] + \omega_0 \sum_{i,j,s} g(i,j)\hat{n}_{js}(d^{\dagger}_i + d_i) + \omega_0 \sum_i [d^{\dagger}_i d_i + 1/2] + U\sum_j \hat{n}_{j\uparrow}\hat{n}_{j\downarrow} \tag{4.136}$$

in the site representation for electrons and phonons (section 4.1), where

$$g(i,j) = g_0[\delta_{i,j} + \delta_{i,j+1}]$$

and i, j are integers sorting the ions and the Wannier sites, respectively. This model is referred to as the *extended* Holstein–Hubbard model (EHHM) [107]. We can view the EHHM as the simplest model with a longer range than the Holstein interaction. In comparison with the Fröhlich model of section 4.3.3, the EHHM lacks a long-range tail in the e–ph interaction but reveals similar physical properties. In the momentum representation, the model is a one-dimensional case of the generic Hamiltonian (4.7), with

$$\gamma(q) = g_0\sqrt{2}(1 + e^{iqa}) \tag{4.137}$$

and $\omega(q) = \omega_0$. Using equations (4.34), (4.39) and (4.32), we obtain

$$E_p = \frac{g_0^2\omega_0 a}{\pi}\int_{-\pi/a}^{\pi/a} dq\,[1+\cos qa] = 2g_0^2\omega_0 \tag{4.138}$$

for the polaron level shift,

$$g^2 = \frac{g_0^2 a}{\pi}\int_{-\pi/a}^{\pi/a} dq\,[1-\cos^2 qa] = g_0^2 \tag{4.139}$$

for the mass enhancement exponent and

$$v(0) = U - 4g_0^2\omega_0 \tag{4.140}$$

$$v(a) = -\frac{2g_0^2\omega_0 a}{\pi}\int_{-\pi/a}^{\pi/a} dq\,[1+\cos qa]\cos qa = -2g_0^2\omega_0$$

for the on-site and inter-site polaron–polaron interactions, respectively. Hence, the EHHM has the numerical coefficient $\gamma = 1/2$, and the polaron mass

$$m^*_{\text{EHP}} \propto \exp\left(\frac{E_p}{2\omega_0}\right) \tag{4.141}$$

scales as the square root of the small Holstein polaron mass, $m^*_{\text{SHP}} \propto \exp(E_p/\omega_0)$. In the case when $U < 2g_0^2\omega_0$, the on-site bipolaron has the lowest energy because $|v(0)| > |v(a)|$. In this regime the bipolaron binding energy is

$$\Delta = 4g_0^2\omega_0 - U. \tag{4.142}$$

Using expression (4.132) for the bipolaron hopping integral, we obtain the bipolaron mass as

$$m^{**}_{\text{EHB}} \propto \exp\left(\frac{2E_p}{\omega_0}\right) \tag{4.143}$$

if $\Delta \gg \omega_0$. It scales as $(m^*_{\text{EHP}}/m)^4$ but is much smaller than the on-site bipolaron mass in the Holstein model, $m^{**}_{\text{SHB}} \propto \exp(4E_p/\omega_0)$, which scales as $(m^*_{\text{SHP}}/m)^4$. In the opposite regime, when $U > 2g_0^2\omega_0$, the inter-site bipolaron has the lowest energy. Its binding energy

$$\Delta = 2g_0^2\omega_0 \tag{4.144}$$

does not depend on U. Differing from the on-site singlet bipolaron, the inter-site bipolaron has four spin states, one singlet $S = 0$ and three triplet states, $S = 1$, with different z-components of the total spin, $S_z = 0, \pm 1$. In the chain model (figure 4.7), the inter-site bipolaron also tunnels only in the second order in $T(a)$, when one of the electrons within the pair hops to the left (right) and then the other follows. This tunnelling involves the multi-phonon correlation function $\Phi^{j+2,j+1}_{j+1,j}$ (equation (4.126)):

$$\Phi^{j+2,j+1}_{j+1,j} = \mathrm{e}^{-2g_0^2}.$$

Hence, the inter-site bipolaron mass enhancement is

$$m^{**}_{\mathrm{EHB}} \propto T^{-2}(a) \exp\left(\frac{E_p}{\omega_0}\right) \propto \left(\frac{m^*_{\mathrm{EHP}}}{m}\right)^2 \tag{4.145}$$

in the infinite Hubbard U limit ($U \to \infty$). We see that the inter-site bipolaron in the chain model is lighter than the on-site bipolaron but still remains much heavier than the polaron.

4.6.3 Superlight inter-site bipolarons

Any realistic theory of doped ionic insulators must include both the long-range Coulomb repulsion between carriers and the strong long-range electron–phonon interaction. From a theoretical standpoint, the inclusion of the long-range Coulomb repulsion is critical in ensuring that the carriers would not form clusters. Indeed, in order to form stable bipolarons, the e–ph interaction has to be strong enough to overcome the Coulomb repulsion at short distances. Since the e–ph interaction is long range, there is a potential possibility for clustering. The inclusion of the Coulomb repulsion V_c makes the clusters unstable. More precisely, there is a certain window of V_c/E_p inside which the clusters are unstable but bipolarons form nonetheless. In this parameter window, bipolarons repel each other and propagate in a narrow band. At a weaker Coulomb interaction, the system is a charge-segregated insulator and at a stronger Coulomb repulsion, the system is the Fermi liquid or the Luttinger liquid, if it is one-dimensional.

Let us now apply a generic 'Fröhlich–Coulomb' Hamiltonian, which explicitly includes the infinite-range Coulomb and electron–phonon interactions, to a particular lattice structure [108]. The implicitly present infinite Hubbard U prohibits double occupancy and removes the need to distinguish the fermionic spin. Introducing spinless fermion operators $c_{\boldsymbol{n}}$ and phonon operators $d_{\boldsymbol{m}\nu}$, the Hamiltonian is written as

$$\begin{aligned} H = & \sum_{\boldsymbol{n} \neq \boldsymbol{n}'} T(\boldsymbol{n} - \boldsymbol{n}') c^{\dagger}_{\boldsymbol{n}} c_{\boldsymbol{n}'} + \sum_{\boldsymbol{n} \neq \boldsymbol{n}'} V_c(\boldsymbol{n} - \boldsymbol{n}') c^{\dagger}_{\boldsymbol{n}} c_{\boldsymbol{n}} c^{\dagger}_{\boldsymbol{n}'} c_{\boldsymbol{n}'} \\ & + \omega_0 \sum_{\boldsymbol{n} \neq \boldsymbol{m}, \nu} g_{\nu}(\boldsymbol{m} - \boldsymbol{n})(\boldsymbol{e}_{\nu} \cdot \boldsymbol{e}_{\boldsymbol{m}-\boldsymbol{n}}) c^{\dagger}_{\boldsymbol{n}} c_{\boldsymbol{n}} (d^{\dagger}_{\boldsymbol{m}\nu} + d_{\boldsymbol{m}\nu}) \\ & + \omega_0 \sum_{\boldsymbol{m}, \nu} (d^{\dagger}_{\boldsymbol{m}\nu} d_{\boldsymbol{m}\nu} + \tfrac{1}{2}). \end{aligned} \tag{4.146}$$

The e–ph term is written in real space, which is more convenient when working with complex lattices.

In general, the many-body model equation (4.146) is of considerable complexity. However, we are interested in the limit of the strong e–ph interaction. In this case, the kinetic energy is a perturbation and the model can be grossly simplified using the canonical transformation of section 4.3 in the Wannier representation for electrons and phonons:

$$S = \sum_{m \neq n, \nu} g_\nu(\boldsymbol{m} - \boldsymbol{n})(\boldsymbol{e}_\nu \cdot \boldsymbol{e}_{\boldsymbol{m}-\boldsymbol{n}}) c_{\boldsymbol{n}}^\dagger c_{\boldsymbol{n}} (d_{\boldsymbol{m}\nu}^\dagger - d_{\boldsymbol{m}\nu}).$$

The transformed Hamiltonian is

$$\tilde{H} = \mathrm{e}^{-S} H \mathrm{e}^{S} = \sum_{n \neq n'} \hat{\sigma}_{\boldsymbol{nn}'} c_{\boldsymbol{n}}^\dagger c_{\boldsymbol{n}'} + \omega_0 \sum_{\boldsymbol{m}\alpha} (d_{\boldsymbol{m}\nu}^\dagger d_{\boldsymbol{m}\nu} + \tfrac{1}{2})$$
$$+ \sum_{n \neq n'} v(\boldsymbol{n} - \boldsymbol{n}') c_{\boldsymbol{n}}^\dagger c_{\boldsymbol{n}} c_{\boldsymbol{n}'}^\dagger c_{\boldsymbol{n}'} - E_p \sum_{\boldsymbol{n}} c_{\boldsymbol{n}}^\dagger c_{\boldsymbol{n}}. \tag{4.147}$$

The last term describes the energy gained by polarons due to the e–ph interaction. E_p is the familiar polaron level shift:

$$E_p = \omega_0 \sum_{\boldsymbol{m}\nu} g_\nu^2(\boldsymbol{m} - \boldsymbol{n})(\boldsymbol{e}_\nu \cdot \boldsymbol{e}_{\boldsymbol{m}-\boldsymbol{n}})^2 \tag{4.148}$$

which is independent of $\boldsymbol{n}$. The third term on the right-hand side of equation (4.147) is the polaron–polaron interaction:

$$v(\boldsymbol{n} - \boldsymbol{n}') = V_\mathrm{c}(\boldsymbol{n} - \boldsymbol{n}') - V_\mathrm{ph}(\boldsymbol{n} - \boldsymbol{n}') \tag{4.149}$$

where

$$V_\mathrm{ph}(\boldsymbol{n} - \boldsymbol{n}') = 2\omega_0 \sum_{\boldsymbol{m},\nu} g_\nu(\boldsymbol{m} - \boldsymbol{n}) g_\nu(\boldsymbol{m} - \boldsymbol{n}')(\boldsymbol{e}_\nu \cdot \boldsymbol{e}_{\boldsymbol{m}-\boldsymbol{n}})(\boldsymbol{e}_\nu \cdot \boldsymbol{e}_{\boldsymbol{m}-\boldsymbol{n}'}).$$

The phonon-induced interaction, V_ph, is due to displacements of common ions by two electrons. Finally, the transformed hopping operator $\hat{\sigma}_{\boldsymbol{nn}'}$ in the first term in equation (4.147) is given by

$$\hat{\sigma}_{\boldsymbol{nn}'} = T(\boldsymbol{n} - \boldsymbol{n}') \exp\Bigg[\sum_{\boldsymbol{m},\nu} [g_\nu(\boldsymbol{m} - \boldsymbol{n})(\boldsymbol{e}_\nu \cdot \boldsymbol{e}_{\boldsymbol{m}-\boldsymbol{n}})$$
$$- g_\nu(\boldsymbol{m} - \boldsymbol{n}')(\boldsymbol{e}_\nu \cdot \boldsymbol{e}_{\boldsymbol{m}-\boldsymbol{n}'})](d_{\boldsymbol{m}\alpha}^\dagger - d_{\boldsymbol{m}\alpha}) \Bigg]. \tag{4.150}$$

This term is a perturbation at large λ. Here we consider a particular lattice structure (ladder), where bipolarons already tunnel in the first order in $T(\boldsymbol{n})$, so that $\hat{\sigma}_{\boldsymbol{nn}'}$ can be averaged over phonons. When $T \lesssim \omega_0$, the result is

$$t(\boldsymbol{n} - \boldsymbol{n}') \equiv \langle\langle \hat{\sigma}_{\boldsymbol{nn}'} \rangle\rangle_\mathrm{ph} = T(\boldsymbol{n} - \boldsymbol{n}') \exp[-g^2(\boldsymbol{n} - \boldsymbol{n}')] \tag{4.151}$$

$$g^2(\boldsymbol{n} - \boldsymbol{n}') = \sum_{\boldsymbol{m},\nu} g_\nu(\boldsymbol{m} - \boldsymbol{n})(\boldsymbol{e}_\nu \cdot \boldsymbol{e}_{\boldsymbol{m}-\boldsymbol{n}})$$
$$\times [g_\nu(\boldsymbol{m} - \boldsymbol{n})(\boldsymbol{e}_\nu \cdot \boldsymbol{e}_{\boldsymbol{m}-\boldsymbol{n}}) - g_\nu(\boldsymbol{m} - \boldsymbol{n}')(\boldsymbol{e}_\nu \cdot \boldsymbol{e}_{\boldsymbol{m}-\boldsymbol{n}'})].$$

By comparing equations (4.151) and (4.149), the mass renormalization exponent can be expressed via E_p and V_{ph} as follows:

$$g^2(\boldsymbol{n}-\boldsymbol{n}') = \frac{1}{\omega_0}\left[E_p - \frac{1}{2}V_{\text{ph}}(\boldsymbol{n}-\boldsymbol{n}')\right]. \tag{4.152}$$

Now phonons are 'integrated out' and the polaronic Hamiltonian is

$$\begin{aligned} H_p &= H_0 + H_{\text{pert}} \\ H_0 &= -E_p \sum_{\boldsymbol{n}} c_{\boldsymbol{n}}^\dagger c_{\boldsymbol{n}} + \sum_{\boldsymbol{n}\neq\boldsymbol{n}'} v(\boldsymbol{n}-\boldsymbol{n}') c_{\boldsymbol{n}}^\dagger c_{\boldsymbol{n}} c_{\boldsymbol{n}'}^\dagger c_{\boldsymbol{n}'} \\ H_{\text{pert}} &= \sum_{\boldsymbol{n}\neq\boldsymbol{n}'} t(\boldsymbol{n}-\boldsymbol{n}') c_{\boldsymbol{n}}^\dagger c_{\boldsymbol{n}'}. \end{aligned} \tag{4.153}$$

When V_{ph} exceeds V_{c}, the full interaction becomes negative and the polarons form pairs. The real-space representation allows us to elaborate the physics behind the lattice sums in equations (4.148) and (4.149) more fully. When a carrier (electron or hole) acts on an ion with a force $\boldsymbol{f}$, it displaces the ion by some vector $\boldsymbol{x} = \boldsymbol{f}/s$. Here s is the ion's force constant. The total energy of the carrier–ion pair is $-f^2/(2s)$. This is precisely the summand in equation (4.148) expressed via dimensionless coupling constants. Now consider two carriers interacting with the *same* ion, see figure 4.8(*a*). The ion displacement is $\boldsymbol{x} = (\boldsymbol{f}_1 + \boldsymbol{f}_2)/s$ and the energy is $-f_1^2/(2s) - f_2^2/(2s) - (\boldsymbol{f}_1 \cdot \boldsymbol{f}_2)/s$. Here the last term should be interpreted as an ion-mediated interaction between the two carriers. It depends on the scalar product of $\boldsymbol{f}_1$ and $\boldsymbol{f}_2$ and, consequently, on the relative positions of the carriers with respect to the ion. If the ion is an isotropic harmonic oscillator, as we assume here, then the following simple rule applies. If the angle ϕ between $\boldsymbol{f}_1$ and $\boldsymbol{f}_2$ is less than $\pi/2$ the polaron–polaron interaction will be attractive, if otherwise it will be repulsive. In general, some ions will generate attraction and some repulsion between polarons (figure 4.8(*b*)).

The overall sign and magnitude of the interaction is given by the lattice sum in equation (4.149), the evaluation of which is elementary. One should also note that, according to equation (4.152), an attractive interaction reduces the polaron mass (and, consequently, the bipolaron mass), while repulsive interaction enhances the mass. Thus, the long-range nature of the e–ph interaction serves a double purpose. First, it generates an additional inter-polaron attraction because the distant ions have small angle ϕ. This additional attraction helps to overcome the direct Coulomb repulsion between the polarons. And, second, the Fröhlich interaction makes the bipolarons lighter.

The many-particle ground state of H_0 depends on the sign of the polaron–polaron interaction, the carrier density and the lattice geometry. Here we consider the zig-zag ladder in figure 4.9(*a*), assuming that all sites are isotropic two-dimensional harmonic oscillators. For simplicity, we also adopt the nearest-neighbour approximation for both interactions, $g_\nu(\boldsymbol{l}) \equiv g$, $V_{\text{c}}(\boldsymbol{n}) \equiv V_{\text{c}}$, and for

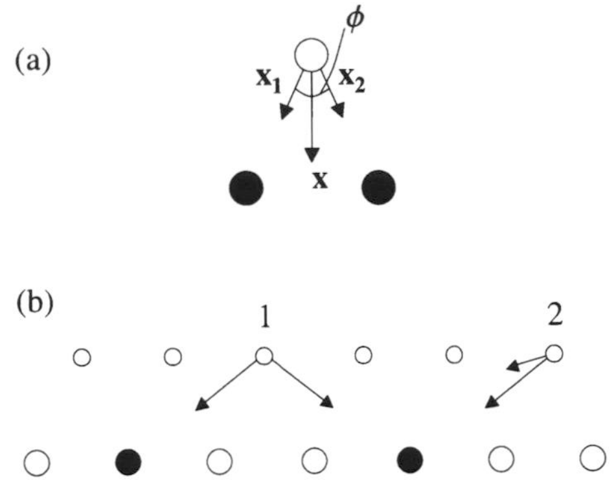

Figure 4.8. The mechanism for the polaron–polaron interaction. (*a*) Together, the two polarons (filled circles) deform the lattice more effectively than separately. An effective attraction occurs when the angle ϕ between x_1 and x_2 is less than $\pi/2$. (*b*) A mixed situation. Ion 1 results in repulsion between two polarons while ion 2 results in attraction.

the hopping integrals, $T(\boldsymbol{m}) = T_{\mathrm{NN}}$ for $l = n = m = a$, and zero otherwise. Hereafter we set the lattice period $a = 1$. There are four nearest neighbours in the ladder, $z = 4$. Then, the *one-particle* polaronic Hamiltonian takes the form

$$H_p = -E_p \sum_n (c_n^\dagger c_n + p_n^\dagger p_n) + \sum_n [t'(c_{n+1}^\dagger c_n + p_{n+1}^\dagger p_n) + t(p_n^\dagger c_n + p_{n-1}^\dagger c_n) + H.c.] \tag{4.154}$$

where c_n and p_n are polaron annihilation operators on the lower and upper sites of the ladder, respectively (figure 4.9(*b*)). Using equations (4.148), (4.149) and (4.152), we find

$$E_p = 4g^2\omega_0 \tag{4.155}$$
$$t' = T_{\mathrm{NN}} \exp\left(-\frac{7E_p}{8\omega_0}\right)$$
$$t = T_{\mathrm{NN}} \exp\left(-\frac{3E_p}{4\omega_0}\right).$$

The Fourier transform of equation (4.154) into momentum space yields

$$H_p = \sum_k (2t'\cos k - E_p)(c_k^\dagger c_k + p_k^\dagger p_k) + t\sum_k [(1+\mathrm{e}^{\mathrm{i}k})p_k^\dagger c_k + H.c.]. \tag{4.156}$$

A linear transformation of c_k and p_k diagonalizes the Hamiltonian, so that the one-particle energy spectrum $E_1(k)$ is found from

$$\det\begin{vmatrix} 2t'\cos k - E_p - E_1(k) & t(1+\mathrm{e}^{\mathrm{i}k}) \\ t(1+\mathrm{e}^{-\mathrm{i}k}) & 2t'\cos k - E_p - E_1(k) \end{vmatrix} = 0. \tag{4.157}$$

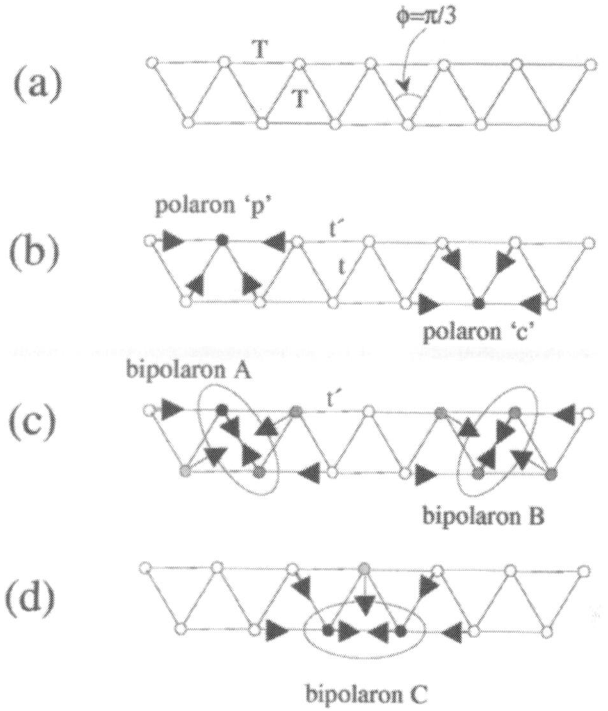

Figure 4.9. One-dimensional zig-zag ladder: (*a*) initial ladder with the bare hopping amplitude $T(a)$; (*b*) two types of polarons with their respective deformations; (*c*) two degenerate bipolaron configurations A and B; and (*d*) a different bipolaron configuration, C, whose energy is higher than that of A and B.

There are two overlapping polaronic bands,

$$E_1(k) = -E_p + 2t' \cos(k) \pm 2t \cos(k/2)$$

with effective mass $m^* = 2/|4t' \pm t|$ near their edges.

Let us now place two polarons on the ladder. The nearest-neighbour interaction (equation (4.149)) is $v = V_c - E_p/2$, if two polarons are on different sides of the ladder, and $v = V_c - E_p/4$, if both polarons are on the same side. The attractive interaction is provided via the displacement of the lattice sites, which are the common nearest neighbours to both polarons. There are two such nearest neighbours for the inter-site bipolaron of type A or B (figure 4.9(*c*)) but there is only one common nearest neighbour for bipolaron C (figure 4.9(*d*)). When $V_c > E_p/2$, there are no bound states and the multi-polaron system is a one-dimensional Luttinger liquid. However, when $V_c < E_p/2$ and, consequently, $v < 0$, the two polarons are bound into an inter-site bipolaron of types A or B.

It is quite remarkable that bipolaron tunnelling in the ladder already appears in the first order with respect to a single-electron tunnelling. This case is different from both the on-site bipolarons discussed in section 4.6.1 and from the inter-site chain bipolaron of section 4.6.2, where the bipolaron tunnelling was of the second order in $T(a)$. Indeed, the lowest-energy configurations A and B are degenerate. They are coupled by H_{pert}. Neglecting all higher-energy configurations, we can project the Hamiltonian onto the subspace containing A, B and empty sites.

The result of such a projection is a bipolaronic Hamiltonian:

$$H_{\mathrm{b}} = (V_{\mathrm{c}} - \tfrac{5}{2}E_p)\sum_n [A_n^\dagger A_n + B_n^\dagger B_n] - t'\sum_n [B_n^\dagger A_n + B_{n-1}^\dagger A_n + H.c.] \quad (4.158)$$

where $A_n = c_n p_n$ and $B_n = p_n c_{n+1}$ are inter-site bipolaron annihilation operators and the bipolaron–bipolaron interaction is dropped (see later). Its Fourier transform yields two *bipolaron* bands,

$$E_2(k) = V_{\mathrm{c}} - \tfrac{5}{2}E_p \pm 2t'\cos(k/2) \quad (4.159)$$

with a combined width $4|t'|$. The bipolaron binding energy in zero order with respect to t, t' is

$$\Delta \equiv 2E_1(0) - E_2(0) = \frac{E_p}{2} - V_{\mathrm{c}}. \quad (4.160)$$

The bipolaron mass near the bottom of the lowest band, $m^{**} = 2/t'$, is

$$m^{**} = 4m^*\left[1 + 0.25\exp\left(\frac{E_p}{8\omega_0}\right)\right]. \quad (4.161)$$

The numerical coefficient $1/8$ ensures that m^{**} remains of the order of m^* even at large E_p. This fact combines with a weaker renormalization of m^*, equation (4.155), providing a *superlight* bipolaron.

In models with strong inter-site attraction, there is a possibility of clusterization. Similar to the two-particle case described earlier, the lowest energy of n polarons placed on the nearest neighbours of the ladder is found as

$$E_n = (2n-3)V_{\mathrm{c}} - \frac{6n-1}{4}E_p \quad (4.162)$$

for any $n \geq 3$. There are *no* resonating states for an n-polaron configuration if $n \geq 3$. Therefore, there is no first-order kinetic energy contribution to their energy. E_n should be compared with the energy $E_1 + (n-1)E_2/2$ of far separated $(n-1)/2$ bipolarons and a single polaron for odd $n \geq 3$, or with the energy of far separated n bipolarons for even $n \geq 4$. 'Odd' clusters are stable if

$$V_{\mathrm{c}} < \frac{n}{6n-10}E_p \quad (4.163)$$

and 'even' clusters are stable if

$$V_{\mathrm{c}} < \frac{n-1}{6n-12}E_p. \quad (4.164)$$

As a result, we find that bipolarons repel each other and single polarons at $V_c > \frac{3}{8}E_p$. If V_c is less than $\frac{3}{8}E_p$, then immobile bound clusters of three and more polarons could form. One should note that at distances much larger than the lattice constant, the polaron–polaron interaction is always repulsive (section 4.4) and the formation of infinite clusters, stripes or strings is impossible. Combining the condition of bipolaron formation and that of the instability of larger clusters, we obtain a window of parameters

$$\tfrac{3}{8}E_p < V_c < \tfrac{1}{2}E_p \tag{4.165}$$

where the ladder is a bipolaronic conductor. Outside the window, the ladder is either charge-segregated into finite-size clusters (small V_c) or it is a liquid of repulsive polarons (large V_c).

4.7 Bipolaronic superconductivity

In the subspace with no single polarons, the Hamiltonian of electrons strongly coupled with phonons is reduced to the bipolaronic Hamiltonian written in terms of creation, $b_{\boldsymbol{m}}^{\dagger} = c_{\boldsymbol{m}\uparrow}^{\dagger}c_{\boldsymbol{m}\downarrow}^{\dagger}$, and annihilation, $b_{\boldsymbol{m}}$, bipolaron operators as

$$H_b = \sum_{\boldsymbol{m}\neq\boldsymbol{m}'} [t(\boldsymbol{m}-\boldsymbol{m}')b_{\boldsymbol{m}}^{\dagger}b_{\boldsymbol{m}'} + \tfrac{1}{2}\bar{v}(\boldsymbol{m}-\boldsymbol{m}')n_{\boldsymbol{m}}n_{\boldsymbol{m}'}] \tag{4.166}$$

where $\bar{v}(\boldsymbol{m}-\boldsymbol{m}')$ is the bipolaron–bipolaron interaction, $n_{\boldsymbol{m}} = b_{\boldsymbol{m}}^{\dagger}b_{\boldsymbol{m}}$, and the position of the middle of the bipolaron band is taken as zero. There are additional (spin) quantum numbers $S = 0, 1$; $S_z = 0, \pm 1$, which should be added to the definition of $b_{\boldsymbol{m}}$ for the case of inter-site bipolarons. Also in some lattice structures (section 4.6.3 and part 2), inter-site bipolarons tunnel via a one-particle hopping rather than via simultaneous two-particle tunnelling of on-site bipolarons. This 'crab-like' tunnelling (figure 4.9) results in a bipolaron bandwidth of the same order as the polaron one. Keeping this in mind, we can apply H_b (equation (4.166)) to both on-site and inter-site bipolarons, and even to more extended non-overlapping pairs, implying that the site index $\boldsymbol{m}$ is the position of the centre of mass of a pair.

4.7.1 Bipolarons and a charged Bose gas

Bipolarons are not perfect bosons. In the subspace of pairs and empty sites, their operators commute as

$$b_{\boldsymbol{m}}b_{\boldsymbol{m}}^{\dagger} + b_{\boldsymbol{m}}^{\dagger}b_{\boldsymbol{m}} = 1 \tag{4.167}$$

$$b_{\boldsymbol{m}}b_{\boldsymbol{m}'}^{\dagger} - b_{\boldsymbol{m}'}^{\dagger}b_{\boldsymbol{m}} = 0 \tag{4.168}$$

for $\boldsymbol{m} \neq \boldsymbol{m}'$. This makes useful the pseudospin analogy [10],

$$b_{\boldsymbol{m}}^{\dagger} = S_{\boldsymbol{m}}^{x} - \mathrm{i}S_{\boldsymbol{m}}^{y} \tag{4.169}$$

and

$$b_m^\dagger b_m = \tfrac{1}{2} - S_m^z \tag{4.170}$$

with the pseudospin $\frac{1}{2}$ operators $S^{x,y,z} = \frac{1}{2}\tau_{1,2,3}$. $S_m^z = \frac{1}{2}$ corresponds to an empty site $\boldsymbol{m}$ and $S_m^z = -\frac{1}{2}$ to a site occupied by the bipolaron. Spin operators preserve the bosonic nature of bipolarons, when they are on different sites and their fermionic internal structure. Replacing bipolarons by spin operators, we transform the bipolaronic Hamiltonian into the anisotropic Heisenberg Hamiltonian:

$$H_b = \sum_{m \neq m'} [\tfrac{1}{2}\bar{v}_{mm'} S_m^z S_{m'}^z + t_{mm'}(S_m^x S_{m'}^x + S_m^y S_{m'}^y)]. \tag{4.171}$$

This Hamiltonian has been investigated in detail as a relevant form for magnetism and also for quantum solids like a lattice model of ^{4}He. However, while in those cases the magnetic field is an independent thermodynamic variable, in our case the total 'magnetization' is fixed,

$$\frac{1}{N}\sum_m \langle\langle S_m^z \rangle\rangle = \frac{1}{2} - n_b \tag{4.172}$$

if the bipolaron density n_b is conserved. The spin-$\frac{1}{2}$ Heisenberg Hamiltonian (4.171) cannot be solved analytically. The complicated commutation rules for bipolaron operators (equations (4.167) and (4.168)) make the problem hard but not in the limit of low atomic density of bipolarons, $n_b \ll 1$ (for a complete phase diagram of bipolarons on a lattice see [10, 109]). In this limit we can reduce the problem to a charged Bose gas on a lattice [110]. Let us transform the bipolaronic Hamiltonian to a representation containing only the Bose operators a_m and $a_m^\dagger$ defined as

$$b_m = \sum_{k=0}^{\infty} \beta_k (a_m^\dagger)^k a_m^{k+1} \tag{4.173}$$

$$b_m^\dagger = \sum_{k=0}^{\infty} \beta_k (a_m^\dagger)^{k+1} a_m^k \tag{4.174}$$

where

$$a_m a_{m'}^\dagger - a_{m'}^\dagger a_m = \delta_{m,m'}. \tag{4.175}$$

The first few coefficients β_k are found by substituting equations (4.173) and (4.174) into equations (4.167) and (4.168):

$$\beta_0 = 1 \qquad \beta_1 = -1 \qquad \beta_2 = \frac{1}{2} + \frac{\sqrt{3}}{6}. \tag{4.176}$$

We also introduce bipolaron and boson Ψ-operators as

$$\Phi(\boldsymbol{r}) = \frac{1}{\sqrt{N}} \sum_{\boldsymbol{m}} \delta(\boldsymbol{r} - \boldsymbol{m}) b_{\boldsymbol{m}} \tag{4.177}$$

$$\Psi(\boldsymbol{r}) = \frac{1}{\sqrt{N}} \sum_{\boldsymbol{m}} \delta(\boldsymbol{r} - \boldsymbol{m}) a_{\boldsymbol{m}}. \tag{4.178}$$

The transformation of the field operators takes the form:

$$\Phi(\boldsymbol{r}) = \left[1 - \frac{\Psi^{\dagger}(\boldsymbol{r})\Psi(\boldsymbol{r})}{N} + \frac{(1/2 + \sqrt{3}/6)\Psi^{\dagger}(\boldsymbol{r})\Psi(\boldsymbol{r})\Psi(\boldsymbol{r})}{N^2} + \cdots\right]\Psi(\boldsymbol{r}). \tag{4.179}$$

Then we write the bipolaronic Hamiltonian as

$$H_{\mathrm{b}} = \int \mathrm{d}\boldsymbol{r} \int \mathrm{d}\boldsymbol{r}'\, \Psi^{\dagger}(\boldsymbol{r}) t(\boldsymbol{r} - \boldsymbol{r}') \Psi(\boldsymbol{r}') + H_{\mathrm{d}} + H_{\mathrm{K}} + H^{(3)} \tag{4.180}$$

where

$$H_{\mathrm{d}} = \tfrac{1}{2} \int \mathrm{d}\boldsymbol{r} \int \mathrm{d}\boldsymbol{r}'\, \bar{v}(\boldsymbol{r} - \boldsymbol{r}') \Psi^{\dagger}(\boldsymbol{r}) \Psi^{\dagger}(\boldsymbol{r}') \Psi(\boldsymbol{r}') \Psi(\boldsymbol{r}) \tag{4.181}$$

is the dynamic part,

$$H_{\mathrm{K}} = \frac{2}{N} \int \mathrm{d}\boldsymbol{r} \int \mathrm{d}\boldsymbol{r}'\, t(\boldsymbol{r} - \boldsymbol{r}') \times [\Psi^{\dagger}(\boldsymbol{r})\Psi^{\dagger}(\boldsymbol{r}')\Psi(\boldsymbol{r}')\Psi(\boldsymbol{r}') + \Psi^{\dagger}(\boldsymbol{r})\Psi^{\dagger}(\boldsymbol{r})\Psi(\boldsymbol{r})\Psi(\boldsymbol{r}')]. \tag{4.182}$$

is the kinematic (hard-core) part due to the 'imperfect' commutation rules, and $H^{(3)}$ includes three- and higher-body collisions. Here

$$t(\boldsymbol{r} - \boldsymbol{r}') = \sum_{\boldsymbol{k}} \epsilon_{\boldsymbol{k}}^{**} \mathrm{e}^{\mathrm{i}\boldsymbol{k}\cdot(\boldsymbol{r}-\boldsymbol{r}')}$$

$$\bar{v}(\boldsymbol{r} - \boldsymbol{r}') = \frac{1}{N} \sum_{\boldsymbol{k}} \bar{v}_{\boldsymbol{k}} \mathrm{e}^{\mathrm{i}\boldsymbol{k}\cdot(\boldsymbol{r}-\boldsymbol{r}')}$$

$\bar{v}_{\boldsymbol{k}} = \sum_{\boldsymbol{m}\neq 0} \bar{v}(\boldsymbol{m}) \exp(\mathrm{i}\boldsymbol{k}\cdot\boldsymbol{m})$ is the Fourier component of the dynamic interaction and

$$\epsilon_{\boldsymbol{k}}^{**} = \sum_{\boldsymbol{m}\neq 0} t(\boldsymbol{m}) \exp(-\mathrm{i}\boldsymbol{k}\cdot\boldsymbol{m}) \tag{4.183}$$

is the bipolaron band dispersion. $H^{(3)}$ contains powers of the field operator higher than four. In the dilute limit ($n_{\mathrm{b}} \ll 1$) only two-particle interactions are essential which include the short-range kinematic and direct density–density repulsions. Because $\bar{v}$ already has a short-range part $v^{(2)}$ (equation (4.129)), the kinematic contribution can be included in the definition of $\bar{v}$. As a result H_b is reduced to the Hamiltonian of interacting hard-core charged bosons tunnelling in a narrow band.

To describe the electrodynamics of bipolarons, we introduce the vector potential $\boldsymbol{A}(\boldsymbol{r})$ using the so-called Peierls substitution [111]:

$$t(\boldsymbol{m}-\boldsymbol{m}') \rightarrow t(\boldsymbol{m}-\boldsymbol{m}')\mathrm{e}^{\mathrm{i}2e\boldsymbol{A}(\boldsymbol{m})\cdot(\boldsymbol{m}-\boldsymbol{m}')}$$

which is a fair approximation when the magnetic field is weak compared with the atomic field, $eHa^2 \ll 1$ (see also section 2.16). It has the following form:

$$t(\mathbf{r}-\mathbf{r}') \rightarrow t(\boldsymbol{r},\boldsymbol{r}') = \sum_{\boldsymbol{k}} \epsilon^{**}_{\boldsymbol{k}-2e\boldsymbol{A}} \mathrm{e}^{\mathrm{i}\boldsymbol{k}\cdot(\boldsymbol{r}-\boldsymbol{r}')} \tag{4.184}$$

in real space. If the magnetic field is weak, we can expand $\epsilon^{**}_{\boldsymbol{k}}$ in the vicinity of $\boldsymbol{k}=0$ to obtain

$$t(\boldsymbol{r},\boldsymbol{r}') \approx -\frac{[\nabla + 2\mathrm{i}e\boldsymbol{A}(\boldsymbol{r})]^2}{2m^{**}}\delta(\boldsymbol{r}-\boldsymbol{r}') \tag{4.185}$$

where

$$\frac{1}{m^{**}} = \left(\frac{\mathrm{d}^2\epsilon^{**}_{\boldsymbol{k}}}{\mathrm{d}k^2}\right)_{k\rightarrow 0} \tag{4.186}$$

is the inverse bipolaron mass. Here we assume a parabolic dispersion near the bottom of the band, $\epsilon^{**}_{\boldsymbol{k}} \propto k^2$. Finally, we arrive at

$$\begin{aligned} H_b \approx & -\int \mathrm{d}\boldsymbol{r}\, \Psi^{\dagger}(\boldsymbol{r}) \left\{ \frac{[\nabla + 2\mathrm{i}e\boldsymbol{A}(\boldsymbol{r})]^2}{2m^{**}} + \mu \right\} \Psi(\boldsymbol{r}) \\ & + \tfrac{1}{2}\int \mathrm{d}\boldsymbol{r}\,\mathrm{d}\boldsymbol{r}'\, \bar{v}(\boldsymbol{r}-\boldsymbol{r}')\Psi^{\dagger}(\boldsymbol{r})\Psi^{\dagger}(\boldsymbol{r}')\Psi(\boldsymbol{r})\Psi(\boldsymbol{r}') \end{aligned} \tag{4.187}$$

where we include the *bipolaron* chemical potential μ. We note that the bipolaron–bipolaron interaction is the Coulomb repulsion, $\bar{v}(\boldsymbol{r}) \sim 1/(\epsilon_0 r)$ at large distances (section 4.4) and the hard-core repulsion is irrelevant in the dilute limit. Hence, equation (4.187) describes a charged Bose gas (CBG) with the effective boson mass m^{**} and charge $2e$.

4.7.2 Bogoliubov equations in the strong-coupling regime

Let us derive equations describing the order parameter and excitations of CBG by the use of the Bogoliubov transformation similar to that in chapter 2. The equation of motion for the field operator, $\psi(\boldsymbol{r},t)$, is derived using the Hamiltonian (4.187) as

$$\begin{aligned} \mathrm{i}\frac{\mathrm{d}}{\mathrm{d}t}\psi(\boldsymbol{r},t) = [H,\psi(\boldsymbol{r},t)] = & \left[-\frac{(\nabla - \mathrm{i}2e\boldsymbol{A})^2}{2m^{**}} - \mu\right]\psi(\boldsymbol{r},t) \\ & + \int \mathrm{d}\boldsymbol{r}'\, \bar{v}(\boldsymbol{r}-\boldsymbol{r}')\psi^{\dagger}(\boldsymbol{r}',t)\psi(\boldsymbol{r}',t)\psi(\boldsymbol{r},t). \end{aligned} \tag{4.188}$$

A large value for the static dielectric constant ($\epsilon_0 \gg 1$) in ionic solids makes the Coulomb repulsion between bipolarons rather weak. If the interaction is weak, we expect that some properties of CBG to be similar to the properties of an ideal Bose gas (appendix B). In particular, the state with zero momentum ($k = 0$) remains macroscopically occupied and the corresponding Fourier component of the field operator $\psi(\boldsymbol{r}, t)$ has an anomalously large matrix element between the states with $N + 1$ and N bosons. Introducing the chemical potential μ we consider each quantum state as a superposition of states with a slightly different total number of bosons. The weight of each state in the superposition is a smooth function of N which is practically unchanged in the window $\pm\sqrt{\bar{N}}$ near the average number $\bar{N}$. Hence, because ψ changes the number of particles only by one, its *diagonal* matrix element is practically the same as the off-diagonal one, calculated for states with fixed $N = \bar{N} + 1$ and $N = \bar{N}$. Following Bogoliubov [4] we separate the large diagonal matrix element ψ_s from ψ treating ψ_s as a number and the rest, $\tilde{\psi}$, as a small fluctuation

$$\psi(\boldsymbol{r}, t) = \psi_s(\boldsymbol{r}, t) + \tilde{\psi}(\boldsymbol{r}, t). \tag{4.189}$$

Substituting equation (4.189) into equation (4.188) and collecting c-*number* terms of ψ_s we obtain a set of Bogoliubov-type equations for the CBG [112]. The *macroscopic* condensate wavefunction $\psi_s(\boldsymbol{r}, t)$, which plays the role of an order parameter, obeys the following equation:

$$\mathrm{i}\frac{\mathrm{d}}{\mathrm{d}t}\psi_s(\boldsymbol{r}, t) = \left[-\frac{(\nabla + \mathrm{i}2e\boldsymbol{A})^2}{2m^{**}} - \mu\right]\psi_s(\boldsymbol{r}, t) + \int \mathrm{d}\boldsymbol{r}'\bar{v}(\boldsymbol{r} - \boldsymbol{r}')n_s(\boldsymbol{r}', t)\psi_s(\boldsymbol{r}, t) \tag{4.190}$$

which is a generalization of the so-called Gross–Pitaevskii (GP) [113] equation applied to neutral bosons. Taking into account the interaction of 'supracondensate' bosons (described by $\tilde{\psi}(\boldsymbol{r}, t)$) with the condensate, and neglecting the interaction between the supra-condensate bosons, from equation (4.188) we also obtain

$$\begin{aligned}\mathrm{i}\frac{\mathrm{d}}{\mathrm{d}t}\tilde{\psi}(\boldsymbol{r}, t) &= \left[-\frac{(\nabla + \mathrm{i}2e\boldsymbol{A})^2}{2m^{**}} - \mu\right]\tilde{\psi}(\boldsymbol{r}, t)\\ &\quad + \int \mathrm{d}\boldsymbol{r}'\,\bar{v}(\boldsymbol{r} - \boldsymbol{r}')[n_s(\boldsymbol{r}, t) + \psi_s^*(\boldsymbol{r}', t')\psi_s(\boldsymbol{r}, t)]\tilde{\psi}(\boldsymbol{r}', t)\\ &\quad + \int \mathrm{d}\boldsymbol{r}'\,\bar{v}(\boldsymbol{r} - \boldsymbol{r}')\psi_s(\boldsymbol{r}', t)\psi_s(\boldsymbol{r}, t)\tilde{\psi}^\dagger(\boldsymbol{r}', t).\end{aligned} \tag{4.191}$$

Here

$$n_s(\boldsymbol{r}, t) = |\psi_s(\boldsymbol{r}, t)|^2 \tag{4.192}$$

is the condensate density. If the Coulomb repulsion of bosons is not very large,

$$r_s = \frac{4m^{**}e^2}{\epsilon_0(4\pi n_b/3)^{1/3}} \lesssim 1 \tag{4.193}$$

the number of bosons $\tilde{n}$ pushed up from the condensate by the repulsion is small (see later). Therefore, the contribution of the terms nonlinear in $\tilde{\psi}$ is negligible in equation (4.191). Applying the linear Bogoliubov transformation of $\tilde{\psi}$ (chapter 2),

$$\tilde{\psi}(\boldsymbol{r}, t) = \sum_n [u_n(\boldsymbol{r}, t)\alpha_n + v_n^*(\boldsymbol{r}, t)\alpha_n^\dagger] \tag{4.194}$$

where now α_n and $\alpha_n^\dagger$ are *Bose* operators, we obtain two coupled Schrödinger equations for the quasi-particle wavefunctions, $u(\boldsymbol{r}, t)$ and $v(\boldsymbol{r}, t)$, as

$$\begin{aligned}
\mathrm{i}\frac{\mathrm{d}}{\mathrm{d}t}u(\boldsymbol{r}, t) = & \left[-\frac{(\nabla + \mathrm{i}2e\boldsymbol{A})^2}{2m^{**}} - \mu\right] u(\boldsymbol{r}, t) \\
& + \int \mathrm{d}\boldsymbol{r}'\, \bar{v}(\boldsymbol{r} - \boldsymbol{r}')[|\psi_s(\boldsymbol{r}', t)|^2 u(\boldsymbol{r}, t) + \psi_s^*(\boldsymbol{r}', t)\psi_s(\boldsymbol{r}, t)u(\boldsymbol{r}', t)] \\
& + \int \mathrm{d}\boldsymbol{r}'\, \bar{v}(\boldsymbol{r} - \boldsymbol{r}')\psi_0(\boldsymbol{r}', t)\psi_0(\boldsymbol{r}, t)v(\boldsymbol{r}', t)
\end{aligned} \tag{4.195}$$

and

$$\begin{aligned}
-\mathrm{i}\frac{\mathrm{d}}{\mathrm{d}t}v(\boldsymbol{r}, t) = & \left[-\frac{(\nabla - \mathrm{i}2e\boldsymbol{A})^2}{2m^{**}} - \mu\right] v(\boldsymbol{r}, t) \\
& + \int \mathrm{d}\boldsymbol{r}'\, \bar{v}(\boldsymbol{r} - \boldsymbol{r}')[|\psi_s(\boldsymbol{r}', t)|^2 v(\boldsymbol{r}, t) + \psi_s(\boldsymbol{r}', t)\psi_s^*(\boldsymbol{r}, t)]v(\boldsymbol{r}', t) \\
& + \int \mathrm{d}\boldsymbol{r}'\, \bar{v}(\boldsymbol{r} - \boldsymbol{r}')\psi_s^*(\boldsymbol{r}', t)\psi_s^*(\boldsymbol{r}, t)u(\boldsymbol{r}', t).
\end{aligned} \tag{4.196}$$

There is also a sum rule,

$$\sum_n [u_n(\boldsymbol{r}, t)u_n^*(\boldsymbol{r}', t) - v_n(\boldsymbol{r}, t)v_n^*(\boldsymbol{r}', t)] = \delta(\boldsymbol{r} - \boldsymbol{r}') \tag{4.197}$$

which retains the Bose commutation relations for new operators. The set of equations (4.190), (4.195) and (4.196) plays the same role in the strong-coupling theory as the Bogoliubov equations for the BCS superconductors (section 2.11).

4.7.3 Excitation spectrum and ground-state energy

For a homogeneous case with $\boldsymbol{A}(\boldsymbol{r}) = 0$, the quasi-particle wavefunctions are plane waves,

$$\begin{aligned}
u_{\boldsymbol{k}}(\boldsymbol{r}, t) &= u_{\boldsymbol{k}}\mathrm{e}^{\mathrm{i}\boldsymbol{k}\cdot\boldsymbol{r} - \mathrm{i}\epsilon_{\boldsymbol{k}}t} \\
v_{\boldsymbol{k}}(\boldsymbol{r}, t) &= v_{\boldsymbol{k}}\mathrm{e}^{\mathrm{i}\boldsymbol{k}\cdot\boldsymbol{r} - \mathrm{i}\epsilon_{\boldsymbol{k}}t}
\end{aligned} \tag{4.198}$$

and the condensate wavefunction is $(\boldsymbol{r}, t)$-independent, $\psi_s(\boldsymbol{r}, t) = \sqrt{n_s} =$ constant. Then a solution of equation (4.190) is

$$\mu = 0 \tag{4.199}$$

because the volume average of $\bar{v}(\boldsymbol{r})$ is zero due to the interaction with the ion background (i.e. because of the global electroneutrality). Then the Bogoliubov equations yield

$$\epsilon_{\boldsymbol{k}} u_{\boldsymbol{k}} = \frac{k^2}{2m^{**}} u_{\boldsymbol{k}} + n_s \bar{v}_{\boldsymbol{k}} [u_{\boldsymbol{k}} + v_{\boldsymbol{k}}] \tag{4.200}$$

$$-\epsilon_{\boldsymbol{k}} v_{\boldsymbol{k}} = \frac{k^2}{2m^{**}} v_{\boldsymbol{k}} + n_s \bar{v}_{\boldsymbol{k}} [u_{\boldsymbol{k}} + v_{\boldsymbol{k}}] \tag{4.201}$$

and

$$|u_{\boldsymbol{k}}|^2 - |v_{\boldsymbol{k}}|^2 = 1. \tag{4.202}$$

Here $\bar{v}_{\boldsymbol{k}}$ is the Fourier transform of $\bar{v}(\boldsymbol{r})$. As a result, we find

$$u_{\boldsymbol{k}}^2 = \frac{1}{2}\left(1 + \frac{\xi_{\boldsymbol{k}}}{\epsilon_{\boldsymbol{k}}}\right) \tag{4.203}$$

$$v_{\boldsymbol{k}}^2 = -\frac{1}{2}\left(1 - \frac{\xi_{\boldsymbol{k}}}{\epsilon_{\boldsymbol{k}}}\right) \tag{4.204}$$

$$u_{\boldsymbol{k}} v_{\boldsymbol{k}} = -\frac{\bar{v}_{\boldsymbol{k}} n_s}{2\epsilon_{\boldsymbol{k}}} \tag{4.205}$$

where $\xi_{\boldsymbol{k}} = k^2/2m^{**} + \bar{v}_{\boldsymbol{k}} n_s$. The quasi-particle energy is

$$\epsilon_{\boldsymbol{k}} = \sqrt{\frac{k^4}{4(m^{**})^2} + \frac{k^2 \bar{v}_{\boldsymbol{k}} n_s}{m^{**}}}. \tag{4.206}$$

Using the Fourier component of the Coulomb interaction yields [114]

$$\epsilon_{\boldsymbol{k}} = \sqrt{\frac{k^4}{4(m^{**})^2} + \omega_{\mathrm{ps}}^2} \tag{4.207}$$

with a gap

$$\omega_{\mathrm{ps}} = \sqrt{\frac{16\pi e^2 n_s}{\epsilon_0 m^{**}}} \tag{4.208}$$

which is the plasma frequency. The quasi-particle spectrum (equation (4.207)) differs qualitatively from the BCS excitation spectrum (section 2.2). The BCS quasi-particles are fermions and their energy is of the order of the BCS gap, $\Delta_{\boldsymbol{k}}$, which is well below the electron plasma frequency, $\Delta_{\boldsymbol{k}} \ll \omega_{\mathrm{e}}$. The quasi-particles in CBG are bosons and their energy is about the (renormalized) plasma frequency

$\omega_{\rm ps}$. The density of bosons pushed up from the condensate by the Coulomb repulsion at $T = 0$ is

$$\tilde{n} = \langle \tilde{\psi}^{\dagger}(\boldsymbol{r})\tilde{\psi}(\boldsymbol{r})\rangle = \sum_{\boldsymbol{k}} v_{\boldsymbol{k}}^2 \tag{4.209}$$

which is small compared with the total density $n_{\rm b}$ as

$$\frac{\tilde{n}}{n_{\rm b}} \approx 0.2 r_s^{3/4}. \tag{4.210}$$

The ground state $|0\rangle$ is a vacuum of quasi-particles, $\alpha_{\boldsymbol{k}}|0\rangle = 0$. The ground-state energy E_0 is obtained by substituting equation (4.194) into the Hamiltonian (4.187) and neglecting terms of higher order than quadratic in $\alpha_{\boldsymbol{k}}$,

$$E_0 \equiv \langle 0|H|0\rangle = \frac{1}{2}\sum_{\boldsymbol{k}}(\epsilon_{\boldsymbol{k}} - \xi_{\boldsymbol{k}}). \tag{4.211}$$

This can be written per particle as

$$\frac{E_0}{n_{\rm b}} = \frac{2^{3/2}}{3^{1/4}\pi}\omega_{\rm ps} r_s^{3/4} \int_0^{\infty} {\rm d}k\, k^2 \left[\sqrt{k^4+1} - k^2 - 1/(2k^2)\right] \approx -0.23\omega_{\rm ps} r_s^{3/4}. \tag{4.212}$$

The negative value of the ground-state energy is due to the opposite charge background. The value of $|E_0|$ is considered as the gain in the total energy due to the condensation of interacting bosons with respect to the ground-state energy ($= 0$) of an ideal Bose gas. Therefore, $|E_0|$ plays the same role as the condensation energy in the BCS superconductor.

These results for three-dimensional charged bosons are readily generalized for any bosons on a lattice beyond the effective mass approximation. For example, the quasi-particle spectrum is given by

$$\epsilon_{\boldsymbol{k}} = \sqrt{(\epsilon_{\boldsymbol{k}}^{**})^2 + 2\epsilon_{\boldsymbol{k}}^{**}\bar{v}_{\boldsymbol{k}} n_s}. \tag{4.213}$$

Hence, if the free-boson dispersion $\epsilon_{\boldsymbol{k}}^{**}$ is anisotropic, the plasma gap is anisotropic as well. In an extreme case of (quasi-)two-dimensional bosons with a parabolic dispersion and a two-dimensional Coulomb repulsion, $\bar{v}_{\boldsymbol{k}} = 8\pi e^2/(\epsilon_0 k)$, the Bogoliubov spectrum is gapless,

$$\epsilon_{\boldsymbol{k}} = E_s\sqrt{k/q_s + k^4/q_s^4}. \tag{4.214}$$

Here $E_s = q_s^2/2m^{**}$ and $q_s = (32\pi e^2 n_s/\epsilon_0)^{1/3}$ is a two-dimensional screening wave-number, n_s is the number of condensed bosons per unit area.

4.7.4 Mixture of two Bose condensates

The quasi-particle spectrum (equation (4.207)) satisfies the Landau criterion of superfluidity (chapter 1), therefore the CBG is a superconductor. The actual spectrum of bipolarons on a lattice is more complicated, because inter-site bipolarons have 'internal' quantum numbers like the spin (section 4.6), orbital momentum and different symmetries of the one-particle Wannier orbitals bound into the bipolaron. These internal degrees of freedom can affect the collective excitations of a bipolaronic superconductor. Here we extend the Bogoliubov-type equations of the previous subsection to the *multi-component* Bose condensate of non-converting charged bosons [115]. Let us consider a two-component (1 and 2) mixture of bosons described by

$$H = \sum_{j=1,2} \int \mathrm{d}\boldsymbol{r}\, \Psi_j^\dagger(\boldsymbol{r}) \left[-\frac{\nabla^2}{2m_j} - \mu_j \right] \Psi_j(\boldsymbol{r}) + \tfrac{1}{2} \sum_{j,j'} \int \mathrm{d}\boldsymbol{r} \int \mathrm{d}\boldsymbol{r}'\, \tilde{v}_{jj'}(\boldsymbol{r}-\boldsymbol{r}') \Psi_j^\dagger(\boldsymbol{r}) \Psi_j(\boldsymbol{r}) \Psi_{j'}^\dagger(\boldsymbol{r}') \Psi_{j'}(\boldsymbol{r}')$$

where m_j is the mass of the boson j.

Using the displacement transformation (equation (4.189)) and the equations of motion for the Heisenberg ψ-operators, the condensate wavefunctions are found from two coupled GP equations:

$$\mathrm{i}\frac{\partial}{\partial t}\psi_{sj}(\boldsymbol{r},t) = \left(-\frac{\nabla^2}{2m_j} - \mu_j \right) \psi_{sj}(\boldsymbol{r},t) + \sum_{j'} \int \mathrm{d}\boldsymbol{r}'\, V_{jj'}(\boldsymbol{r}-\boldsymbol{r}') |\psi_{sj'}(\boldsymbol{r}',t)|^2 \psi_{sj}(\boldsymbol{r},t). \tag{4.215}$$

The supracondensate wavefunctions satisfy four Bogoliubov-type equations:

$$\mathrm{i}\frac{\partial}{\partial t}u_j(\boldsymbol{r},t) = \left(-\frac{\nabla^2}{2m_j} - \mu_j \right) u_j(\boldsymbol{r},t) + \sum_{j'} \int \mathrm{d}\boldsymbol{r}'\, V_{jj'}(\boldsymbol{r}-\boldsymbol{r}') \times [|\psi_{sj'}(\boldsymbol{r}',t)|^2 u_j(\boldsymbol{r},t) + \psi_{sj}^*(\boldsymbol{r}',t)\psi_{sj'}(\boldsymbol{r},t) u_{j'}(\boldsymbol{r}',t) + \psi_{sj}(\boldsymbol{r}',t)\psi_{sj'}(\boldsymbol{r},t) v_{j'}(\boldsymbol{r}',t)]. \tag{4.216}$$

and

$$-\mathrm{i}\frac{\partial}{\partial t}v_j(\boldsymbol{r},t) = \left(-\frac{\nabla^2}{2m_j} - \mu_j \right) v_j(\boldsymbol{r},t) + \sum_{j'} \int \mathrm{d}\boldsymbol{r}'\, V_{jj'}(\boldsymbol{r}-\boldsymbol{r}') \times [|\psi_{sj'}(\boldsymbol{r}',t)|^2 v_j(\boldsymbol{r},t) + \psi_{sj}(\boldsymbol{r}',t)\psi_{sj'}^*(\boldsymbol{r},t) v_{j'}(\boldsymbol{r}',t) + \psi_{sj}^*(\boldsymbol{r}',t)\psi_{sj'}^*(\boldsymbol{r},t) u_{j'}(\boldsymbol{r}',t)]. \tag{4.217}$$

Here we applied the linear transformation of $\tilde{\psi}$,

$$\tilde{\psi}_j(\boldsymbol{r},t) = \sum_n u_{nj}(\boldsymbol{r},t)(\alpha_n + \beta_n) + v_{nj}^*(\boldsymbol{r},t)(\alpha_n^\dagger + \beta_n^\dagger)$$

where α_n, β_n are bosonic operators annihilating quasi-particles in the quantum state n. Solving two GP equations (4.215), we obtain the chemical potentials of a homogeneous system as

$$\mu_1 = \bar{v}n_1 + \bar{w}n_2$$
$$\mu_2 = \bar{u}n_2 + \bar{w}n_1$$

and solving the four Bogoliubov equations, we determine the excitation spectrum, $E(\boldsymbol{k})$, from

$$\mathrm{Det}\begin{bmatrix} \xi_1(\boldsymbol{k}) - E(\boldsymbol{k}) & \bar{v}_{\boldsymbol{k}}\psi_{s1}^2 & \bar{w}_{\boldsymbol{k}}\psi_{s1}\psi_{s2}^* & \bar{w}_{\boldsymbol{k}}\psi_{s1}\psi_{s2} \\ \bar{v}_{\boldsymbol{k}}\psi_{s1}^{*2} & \xi_1(\boldsymbol{k}) + E(\boldsymbol{k}) & \bar{w}_{\boldsymbol{k}}\psi_{s1}^*\psi_{s2}^* & \bar{w}_{\boldsymbol{k}}\psi_{s1}^*\psi_{s2} \\ \bar{w}_{\boldsymbol{k}}\psi_{s1}^*\psi_{s2} & \bar{w}_{\boldsymbol{k}}\psi_{s1}\psi_{s2} & \xi_2(\boldsymbol{k}) - E(\boldsymbol{k}) & \bar{u}_{\boldsymbol{k}}\psi_{s2}^2 \\ \bar{w}_{\boldsymbol{k}}\psi_{s1}^*\psi_{s2}^* & \bar{w}_{\boldsymbol{k}}\psi_{s1}\psi_{s2}^* & \bar{u}_{\boldsymbol{k}}\psi_{s2}^{*2} & \xi_2(\boldsymbol{k}) + E(\boldsymbol{k}) \end{bmatrix} = 0. \tag{4.218}$$

Here $\xi_1(\boldsymbol{k}) = k^2/2m_1 + \bar{v}_{\boldsymbol{k}}n_1$, $\xi_2(\boldsymbol{k}) = k^2/2m_2 + \bar{u}_{\boldsymbol{k}}n_2$ and $\bar{v}_{\boldsymbol{k}}, \bar{u}_{\boldsymbol{k}}, \bar{w}_{\boldsymbol{k}}$ are the Fourier components of $\bar{v}_{11}(\boldsymbol{r})$, $\bar{v}_{22}(\boldsymbol{r})$ and $\bar{v}_{12}(\boldsymbol{r})$, respectively, $\bar{v} \equiv \bar{v}_0$, $\bar{u} \equiv \bar{u}_0$, $\bar{w} \equiv \bar{w}_0$ and $n_j = |\psi_{sj}|^2$ are the condensate densities. There are two branches of excitations with dispersion

$$E(\boldsymbol{k})_{1,2} = 2^{-1/2}\left(\epsilon_1^2(\boldsymbol{k}) + \epsilon_2^2(\boldsymbol{k}) \pm \sqrt{[\epsilon_1^2(\boldsymbol{k}) - \epsilon_2^2(\boldsymbol{k})]^2 + \frac{4k^4}{m_1 m_2}\bar{w}_{\boldsymbol{k}}^2 n_1 n_2}\right)^{1/2} \tag{4.219}$$

where $\epsilon_1(\boldsymbol{k}) = [k^4/(4m_1^2) + k^2\bar{v}_{\boldsymbol{k}}n_1/m_1]^{1/2}$ and $\epsilon_2(\boldsymbol{k}) = [k^4/(4m_2^2) + k^2\bar{u}_{\boldsymbol{k}}n_2/m_2]^{1/2}$ are Bogoliubov modes of two components. If the interaction is purely the Coulomb repulsion, $\bar{v}_{\boldsymbol{k}} = 4\pi q_1^2/k^2$, $\bar{u}_{\boldsymbol{k}} = 4\pi q_2^2/k^2$ and $\bar{w}_{\boldsymbol{k}} = 4\pi q_1 q_2/k^2$, the upper branch is the geometric sum of familiar plasmon modes for $k \to 0$,

$$E_1(\boldsymbol{k}) = \sqrt{\frac{4\pi q_1^2 n_1}{m_1} + \frac{4\pi q_2^2 n_2}{m_2}} \tag{4.220}$$

while the lowest branch is gapless,

$$E_2(\boldsymbol{k}) = \frac{k^2}{2(m_1 m_2)^{1/2}}\sqrt{\frac{q_1^2 n_1 m_1 + q_2^2 n_2 m_2}{q_1^2 n_1 m_2 + q_2^2 n_2 m_1}}. \tag{4.221}$$

Remarkably, this mode does not depend on the interaction at *any* charges of the components, if $m_1 = m_2$. It corresponds to a low-frequency oscillation in which two condensates move in anti-phase with one another, in contrast to the usual optical high-frequency plasmon (equation (4.220)) in which the components oscillate inphase. The mode is similar to the acoustic plasmon (AP) mode in the electron–ion [116] and electron–hole [117] plasmas. However, differing from these normal-state APs with a linear dispersion, the AP of Bose mixtures has a quadratic dispersion in the long-wavelength limit. We conclude that while the

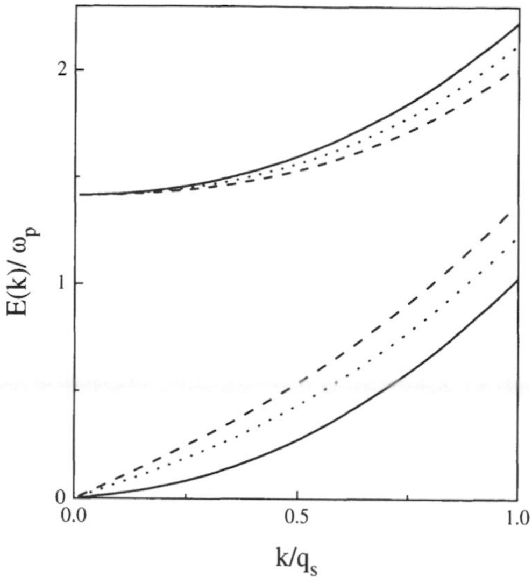

Figure 4.10. Excitation energy spectrum of the two-component Bose–Einstein condensate with the long-range repulsive and hard-core interactions. In this plot $n_1 = n_2 = n$, $m_1 = m_2$ and $\bar{u} = \bar{v} = \omega_p$, $q_1 = q_2 = e$. Different curves correspond to the different values of $\bar{w}/\bar{v} = 0.1$ (dashes), 0.5 (dots), and 0.95 (full line), respectively. Excitation energy is measured in units of the plasma frequency $\omega_p = 4\pi ne^2/m$ and momentum k is measured in inverse screening length $q_s = (16\pi e^2 nm)^{1/4}$.

CBG is a superfluid (according to the Landau criterion), their mixture is not. In the case of bipolarons, the interaction includes both the long-range repulsion and the hard-core interaction (section 4.7.1). Combining both interactions, i.e. taking $\bar{v}_k, \bar{u}_k, \bar{w}_k \propto$ constant $+ 1/k^2$, transforms the lowest AP mode into the Bogoliubov sound with a linear dispersion at $k \to 0$, figure 4.10. Hence, the two-component condensate of bipolarons is a superconductor.

4.7.5 Critical temperature and isotope effect

Polarons and bipolarons differ from electrons in many ways. One of the differences is their effective mass. Electrons change their mass in solids due to interactions with ions, spins and between themselves. Because λ does not depend on the ion mass, the renormalized mass of electrons is independent of the ion mass M in ordinary metals, where the Migdal adiabatic approximation is believed to be valid (chapter 3). However, if the interaction between electrons and ion vibrations is strong and/or the adiabatic approximation is not applied, electrons form polarons and their effective mass m^* depends on M as $m^* = m\exp(A/\omega_0)$ [82] (section 4.3.1), where m is the band mass in the absence of the electron–phonon interaction, A is a constant independent of M and ω_0 is the characteristic phonon frequency, which depends on M as $\omega_0 \propto M^{-1/2}$. Hence, there is a substantial isotope effect on the carrier mass in polaronic and bipolaronic systems, in contrast to the zero isotope effect in ordinary metals.

The isotope exponent in m^* is defined as $\alpha_{m^*} = \sum_i \mathrm{d}\ln m^*/\mathrm{d}\ln M_i$, where M_i is the mass of the ith ion in a unit cell. Using this definition and the expression for the polaron mass m^* mentioned above, one readily finds

$$\alpha_{m^*} = \tfrac{1}{2}\ln(m^*/m). \qquad (4.222)$$

The same isotope exponent is predicted for inter-site bipolarons, when their mass m^{**} is proportional to the polaron mass (section 4.6.3). The isotope effect on the polaron/bipolaron mass leads to an anomalous isotope effect on the superconducting critical temperature of polaronic and bipolaronic superconductors [82]. T_c of bipolarons is the Bose–Einstein condensation (BEC) temperature of CBG. It is approximately the Bose–Einstein condensation temperature of ideal bosons (appendix B) as long as the Coulomb repulsion is weak ($r_s \lesssim 1$):

$$T_c \approx \frac{3.31 n_b^{2/3}}{m^{**}} \qquad (4.223)$$

where n_b is the volume density of bipolarons. Corrections to equation (4.223), caused by the Coulomb repulsion, are numerically small even at $r_s = 1$ [118,119]. Hence, the isotope exponent α in T_c of inter-site bipolarons is the same as $\alpha_{m^{**}}$,

$$\alpha \equiv -\frac{\mathrm{d}\ln T_c}{\mathrm{d}\ln M} = \alpha_{m^{**}}. \qquad (4.224)$$

In the crossover region from the normal polaron Fermi liquid to bipolarons, one can apply the BCS-like expression (4.113) for T_c of non-adiabatic carriers which we write as

$$T_c \propto D \exp\left(-g^2 - \frac{Z'}{\lambda - \mu_c}\right). \qquad (4.225)$$

Here only $g^2 \propto \sqrt{M}$ depends on M. When $\lambda - \mu_c \gg Z' = \mathrm{e}^{-g^2}$, this expression also describes the isotope exponent in T_c of inter-site bipolarons (equation (4.223)) if their effective mass $m^{**} \propto m^*$. In this case we can apply equation (4.225) to the whole region of the phase diagram (figure 4.6) including the BCS-like crossover regime and the Bose–Einstein condensation. Differentiating equation (4.225) with respect to M yields

$$\alpha = \alpha_{m^*}\left(1 - \frac{Z'}{\lambda - \mu_c}\right). \qquad (4.226)$$

We see that the isotope exponent is negative in polaronic superconductors, where $\lambda - \mu_c < Z'$ but it is positive in bipolaronic superconductors, where $\lambda - \mu_c > Z'$. Equation (4.225), interpolating between BCS- and BEC-type superconductivity, allows us to understand the origin of high values of T_c in comparison with the weak-coupling BCS theory (equation (2.48)). High T_c originates in the polaron narrowing of the band. The exponentially enhanced DOS of the narrow polaronic band effectively eliminates the small exponential factor in equation (4.225). But at a very strong e–ph coupling, T_c drops again because the carriers become very heavy. Therefore, we conclude that the highest T_c is in the crossover region from polaronic to bipolaronic superconductivity where $\lambda - \mu_c \approx Z'$. The polaron half-bandwidth $Z'D$ is normally less or about the phonon frequency ω_0, so that the maximum value of T_c is estimated to be $T_c \lesssim \omega_0/3$ [13]. In novel oxygen and carbon-based superconductors (part 2), the characteristic optical phonon frequency is about 500 to 2000 K. That is why T_c is remarkably high in these compounds.

4.7.6 Magnetic field expulsion

The linear response function is defined as

$$j_\alpha(\boldsymbol{q},\omega) = \sum_{\beta=x,y,z} K^{\alpha\beta}(\boldsymbol{q},\omega) a_\beta(\boldsymbol{q},\omega) \tag{4.227}$$

where $j_\alpha(\boldsymbol{q},\omega)$ and $a_\beta(\boldsymbol{q},\omega)$ are the Fourier transforms of the current and of the vector potential, respectively. To calculate the response, we need to solve equation (4.190) with $\mu = 0$ in the first order with respect to $\boldsymbol{A}(\boldsymbol{r},t)$,

$$\mathrm{i}\frac{\mathrm{d}}{\mathrm{d}t}\psi_s(\boldsymbol{r},t) = -\frac{[\nabla + \mathrm{i}2e\boldsymbol{A}(\boldsymbol{r},t)]^2}{2m^{**}}\psi_s(\boldsymbol{r},t) + \int \mathrm{d}\boldsymbol{r}'\, \bar{v}(\boldsymbol{r}-\boldsymbol{r}')|\psi_s^*(\boldsymbol{r}',t)|^2\psi_s(\boldsymbol{r},t).$$

Using a perturbed wavefunction,

$$\psi_s(\boldsymbol{r},t) = \sqrt{n_s} + \phi(\boldsymbol{r},t) \tag{4.228}$$

and keeping only the terms linear in $\boldsymbol{A}(\boldsymbol{r},t)$, we obtain for the Fourier transform $\phi(\boldsymbol{q},\omega)$:

$$\omega\phi(\boldsymbol{q},\omega) = \frac{q^2}{2m^{**}}\phi(\boldsymbol{q},\omega) + n_s\bar{v}_{\boldsymbol{q}}\{\phi(\boldsymbol{q},\omega) + \phi^*(-\boldsymbol{q},-\omega)\} - \frac{2e\sqrt{n_s}}{2m^{**}}\boldsymbol{q}\cdot\boldsymbol{a}(\boldsymbol{q},\omega). \tag{4.229}$$

The solution is

$$\phi(\boldsymbol{q},\omega) + \phi^*(-\boldsymbol{q},-\omega) = -\frac{2e\sqrt{n_s}}{m^{**}}\frac{\omega}{\omega^2 - \epsilon_{\boldsymbol{q}}^2}\boldsymbol{q}\cdot\boldsymbol{a}(\boldsymbol{q},\omega) \tag{4.230}$$

$$\phi(\boldsymbol{q},\omega) - \phi^*(-\boldsymbol{q},-\omega) = -\frac{4e\sqrt{n_s}}{q^2}\frac{\epsilon_{\boldsymbol{q}}^2}{\omega^2 - \epsilon_{\boldsymbol{q}}^2}\boldsymbol{q}\cdot\boldsymbol{a}(\boldsymbol{q},\omega). \tag{4.231}$$

The expectation value of the current is given by

$$\boldsymbol{j}(\boldsymbol{r},t) = \frac{\mathrm{i}e}{m^{**}}[\psi_s^*(\boldsymbol{r},t)\nabla\psi_s(\boldsymbol{r},t) - \psi_s(\boldsymbol{r},t)\nabla\psi_s^*(\boldsymbol{r},t)] - \frac{4e^2n_s}{m^{**}}\boldsymbol{A}(\boldsymbol{r},t). \tag{4.232}$$

Using the perturbed wavefunction, we obtain the Fourier transform of the current as

$$\boldsymbol{j}(\boldsymbol{q},\omega)=\frac{e\sqrt{n_s}}{m^{**}}\boldsymbol{q}[\phi(\boldsymbol{q},\omega)-\phi^{*}(-\boldsymbol{q},-\omega)]-\frac{4e^2n_s}{m^{**}}\boldsymbol{a}(\boldsymbol{q},\omega) \tag{4.233}$$

and

$$K^{\alpha\beta}(\boldsymbol{q},\omega)=\frac{4e^2n_s}{m^{**}}\left[\delta^{\alpha\beta}\frac{\omega^2}{\epsilon_{\boldsymbol{q}}^2-\omega^2}+(q^\alpha q^\beta-\delta^{\alpha\beta}q^2)\frac{\epsilon_{\boldsymbol{q}}^2}{q^2(\epsilon_{\boldsymbol{q}}^2-\omega^2)}\right]. \tag{4.234}$$

This response function has been split into a longitudinal K_{l} $(\propto\delta^{\alpha\beta})$ and a transverse K_{t} $(\propto(q^\alpha q^\beta-\delta^{\alpha\beta}q^2))$ part. The longitudinal response to the field $(\boldsymbol{D}\parallel\boldsymbol{q})$ is expressed in terms of the so-called external conductivity σ_{ex}:

$$\boldsymbol{j}_{\mathrm{l}}(\boldsymbol{q},\omega)=\sigma_{\mathrm{ex}}(\boldsymbol{q},\omega)\boldsymbol{D}(\boldsymbol{q},\omega) \tag{4.235}$$

where $\boldsymbol{D}$ is the external electric field. Using equation (4.234), we find

$$\sigma_{\mathrm{ex}}(\boldsymbol{q},\omega)=\frac{K_{\mathrm{l}}}{\mathrm{i}\omega}=\frac{\mathrm{i}\epsilon_0\omega\omega_{\mathrm{ps}}^2}{4\pi(\omega^2-\epsilon_{\boldsymbol{q}}^2)}. \tag{4.236}$$

The conductivity sum rule (1.18) is satisfied as

$$\int_0^\infty \mathrm{d}\omega\,\mathrm{Re}\,\sigma_{\mathrm{ex}}(\omega)=\frac{\epsilon_0\omega_{\mathrm{ps}}^2}{8}. \tag{4.237}$$

The conductivity in the transverse electromagnetic field $(\boldsymbol{D}\perp\boldsymbol{q})$ is given by

$$\sigma_{\mathrm{t}}=\frac{\mathrm{i}}{4\pi\lambda_{\mathrm{H}}^2\omega} \tag{4.238}$$

where

$$\lambda_{\mathrm{H}}=\left[\frac{m^{**}}{16\pi e^2n_s}\right]^{1/2}. \tag{4.239}$$

This expression combined with the Maxwell equation describes the Meissner–Ochsenfeld effect in the CBG with the magnetic field penetration depth λ_{H}.

4.7.7 Charged vortex and lower critical field

The CBG is an extreme type II superconductor, as shown later. Hence, we can analyse a single vortex in the CBG and calculate the lower (first) critical field H_{c1} by solving a stationary equation for the macroscopic condensate wavefunction [120]:

$$\left\{-\frac{[\nabla+2\mathrm{i}e\boldsymbol{A}(\boldsymbol{r})]^2}{2m^{**}}-\mu+\frac{4e^2}{\epsilon_0}\int\mathrm{d}\boldsymbol{r}'\,\frac{|\psi_s(\boldsymbol{r}',t)|^2-n_{\mathrm{b}}}{|\boldsymbol{r}-\boldsymbol{r}'|}\right\}\psi_s(\boldsymbol{r},t)=0. \tag{4.240}$$

Subtracting n_b in the integral of equation (4.240), we explicitly take into account the Coulomb interaction with a homogeneous charge background with the same density as the density of charged bosons.

The integra-differential equation (4.240) is quite different from the Ginsburg–Landau equation (1.43). While the CBG shares quantum coherence with the BCS superconductors owing to the Bose–Einstein condensate (BEC), the long-range (non-local) interaction leads to some peculiarities. In particular, the vortex is charged in the CBG and the coherence length occurs just the same as the screening radius.

Indeed, introducing dimensionless quantities $f = |\psi_s|/n_b^{1/2}$, $\boldsymbol{\rho} = \boldsymbol{r}/\lambda(0)$ and $\boldsymbol{h} = 2e\xi(0)\lambda(0)\nabla \times \boldsymbol{A}$ for the order parameter, length and magnetic field, respectively, equation (4.240) and the Maxwell equations take the following form:

$$\frac{1}{\kappa^2\rho}\frac{\mathrm{d}}{\mathrm{d}\rho}\rho\frac{\mathrm{d}f}{\mathrm{d}\rho} - \frac{1}{f^3}\left(\frac{\mathrm{d}h}{\mathrm{d}\rho}\right)^2 - \phi f = 0 \tag{4.241}$$

$$\frac{1}{\kappa^2\rho}\frac{\mathrm{d}}{\mathrm{d}\rho}\rho\frac{\mathrm{d}\phi}{\mathrm{d}\rho} = 1 - f^2 \tag{4.242}$$

$$\frac{1}{\rho}\frac{\mathrm{d}}{\mathrm{d}\rho}\frac{\rho}{f^2}\frac{\mathrm{d}h}{\mathrm{d}\rho} = h. \tag{4.243}$$

The new feature compared with the GL equations of chapter 1 is the electric field potential determined as

$$\phi = \frac{1}{2e\phi_c}\int \mathrm{d}\boldsymbol{r}'\,\bar{v}(\boldsymbol{r}-\boldsymbol{r}')[|\psi_s(\boldsymbol{r}')|^2 - n_B] \tag{4.244}$$

with a new fundamental unit $\phi_c = em^{**}\xi(0)^2$. The potential is calculated using the Poisson equation (4.242). At $T = 0$, the coherence length is the same as the screening radius,

$$\xi(0) = (2^{1/2}m^{**}\omega_{ps})^{-1/2} \tag{4.245}$$

and the London penetration depth is

$$\lambda(0) = \left(\frac{m^{**}}{16\pi n_B e^2}\right)^{1/2}. \tag{4.246}$$

There are now six boundary conditions in a single-vortex problem. Four of them are the same as in the BCS superconductor (section 1.6.3), $h = \mathrm{d}h/\rho = 0$, $f = 1$ for $\rho = \infty$ and the flux quantization condition, $\mathrm{d}h/\mathrm{d}\rho = -pf^2/\kappa\rho$ for $\rho = 0$, where p is an integer. The remaining two conditions are derived from the global charge neutrality, $\phi = 0$ for $\rho = \infty$ and

$$\phi(0) = \int_0^\infty \rho\ln(\rho)(1 - f^2)\,\mathrm{d}\rho \tag{4.247}$$

for the electric field at the origin, $\rho = 0$. We note that the chemical potential is zero at any point in the thermal equilibrium.

CBG is an extreme type II superconductor with a very large Ginsburg–Landau parameter ($\kappa = \lambda(0)/\xi(0) \gg 1$). Indeed, with the material parameters typical for oxides, such as $m^{**} = 10m_e$, $n_b = 10^{21}$ cm^{-3} and $\epsilon_0 = 10^3$, we obtain $\xi(0) \simeq 0.48$ nm, $\lambda(0) \simeq 265$ nm, and the Ginsburg–Landau ratio $\kappa \simeq 552$. Owing to a large dielectric constant, the Coulomb repulsion remains weak even for heavy bipolarons, $r_s \simeq 0.46$. If $\kappa \gg 1$, equation (4.243) is reduced to the London equation with the familiar solution $h = pK_0(\rho)/\kappa$, where $K_0(\rho)$ is the Hankel function of imaginary argument of zero order (section 1.6.3). For the region $\rho \le p$, where the order parameter and the electric field differ from unity and zero, respectively, we can use the flux quantization condition to 'integrate out' the magnetic field in equation (4.241). This leaves us with two parameter-free equations written for $r = \kappa\rho$:

$$\frac{1}{r}\frac{\mathrm{d}}{\mathrm{d}r}r\frac{\mathrm{d}f}{\mathrm{d}r} - \frac{p^2 f}{r^2} - \phi f = 0 \tag{4.248}$$

and

$$\frac{1}{r}\frac{\mathrm{d}}{\mathrm{d}r}r\frac{\mathrm{d}\phi}{\mathrm{d}r} = 1 - f^2. \tag{4.249}$$

They are satisfied by regular solutions of the form $f = c_p r^{\mathrm{p}}$ and $\phi = \phi(0) + (r^2/4)$, when $r \to 0$. The constants c_p and $\phi(0)$ are determined by complete numerical integration of equations (4.248) and (4.249). The numerical results for $p = 1$ are shown in figures 4.11 and 4.12, where $c_1 \simeq 1.5188$ and $\phi(0) \simeq -1.0515$.

In the region $p \ll r < p\kappa$, the solutions are $f = 1 + (4p^2/r^4)$ and $\phi = -p^2/r^2$. In this region, f differs qualitatively from the BCS order parameter, $f_{\mathrm{BCS}} = 1 - (p^2/r^2)$ (see also figure 4.11). The difference is due to a local charge redistribution caused by the magnetic field in the CBG. Being quite different from the BCS superconductor, where the total density of electrons remains constant across the sample, the CBG allows for flux penetration by redistributing the density of bosons within the coherence volume. This leads to an increase of the order parameter compared with the homogeneous case ($f = 1$) in the region close to the vortex core. Inside the core the order parameter is suppressed (figure 4.11) as in the BCS superconductor. The resulting electric field (figure 4.12) (together with the magnetic field) acts as an additional centrifugal force increasing the steepness (c_p) of the order parameter compared with the BCS superfluid, where $c_1 \simeq 1.1664$, figure 4.11(*a*).

The breakdown of the local charge neutrality is due to the absence of any equilibrium *normal*-state solution in the CBG below the $H_{c2}(T)$ line. Both superconducting ($\Delta_k \neq 0$) and normal ($\Delta_k = 0$) solutions are allowed at any temperature in the BCS superconductors (chapter 2). Then the system decides which of two phases (or their mixture) is energetically favourable but the local charge neutrality is respected. In contrast, there is no equilibrium normal-state solution (with $\psi_s = 0$) in CBG below the $H_{c2}(T)$-line because it does not respect the density sum rule. Hence, there are no different phases to mix and the only

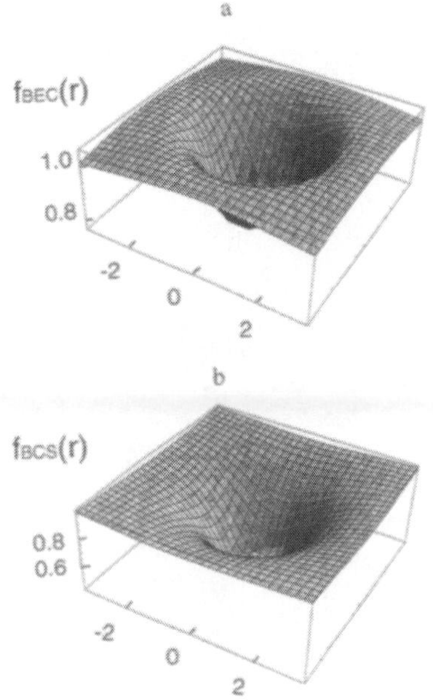

Figure 4.11. Vortex core profile in the charged Bose gas, f_{BEC}, (*a*) compared with the vortex in the BCS superconductor, f_{BCS}, (*b*).

way to acquire a flux in the thermal equilibrium is to redistribute the local density of bosons at the expense of their Coulomb energy. This energy determines the vortex free energy $\mathcal{F} = E_v - E_0$, which is the difference between the energy of the CBG with, E_v, and without, E_0, a magnetic flux:

$$\mathcal{F} = \int \mathrm{d}\boldsymbol{r} \left\{ \frac{1}{2m^*} |[\nabla + 2\mathrm{i}e\boldsymbol{A}(\boldsymbol{r})]\psi_s(\boldsymbol{r})|^2 + e\phi_{\mathrm{c}}\phi[|\psi_s(\boldsymbol{r})|^2 - n_{\mathrm{b}}] + \frac{(\nabla \times \boldsymbol{A})^2}{8\pi} \right\}. \tag{4.250}$$

Using equations (4.241), (4.242) and (4.243), this can be written in dimensionless form:

$$F = 2\pi \int_0^\infty [h^2 - \tfrac{1}{2}\phi(1 + f^2)]\rho \,\mathrm{d}\rho. \tag{4.251}$$

In the large κ limit, the main contribution comes from the region $p/\kappa < \rho < p$, where $f \simeq 1$ and $\phi \simeq -p^2/(\kappa^2\rho^2)$. The energy is, thus, the same as that in the BCS superconductor, $F \simeq 2\pi p^2 \ln(\kappa)/\kappa^2$ (section 1.6.3). It can be seen that the most stable solution is the formation of the vortex with one flux quantum, $p = 1$, and the lower critical field is the same as in the BCS superconductor, $h_{\mathrm{c1}} \approx \ln\kappa/(2\kappa)$. However, differing from the BCS superconductor, where the Ginsburg–Landau phenomenology is microscopically justified in the temperature region close to T_{c} (section 2.16), the CBG vortex structure is derived here in the

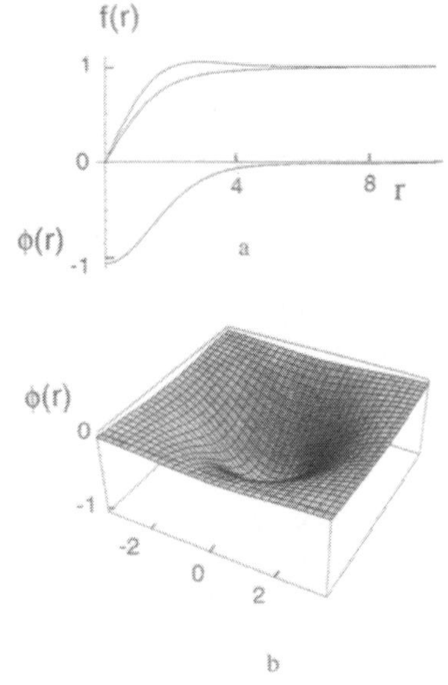

Figure 4.12. Electric field potential ϕ as a function of the distance (measured in units of $\xi(0)$) (lower curve) together with the CBG (upper curve) and BCS order parameters (*a*); its profile is shown in (*b*).

low-temperature region. In fact, the zero-temperature solution (figure 4.11) is applied in a wide temperature region well below the Bose–Einstein condensation temperature, where the depletion of the condensate remains small. The actual size of the charged core is about 4ξ (figure 4.12).

4.7.8 Upper critical field in the strong-coupling regime

If we 'switch off' the Coulomb repulsion between bosons, an ideal CBG cannot be bose-condensed at finite temperatures in a homogeneous magnetic field because of one-dimensional particle motion at the lowest Landau level [7]. However, an interacting CBG condenses in a field lower than a certain critical value $H_{c2}(T)$ [121]. Collisions between bosons and/or with impurities and phonons make the motion three dimensional, and eliminate the one-dimensional singularity of the density of states, which prevents BEC of the ideal gas in the field. As we show later, the upper critical field of the CBG differs significantly from the $H_{c2}(T)$ of BCS superconductors (section 1.6.4). It has an unusual positive curvature near T_c ($H_{c2}(T) \propto (T_c - T)^{3/2}$) and diverges at $T \to 0$, if there is no localization due to a random potential. The localization can drastically change the low-temperature behaviour of $H_{c2}(T)$ (part 2).

In line with the conventional definition (section 1.6.4), $H_{c2}(T)$ is a field, where the first non-zero solution of the linearized stationary equation for the macroscopic condensate wavefunction occurs:

$$\left[-\frac{1}{2m^{**}}[\nabla - 2ie\boldsymbol{A}(\boldsymbol{r})]^2 + V_{\text{scat}}(\boldsymbol{r})\right]\psi_s(\boldsymbol{r}) = \mu\psi_s(\boldsymbol{r}). \tag{4.252}$$

Here we include the 'scattering' potential $V_{\text{scat}}(\boldsymbol{r})$ caused, for example, by impurities. Let us first discuss non-interacting bosons, $V_{\text{scat}}(\boldsymbol{r}) = 0$. Their energy spectrum in a homogeneous magnetic field is (section 1.6.4)

$$\varepsilon_n = \omega(n + 1/2) + \frac{k_z^2}{2m^{**}} \tag{4.253}$$

where $\omega = 2eH_{c2}/m^{**}$ and $n = 0, 1, 2, \ldots, \infty$. BEC occurs when the chemical potential 'touches' the lowest band edge from below, i.e. $\mu = \omega/2$ (appendix B). Hence, quite different from the GL equation (1.90), the Schrödinger equation (4.252) does not allow for a direct determination of H_{c2}, In fact, it determines the value of the chemical potential. Then using this value, the upper critical field is found from the density sum rule,

$$\int_{E_c}^{\infty} f(\epsilon)N(\epsilon, H_{c2})\,d\epsilon = n_b \tag{4.254}$$

where $N(\epsilon, H_{c2})$ is the DOS of the Hamiltonian (4.252), $f(\epsilon) = [\exp(\epsilon - \mu)/T - 1]^{-1}$ is the Bose–Einstein distribution function and E_c is the lowest band edge. For ideal bosons, we have $\mu = E_c = \omega/2$ and

$$N(\epsilon, H_{c2}) = \frac{\sqrt{2}(m^{**})^{3/2}\omega}{4\pi^2}\,\mathrm{Re}\sum_{n=0}^{\infty}\frac{1}{\sqrt{\epsilon - \omega(n + 1/2)}}. \tag{4.255}$$

Substituting equation (4.255) into equation (4.254) yields

$$\frac{\sqrt{2}(m^{**})^{3/2}\omega}{4\pi^2}\int_0^{\infty}\frac{dx}{x^{1/2}}\frac{1}{\exp(x/T) - 1} = n_b - \tilde{n}(T) \tag{4.256}$$

where

$$\tilde{n}(T) = \frac{\sqrt{2}(m^{**})^{3/2}\omega}{4\pi^2}\int_0^{\infty}\frac{dx}{\exp(x/T) - 1}\sum_{n=1}^{\infty}\frac{1}{\sqrt{x - \omega n}} \tag{4.257}$$

is the number of bosons occupying the levels from $n = 1$ to $n = \infty$. This number is practically the same as in zero field, $\tilde{n}(T) = n_b(T/T_c)^{3/2}$ (see equation (B.61)), if $\omega \ll T_c$. In contrast, the number of bosons on the lowest level, $n = 0$, is given by a divergent integral on the left-hand side of equation (4.256). Hence the only solution to equation (4.256) is $H_{c2}(T) = 0$.

The scattering of bosons effectively removes the one-dimensional singularity in $N_0(\epsilon, H_{c2}) \propto \omega(\epsilon - \omega/2)^{-1/2}$ leading to a finite DOS near the bottom of the lowest level,

$$N_0(\epsilon, H_{c2}) \propto \frac{H_{c2}}{\sqrt{\Gamma_0(H_{c2})}}. \tag{4.258}$$

Using the Fermi–Dirac golden rule, the collision broadening of the lowest level $\Gamma_0(H_{c2})$ is proportional to the same DOS:

$$\Gamma_0(H_{c2}) \propto N_0(\epsilon, H_{c2}) \tag{4.259}$$

so that Γ_0 scales with the field as $\Gamma_0(H_{c2}) \propto H_{c2}^{2/3}$. Then the number of bosons at the lowest level is estimated to be

$$n_0 = \frac{\sqrt{2}(m^{**})^{3/2}\omega}{4\pi^2} \int_{\Gamma_0}^{\infty} \frac{dx}{x^{1/2}} \frac{1}{\exp(x/T) - 1} \propto T H_{c2}^{2/3} \tag{4.260}$$

as long as $T \gg \Gamma_0$. Here we apply the one-dimensional DOS but cut the integral at Γ_0 from below. Finally we arrive at

$$H_{c2}(T) = H_0(t^{-1} - t^{1/2})^{3/2} \tag{4.261}$$

where $t = T/T_c$ and H_0 is a temperature-independent constant. The scaling constant, H_0, depends on the scattering mechanism. If we write $H_0 = \Phi_0/(2\pi\xi_0^2)$, then the characteristic length is

$$\xi_0 \approx \left(\frac{l}{n_b}\right)^{1/4} \tag{4.262}$$

where l is the zero-field mean-free path of low-energy bosons. The upper critical field has a nonlinear behaviour:

$$H_{c2}(T) \propto (T_c - T)^{3/2}$$

in the vicinity of T_c and diverges at low temperatures as

$$H_{c2}(T) \propto T^{-3/2}.$$

These simple scaling arguments are fully confirmed by DOS calculations with impurity [121] and boson–boson [122] scattering. For impurities, the energy spectrum of the Hamiltonian (4.252) consists of discrete levels (i.e. localized states) and a continuous spectrum (extended states). The density of extended states $\tilde{N}(\epsilon, H_{c2})$ and their lowest energy E_c (the so-called mobility edge, where $\tilde{N}(E_c, H_{c2}) = 0$) can be found, if the electron self-energy $\Sigma(\epsilon)$ is known. We calculate $\Sigma(\epsilon)$ in the non-crossing approximation of section 3.3 considering the impurity scattering as the elastic limit of the phonon scattering:

$$\Sigma_\nu(\epsilon) = \sum_{\nu'} \frac{V_{\nu\nu'}^2}{\epsilon - \epsilon_{\nu'} - \Sigma_{\nu'}(\epsilon)}. \tag{4.263}$$

To obtain analytical results, we choose the matrix elements of the scattering potential as $V_{\nu\nu'} = V\delta_{nn'}$. Here $\nu \equiv (n, k_x, k_z)$ are the quantum numbers of a charge particle in the magnetic field. Then the DOS is found to be

$$\tilde{N}(\epsilon, H_{c2}) = \frac{1}{\pi V^2} \sum_n \operatorname{Im} \Sigma_n(\epsilon). \tag{4.264}$$

Integrating over k_z in equation (4.263) yields a cubic algebraic equation for the self-energy $\Sigma_n(\epsilon)$ of the nth level:

$$\Sigma_n(\epsilon) = \frac{dV^2}{2\pi}\int_{-\infty}^{\infty} dk_z \frac{1}{\epsilon - \omega(n+1/2) - k_z^2/(2m^{**}) - \Sigma_n(\epsilon)}$$
$$= \frac{eH_{c2}V^2\sqrt{m^{**}}}{\pi\sqrt{2}} \frac{i}{\sqrt{\epsilon - \omega(n+1/2) - \Sigma_n(\epsilon)}}. \qquad (4.265)$$

Solving this equation, we obtain

$$\tilde{N}(\epsilon, H) = \frac{eH_{c2}}{4\pi^2}\sqrt{\frac{6m^{**}}{\Gamma_0}} \times \sum_{n=0}^{\infty}\left[\left(\frac{\tilde{\epsilon}_n^3}{27} + \frac{1}{2} + \sqrt{\frac{\tilde{\epsilon}_n^3}{27} + \frac{1}{4}}\right)^{1/3} - \left(\frac{\tilde{\epsilon}_n^3}{27} + \frac{1}{2} - \sqrt{\frac{\tilde{\epsilon}_n^3}{27} + \frac{1}{4}}\right)^{1/3}\right] \qquad (4.266)$$

and the mobility edge

$$E_c = \frac{\omega}{2} - \frac{3\Gamma_0}{2^{2/3}}. \qquad (4.267)$$

Here

$$\Gamma_0 = \frac{1}{2}\left(\frac{2V^2eH\sqrt{m^{**}}}{\pi}\right)^{2/3}$$

is the collision broadening of levels and $\tilde{\epsilon}_n = [\epsilon - \omega(n+1/2]/\Gamma_0$.

With this DOS in the sum rule (equation (4.254)), we obtain the same $H_{c2}(T)$ for $T \gg \Gamma_0$ as through the scaling, equation (4.261), and

$$\xi_0 \approx 0.8\left(\frac{l}{n_b}\right)^{1/4}. \qquad (4.268)$$

The zero-field mean free path l is expressed via microscopic parameters as $l = \pi/(Vm^{**})^{-2}$. The 'coherence' length ξ_0 of the CBG (equation (4.268)) depends on the mean free path l and the inter-particle distance $n_b^{-1/3}$. It has little to do with the size of the bipolaron and could be as large as the coherence length of weak-coupling BCS superconductors.

Thus $H_{c2}(T)$ of strongly-coupled superconductors has a '3/2' curvature near T_c which differs from the linear weak-coupling $H_{c2}(T)$ of section 1.6 (figure 4.13). The curvature is a universal feature of the CBG, which does not depend on a particular scattering mechanism and on any approximations. Another interesting feature of strongly-coupled superconductors is a breakdown of the Pauli paramagnetic limit given by $H_p \simeq 1.84T_c$ in the weak-coupling theory (section 1.6.4). The $H_{c2}(T)$ of bipolarons exceeds this limit because the singlet

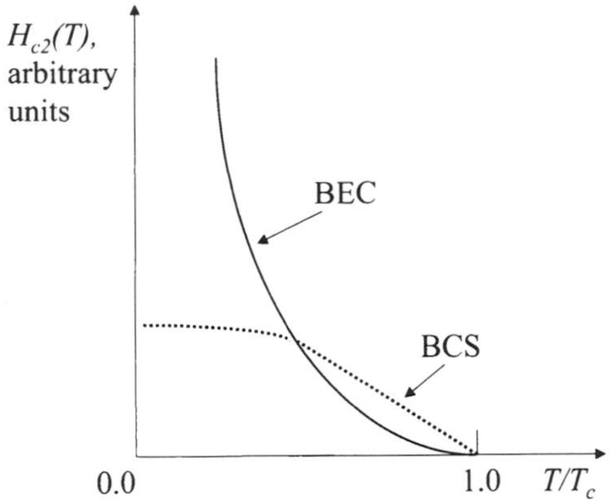

Figure 4.13. BEC critical field of a CBG compared with the $H_{c2}(T)$ of BCS superconductors.

bipolaron binding energy Δ is much larger than their T_c. Bosons are condensed at $T = 0$ no matter what their energy spectrum is. Hence, in the CBG model, $H_{c2}(0) = \infty$ (figure 4.13). For composed bosons, like bipolarons, the pair-breaking limit is given by $\mu_B H_{c2}(0) \approx \Delta$, so that $H_{c2}(0) \gg H_p$.

4.7.9 Symmetry of the order parameter

The anomalous Bogoliubov average

$$F_{ss'}(\boldsymbol{r}_1, \boldsymbol{r}_2) = \langle\langle \Psi_s(\boldsymbol{r}_1) \Psi_{s'}(\boldsymbol{r}_2) \rangle\rangle$$

is the superconducting order parameter both in the weak- and strong-coupling regimes. It depends on the relative coordinate $\boldsymbol{\rho} = \boldsymbol{r}_1 - \boldsymbol{r}_2$ of two electrons of the pair and on the centre-of-mass coordinate $\boldsymbol{R} = (\boldsymbol{r}_1 + \boldsymbol{r}_2)/2$. Hence, its Fourier transform, $f(\boldsymbol{k}, \boldsymbol{K})$, depends on the relative momentum $\boldsymbol{k}$ and on the centre-of-mass momentum, $\boldsymbol{K}$. In the BCS theory, $\boldsymbol{K} = 0$ (in a homogeneous superconductor) and the Fourier transform of the order parameter is proportional to the gap in the quasi-particle excitation spectrum, $f(\boldsymbol{k}, \boldsymbol{K}) \propto \Delta_{\boldsymbol{k}}$ (section 2.2). Hence, the symmetry of the order parameter and the symmetry of the gap are the same in the weak-coupling regime. Under the rotation of the coordinate system, $\Delta_{\boldsymbol{k}}$ changes its sign if the Cooper pairing is unconventional (section 2.10). In this case, the BCS quasi-particle spectrum is gapless.

In the bipolaron theory, the symmetry of the BEC is not necessarily the same as the 'internal' symmetry of a pair [123]. While the latter describes the transformation of $f(\boldsymbol{k}, \boldsymbol{K})$ with respect to the rotation of $\boldsymbol{k}$, the former ('external') symmetry is related to the rotation of $\boldsymbol{K}$. Therefore, it depends on the bipolaron band dispersion but not on the symmetry of the bound state. As an example, let us consider a tight-binding bipolaron spectrum comprising two bands on a square

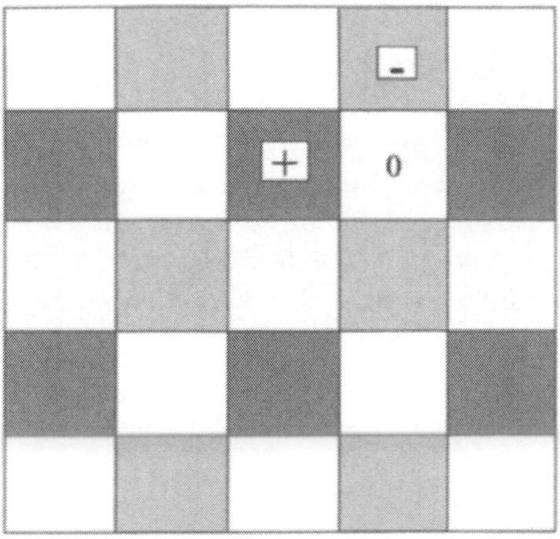

Figure 4.14. d-wave condensate wavefunction, in the Wannier representation. The order parameter has different signs in the shaded cells and is zero in the blank cells.

lattice with the period $a = 1$:

$$E_{\mathbf{K}}^{x} = t\cos(K_x) - t'\cos(K_y) \tag{4.269}$$
$$E_{\mathbf{K}}^{y} = -t'\cos(K_x) + t\cos(K_y).$$

They transform into one another under $\pi/2$ rotation. If $t, t' > 0$, 'x' bipolaron band has its minima at $\mathbf{K} = (\pm\pi, 0)$ and the y-band at $\mathbf{K} = (0, \pm\pi)$. These four states are degenerate, so that the condensate wavefunction $\psi_s(\boldsymbol{m})$ in the site space, $\boldsymbol{m} = (m_x, m_y)$, is given by

$$\psi_s(\boldsymbol{m}) = N^{-1/2} \sum_{\mathbf{K}=(\pm\pi,0),(0,\pm\pi)} b_{\mathbf{K}} \mathrm{e}^{-\mathrm{i}\mathbf{K}\cdot\boldsymbol{m}}. \tag{4.270}$$

where $b_{\mathbf{K}} = \pm\sqrt{n_s}$ are c-numbers at $T = 0$. The superposition (4.270) respects the time-reversal and parity symmetries, if

$$\psi_s^{\pm}(\boldsymbol{m}) = \sqrt{n_s}[\cos(\pi m_x) \pm \cos(\pi m_y)]. \tag{4.271}$$

The two order parameters (equation (4.271)) are physically identical because they are related by the translation transformation, $\psi_s^{+}(m_x, m_y) = \psi_s^{-}(m_x, m_y + 1)$. Both have a d-wave symmetry-changing sign, when the lattice is rotated by $\pi/2$ (figure 4.14). The d-wave symmetry is entirely due to the bipolaron energy dispersion with four minima at $\mathbf{K} \neq 0$. When the bipolaron spectrum is not degenerate and its minimum is located at the Γ point of the Brillouin zone, the condensate wavefunction is s-wave with respect to the centre-of-mass coordinate. The symmetry of the gap has little to do with the symmetry of the order parameter in the strong-coupling regime. The one-particle excitation gap is half of the bipolaron binding energy, $\Delta/2$, and does not depend on any momentum in zero order of the polaron bandwidth, i.e. it has an 's'-wave symmetry. In fact, due to a finite dispersion of polaron and bipolaron bands (sections 4.3 and 4.6), the one-particle gap is an *anisotropic* s-wave. A multi-band electron structure can include bands only weakly coupled with phonons which could overlap with the bipolaronic band (see also part 2). In this case, the CBG coexists with the Fermi gas just as ^{4}He bosons coexist with ^{3}He fermions in the mixture of He-4 and He-3. The one-particle excitation spectrum of such mixtures is gapless.

PART 2

APPLICATIONS TO HIGH-T_c SUPERCONDUCTORS

Chapter 5

Competing interactions in unconventional superconductors

5.1 High-T_c superconductors: different concepts

Nowadays there are many complex high-T_c superconductors including copper [1, 127] and a few other oxides, doped fullerenes M_xC_{60} [128, 129] (M is an alkali metal) and the more recently discovered magnesium diborade MgB_2 [130] with a critical temperature above 30 K, figure 5.1.

These discoveries have broken all constraints on the maximum T_c predicted by conventional theory of low-temperature superconducting metals and their alloys. As highlighted by Simon [132], the canonical theory has not provided a 'materials' aspect in the search for new high-T_c superconductors. Nevertheless, in a phenomenological sense, any superconductivity can just be treated as a consequence of the formation of pairs and their condensation—even within the BCS framework, as was noted by Ginzburg [133] in 1968 (chapter 1). At any value of the e–ph interaction superconductivity is a correlated many-body state of pairs which is well described by BCS theory in the weak-coupling regime $\lambda \ll 1$, where pairing takes place in the momentum space due to the Pauli exclusion principle (*collective* pairing) and by the bipolaron theory in the strong coupling regime, $\lambda \gtrsim 1$, where pairing is *individual* (upper half-plane in figure 5.2). Hence, knowledge of the carrier-pairing mechanism and of the nature of the normal state is central to an understanding of high-T_c supeconductivity.

In general, the bosonic field, which 'glues' two carriers together, can not only be 'phononic', as in BCS and bipolaron theory, but also 'excitonic' [133, 134], 'plasmonic' [135, 136] and/or of magnetic origin [137, 138]. BCS theory like any mean-field theory is rather universal, so that it describes the cooperative quantum phenomenon of superconductivity well even with these non-phononic attraction mechanisms, if the coupling is weak (see left-hand side of figure 5.2). If the coupling is strong, magnetic interaction could result in the spin–bipolaron formation as suggested by Mott [139]. The main motivation behind these concepts

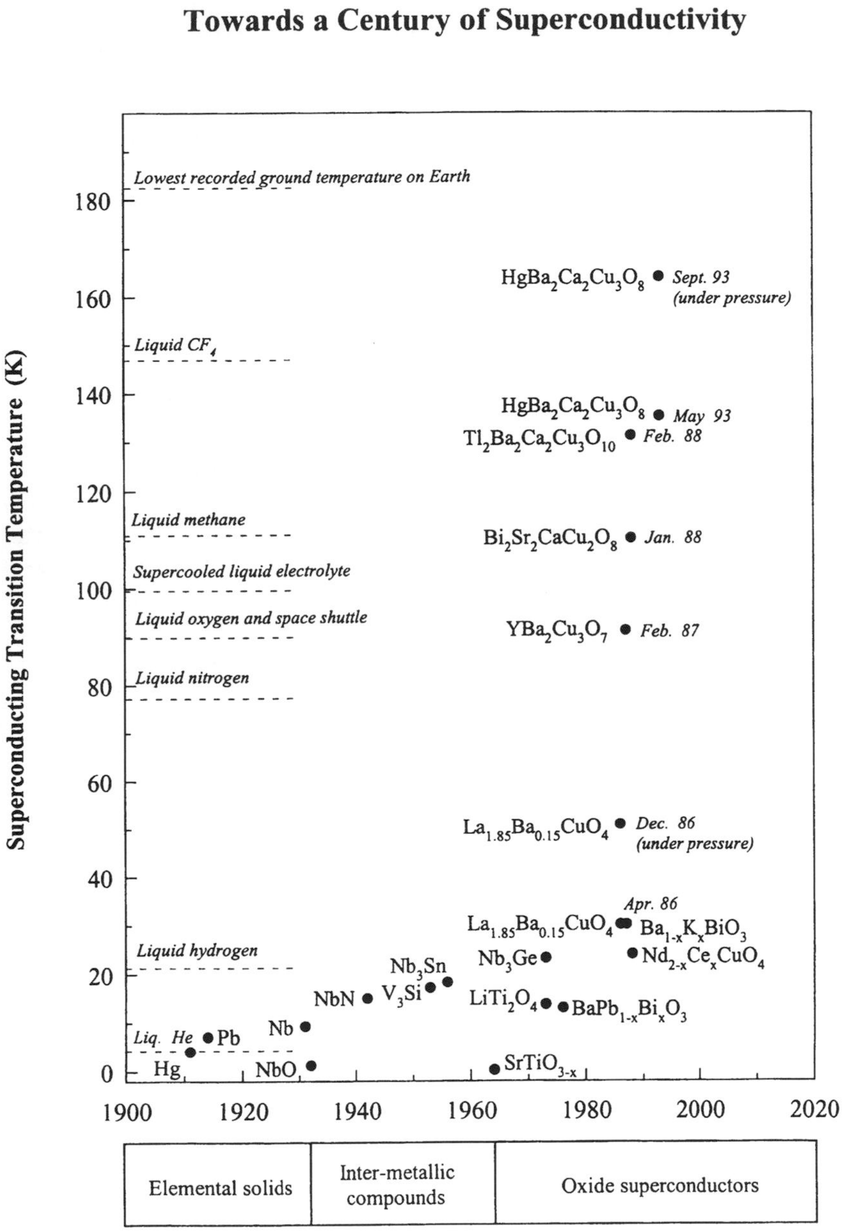

Figure 5.1. Towards a century of superconductivity. A plot of the evolution of the superconducting transition temperature, from 1911, to the present situation. We also include the characteristic temperatures of a variety of cryogenic liquids, as well as the lowest recorded ground temperature on Earth (−89.2 °C) [131].

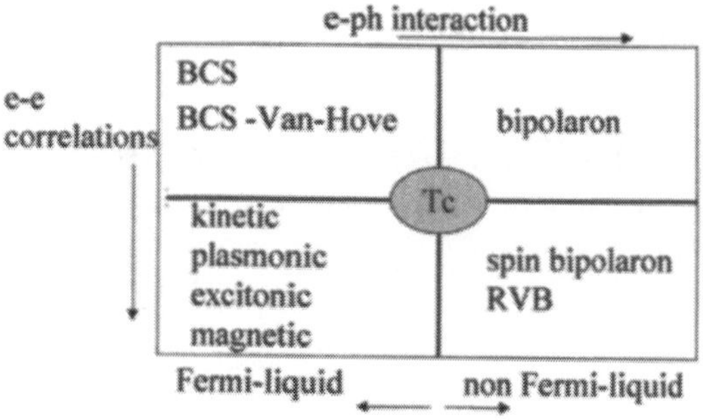

Figure 5.2. BSC-like theories of high-T_c superconductors with the electron–phonon (top) and electron–electron (bottom) interactions (left-hand side). Right-hand side: non-Fermi liquid theories with electron–phonon (top) and electron–electron (bottom) correlations.

is that high T_c might be achieved by replacing low-frequency acoustic phonons in conventional BCS theory by higher-frequency bosonic modes, such as the electron plasmon (chapter 3) or by spin waves (pseudomagnons). However, the Coulomb pseudopotential creates a serious problem for any of these non-phononic mechanisms. With the increasing energy scale of the bosonic field, the retardation of the effective attraction vanishes and the Tolmachev logarithm (section 3.5) cannot be applied. Then the direct Coulomb repulsion would definitely prevent any pairing in these models because the exchange electron–electron interaction is always weaker than direct repulsion. The problem is particularly grave in novel superconductors, where the carrier density is low, and the screening of the direct Coulomb repulsion is poor.

Therefore, some authors [140, 141] dogmatized that the interaction in novel superconductors is essentially repulsive and unretarded and that it also provides high T_c without any phonons. A motivation for this concept can be found in the earlier work by Kohn and Luttinger [57], who showed that the Cooper pairing of repulsive fermions is possible. But the same work clearly showed that the T_c of repulsive fermions is extremely low, well below the mK scale (section 3.6). Nevertheless, BCS and BCS-like theories (including the Kohn–Luttinger consideration) rely heavily on the Fermi-liquid model of the *normal* state. This model fails in many high-temperature superconductors (chapter 6). There are no obvious *a priori* reasons for discarding the dogma, if the normal state is not the Fermi liquid. There is little doubt that strong on-site repulsive correlations (Hubbard U) are an essential feature of the cuprates. Indeed all undoped parent compounds are *insulators* with an insulating gap of about 2 eV or so. But if the repulsive correlations are weak, one would expect a metallic behaviour for the half-filled d-band of copper in cuprates or, at most, a much smaller gap caused by lattice and spin distortions (i.e. due to charge and/or spin density waves [142]). Therefore, it is the strong on-site repulsion of d-electrons in cuprates which results in their parent 'Mott' insulating state (section 5.2).

Differing from conventional band-structure insulators with completely filled or empty Bloch bands, the Mott insulator arises from a potentially metallic half-filled band due to the Coulomb blockade of electron tunnelling to neighbouring sites, if $U > zT(a)$ [143]. The insulator is antiferromagnetic with one hole and spin-$\frac{1}{2}$ per site. In using this model, we have to realize that the insulating properties of the Mott insulator do not depend on the ordering of the spins; they persist above the Néel temperature and arise because the on-site Coulomb repulsion is larger than the half-bandwidth.

When on-site correlations are strong and dimensionality is low, there is an alternative to the usual Fermi-liquid description proposed by Anderson [140]. In Anderson's resonating-valence-bond (RVB) theory, the ground state supports 'topological solitons' (the so-called spinons and holons), such as occur in one-dimensional models like the one-dimensional Hubbard model (see later). The main idea is that an electron injected into a two-dimensional layer decays into a singlet charge e component (holon) and a spin-$\frac{1}{2}$ component (spinon) and, conversely, must form again in order to come out. This is the case for one-dimensional repulsive electrons, which form the so-called Luttinger liquid in one dimension. Bose quasi-particles imply a condensate. However, there is no experimental evidence for a charge e superfluid. Therefore, in the so-called *interlayer* RVB extension of the model [144], it was suggested that the superconductivity of copper-based high-T_c materials is due to holon-pair tunnelling between the copper–oxygen planes. There is no single-particle coherent tunnelling between two spinon–holon planes above T_c. However, there is a coherent two-particle tunnelling between them below T_c. Then the corresponding kinetic energy should be responsible for the BCS-like pairing at temperatures below $T_c \approx t_\perp^2/t$ and for the plasma-like gap, observed in the c-axis conductivity. Here t and $t_\perp$ are the in-plane and out-of-plane renormalized hopping integrals, respectively. Anderson argued that the existence of the upper Hubbard band (section 5.2) would necessarily lead to the Luttinger liquid even in two dimensions, as opposed to the Fermi liquid. While the interlayer RVB model was found to be incompatible with experiments [145], the basic idea of spin and charge separation had been worked out in great detail [146]. The microscopic Hubbard Hamiltonian [147]

$$H = T(a) \sum_{\langle \boldsymbol{mn} \rangle, s} [c^\dagger_{\boldsymbol{m}s} c_{\boldsymbol{n}s} + H.c.] + U \sum_{\boldsymbol{m}} \hat{n}_{\boldsymbol{m}\uparrow} \hat{n}_{\boldsymbol{m}\downarrow} \tag{5.1}$$

was proposed to justify the RVB concept, where $\langle \boldsymbol{mn} \rangle$ are the nearest neigbours. The Hamiltonian describes the antiferromagnetic Mott insulator at half filling when $U > D$. In the strong correlation limit, $U \gg T(a)$, the doubly occupied sites take a large Coulomb energy and the Hubbard Hamiltonian can be reduced to the so-called t–J model [148]

$$\tilde{H} = T(a) \sum_{\langle \boldsymbol{mn} \rangle, s} [\tilde{c}^\dagger_{\boldsymbol{m}s} \tilde{c}_{\boldsymbol{n}s} + H.c.] + J \sum_{\langle \boldsymbol{mn} \rangle} (\hat{\boldsymbol{S}}_{\boldsymbol{m}} \hat{\boldsymbol{S}}_{\boldsymbol{n}} - \tfrac{1}{4} \hat{n}_{\boldsymbol{m}} \hat{n}_{\boldsymbol{n}}) \tag{5.2}$$

where the projected electron operators $\tilde{c}_{m\uparrow} = c_{m\uparrow}(1 - \hat{n}_{m\downarrow})$ act in the subspace without double occupancy, $\hat{n}_m = \hat{n}_{m\uparrow} + \hat{n}_{m\downarrow}$,

$$\hat{S}_m^{\alpha} = \frac{1}{2} \sum_{ss'} \tilde{c}_{ms}^{\dagger} (\tau_{\alpha})_{ss'} \tilde{c}_{ms'}$$

are three components of the on-site spin-$\frac{1}{2}$ operator ($\alpha = 1, 2, 3$) and τ_{α} are the Pauli matrices as in equation (3.73). The second term in $\tilde{H}$ describes the spin-$\frac{1}{2}$ Heisenberg antiferromagnet with the exchange energy

$$J = \frac{4T^2(a)}{U} \tag{5.3}$$

for the nearest neigbours. The exchange energy leads to spin polarons similar to lattice polarons but dressed with magnetic fluctuations rather than with phonons. However, the non-fermionic commutation relations for the projected electron operators lead to an additional *kinematic* interaction between carriers doped into the Mott insulator. As a result an analytical solution of the t–J model is a very hard problem but not in the dilute limit. In this limit a doped hole dressed by antiferromagnetic spin fluctuations propagates coherently in a narrow band of the order of $J \ll T(a)$ like a small polaron. One could believe that the same spin fluctuations, which dress a single hole, mediate a superconducting pairing of spin polarons due to an effective attractive interaction of the order of J. To be consistent with the Néel temperature of parent undoped cuprates and with their spin-wave excitation spectrum, J should be of the order of 0.1 eV. The magnetic singlet pairing might be effective in the *d-channel* (section 2.10), where the direct Coulomb repulsion is diminished due to the d-wave symmetry of the BCS order parameter $\Delta_{\mathbf{k}}$. The possibility of d-wave BCS-like superconductivity in the two-dimensional Hubbard model near half-filling was suggested by Scalapino *et al* [149] concurrently with the discovery of novel superconductors. There is now a variety of phase-sensitive experiments, which support the unconventional symmetry of the order parameter in some cuprates, while other experiments appear to contradict this symmetry (see section 8.2).

Independent of any experimental evidence, the Hubbard U and t–J models share an inherent difficulty in determining the order. While some groups claimed that they describe high-T_c superconductors at finite doping, other authors could not find any superconducting instabilty without an additonal (i.e. e–ph) interaction [150]. Therefore, it has been concluded that models of this kind are highly conflicting and confuse the issue by exaggerating the magnetism rather than clarifying it [151]. In our view [12], the problem with the RVB-like concepts of high-T_c is that for them to be valid, the diffractive scattering between holes of cuprate superconductors needs to be absent. A known case where diffractive scattering is absent is that of one-dimensional interacting fermion systems. But in one-dimensional interacting fermion systems the scattering is non-diffractive due to topological reasons. In two and higher dimensions, one does not have such a topological constraint. On the experimental side, there are measurements of out-of-plane resistivity ρ_c [152, 153] which show the metallic-like temperature dependence of ρ_c and the mean-free path in the c-direction comparable with the interplane spacing in a number of high-T_c cuprates. Well-developed interlayer tunnelling invalidates these and any other low-dimensional concepts (see figure 5.2), which definitely fail to account for $d\rho_c/dT > 0$ and high-T_c of highly oxygenated cuprates, doped fullerenes and MgB_2.

Another problem of the microscopic Hamiltonians (equations (5.1) and (5.2)), which are behind the RVB concept, is that they neglect the long-range Coulomb and electon-phonon interactions, which are essential in novel superconductors (section 5.2). As a result, some exact solutions of various limits and numerical studies on t–J and Hubbard models led to the conclusion that the electronic structure in cuprates is much more prone to inhomogeneity and intermediate-scale structures such as stripes of hole-rich domains separated by insulating antiferromagnetic domains. Moreover, it has been proposed that the stripes are essential to a high-T_c mechanism [141], especially in the underdoped regime. In this scenario, stripe formation permits hole delocalization in one direction, but hole motion transverse to the stripe is still restricted. It is thus favourable for the holes to pair so that the pairs can spread out somewhat into the antiferromagnetic neighbourhood of the stripe, where their interaction is allegedly attractive due to spin fluctuations. The proximity effect in conventional superconductor–normal metal structures is a prototypical example of such a mechanism of pairing: when the BCS superconductor and a normal metal are placed in contact with each other, electrons in the metal pair even if the interaction between them is repulsive. As we discuss in section 8.6, this 'stripe' scenario is incompatible with Coulomb's law: there are no stripes if the long-range Coulomb repulsion is properly taken into account. On the experimental side, recent neutron [154] and X-ray [155] spectroscopic studies did not find any bulk charge/spin segregation in the *normal* state of a few cuprates suggesting an absence of in-plane carrier density modulations in these materials above T_c.

Strong repulsive correlations between holes could provide another novel mechanism for Cooper pairing, i.e. the *kinetic-energy-driven mechanism of superconductivity* proposed by Hirsch [156] (see figure 5.2). Such a mechanism does not require any dynamic attraction between holes. The qualitative explanation is as follows: electrons in metals are 'dressed' by a cloud of other electrons with which they interact and form the so-called *electronic* polarons. The dressing causes an increase in the electron's effective mass and when the dressing is large, the metal is a poor conductor. If, however, the electrons manage to 'undress' when the temperature is lowered, their effective mass will be reduced and electricity will flow easily. A model Hamiltonian, which describes the 'undressing', is that of small polarons with a nonlinear interaction with a background bosonic degree of freedom which gives rise to an effective mass enhancement that depends on the local charge occupation. The undressing process can only occur if the carriers are 'holes' rather than electrons and when two hole carriers form a pair. High-precision optical conductivity measurements [157] observed a temperature-dependent change in the conductivity sum rule (1.18) in the optimally doped Bi-cuprate integrated up frequencies a few orders of magnitude greater than T_c. According to [157], this observation implies a change in the kinetic energy anticipated in the theory of 'hole' superconductivity [156] and in the bipolaron theory [13].

Less exotic concepts are based on the Fermi-liquid and BCS-like approach but take into account the van Hove singularities of the electron density of states [158], and/or a singular interaction between holes due to a closeness to a quantum phase transition of any type [138, 142, 159, 160]. In particular, Pines and his co-workers argued that the magnetic interaction between planar quasi-particles in cuprate superconductors leads to a new quantum state of matter, the nearly antiferromagnetic Fermi liquid (NAFL). It possesses a well-defined Fermi surface but it is not the ordinary Landau–Fermi liquid, because of a singular interaction between electrons with their Fourier component peaked at some momentum caused by soft magnetic fluctuations. The model clearly contradicts the neutron data by Smith *et al* [161], obtained on the high flux reactor at Grenoble. The quasi-elastic peak and diffuse magnetic scattering are virtually absent in the metallic phase of $YBa_2Cu_3O_{7-\delta}$ in a wide energy range up to 30 meV. It follows from these data that the local magnetic moments are practically absent from optimally doped cuprates. Some authors attribute high T_c and non-Fermi liquid features of the cuprates to the proximity of the Fermi level to a van Hove singularity in the density of states. However, other authors [162] conclude in the framework of the Eliashberg theory (chapter 3), that this model yields $T_c \simeq 10$ K for both $La_{2-x}Sr_xCuO_4$ and $YBa_2Cu_3O_{7-\delta}$, when its parameters are constrained by neutron-scattering and transport measurements. Hence, the van Hove singularity scenario enhances T_c much less effectively than weak-coupling *ad hoc* calculations would suggest.

There are several semi-phenomelogical concepts for the behaviour of a non-Fermi liquid of cuprates. In particular, it is conceivable that novel superconductors are *marginal* Fermi liquids (MFL) [163] in the sense that the energy interval near the Fermi level in which the Landau quasi-particles exist is small compared with T_c. Outside the interval a linear energy dependence of the quasi-particle scattering rate has been postulated, which allows for a description of optical responce in some cuprates. A linear frequency dependence of the imaginary part of the electron self-energy does not follow from the e–ph or Coulomb interactions (see chapter 3), where it is cubic or quadratic, respectively. Therefore, an MFL needs a singular interaction like NAFL, which might be caused by a closeness to a quantum phase transition of any type.

Other authors [141,164] dismiss any real-space pairing claiming that pairing is collective even in underdoped cuprates. They believe in a large Fermi surface with the number of holes $(1 + x)$ rather than x in superconducting cuprates, where x is the doping level like in $La_{2-x}Sr_xCuO_4$. As an alternative to a three-dimensional Bose–Einstein condensation of bipolarons (chapter 4), these authors suggest a collective pairing (i.e. the *Cooper* pairs in the momentum space) at some temperature $T^* > T_c$ but without phase ordering. In this concept, the phase coherence and superconducting critical temperature T_c are determined by the superfluid density, which is proportional to doping x due to a low dimensionality, rather than to the total density of carriers $(1 + x)$. However, a large Fermi surface is clearly incompatible with a great number of thermodynamic, magnetic and kinetic measurements, which show that only holes *doped* into a parent insulator are carriers in the *normal* state (chapter 6). On theoretical grounds this preformed *Cooper*-pair (or phase-fluctuation) scenario contradicts a theorem [165], which proves that the number of supercarriers (at $T = 0$) and normal-state carriers is the same in any *clean* superfluid. It also contradicts a parameter-free estimate of the Fermi energy and T_c in the cuprates, as explained in sections 5.3 and 7.1, respectively.

For although high-temperature superconductivity has not yet been targeted as '*the shame and despair of theoretical physics*'—a label attributed to superconductivity during the first half-century after its discovery—the parlous state of current theoretical constructions has led to a current consensus that there is no consensus on the theory of high-T_c superconductivity. Our view, which we discuss in the rest of the book, is that the extension of BCS theory towards the strong interaction between electrons and ion vibrations (chapter 4) describes the phenomenon naturally and that high-temperature superconductivity exists in the crossover region of the electron–phonon interaction strength from the BCS-like to bipolaronic superconductivity as was predicted before [11], and explored in greater detail after discovery [13, 64, 166–169]. Experimental [170–177] and theoretical [64, 84, 178–182] evidence for an exceptionally strong electron–phonon interaction in high temperature superconductors is now so overwhelming, that even advocates of a non-phononic mechanism [141] accept this fact.

5.2 Band structure and essential interactions in cuprates

In this book we take the view that cuprates and related transition metal oxides are charge transfer Mott insulators at *any* level of doping [183]. As established in a few site-selective experiments [184] and in first-principle (the so-called 'LDA+U') band-structure calculations, their one-particle DOS is schematically represented by figure 5.3. Here a d-band of a transition metal is split into the lower and upper Hubbard bands due to on-site repulsive interaction (the Hubbard U), while the first band to be doped is the oxygen band lying within the Hubbard gap. The oxygen band is less correlated and completely filled in parent insulators. As a result, a single oxygen hole has well-defined quasi-particle properties in the absence of interactions with phonons and fluctuations of d-band spins. A strong coupling with phonons, unambiguously established for many oxides, should lead to a high-energy spectral weight in the spectral function of an oxygen hole at energies about twice the Franck–Condon (polaron) level shift, E_p, and to a band-narrowing effect (section 4.3).

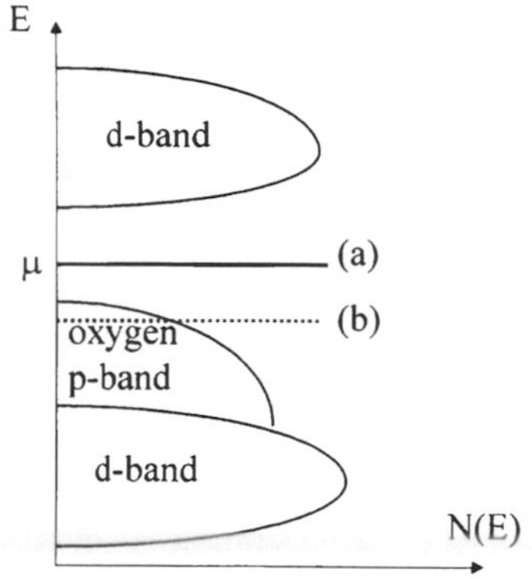

Figure 5.3. Density of states in cuprates. The chemical potential is inside the charge transfer gap due to bipolaron formation (a) and bipolarons coexist with thermally excited *non-degenerate* polarons. It might enter the oxygen band in overdoped cuprates (b), where bipolarons may coexist with unpaired *degenerate* polarons.

The e–ph interaction also binds holes into inter-site oxygen bipolarons the size of a lattice constant (section 4.4). The bipolaron density remains relatively low (below 0.15 per cell) at any relevant level of doping and the residual repulsive interaction of bipolarons is strongly suppressed by the lattice polarization owing to a large static dielectric constant (section 4.4). That is why bipolarons in oxides are fairly well described by the charged Bose-gas model of section 4.7. One of their roles in the one-particle DOS is to pin the chemical potential inside the charge transfer gap, half the bipolaron binding energy above the oxygen band edge, shifted by E_p, figure 5.3. This binding energy as well as the singlet–triplet bipolaron exchange energy are thought to be the origin of normal-state pseudogaps, as first proposed by us [185] (sections 6.4 and 6.5). In overdoped samples, carriers screen part of the e–ph interaction with low-frequency phonons. Hence, the binding energy and the hole spectral function depend on doping. In particular, the bipolaron and polaron bands could overlap because the bipolaron binding energy becomes smaller [186], so the chemical potential could enter the oxygen band in the overdoped cuprates, figure 5.3(b). Then a Fermi-level crossing could be seen in the angle-resolved photoemission spectra (ARPES) (section 8.4).

To assess quantitatively the role of the Fröhlich e–ph interaction, we apply an expression for E_p, which depends only on the experimentally measured high-frequency, ϵ_∞, and static, ϵ_0, dielectric constants of the host insulator as [187]

$$E_p = \frac{1}{2\kappa} \int_{\mathrm{BZ}} \frac{\mathrm{d}^3 q}{(2\pi)^3} \frac{4\pi e^2}{q^2} \tag{5.4}$$

where $\kappa^{-1} = \epsilon_\infty^{-1} - \epsilon_0^{-1}$. Here the size of the integration region, which is the Brillouin zone (BZ) is determined by the lattice constants a, b, c. As shown in

Table 5.1. Polaron shift E_p due to the Fröhlich interaction. Data taken from the 1997 *Handbook of Optical Constants of Solids* ed E D Palik (New York: Academic). The value $\epsilon_\infty = 5$ for WO_3 is an estimate.

System	ϵ_∞	ϵ_0	$a \times b \times c$ (Å^3)	E_p (eV)
$SrTiO_3$	5.2	310	3.905^3	0.852
$BaTiO_3$	5.1–5.3	1499	$3.992^2 \times 4.032$	0.842
$BaBiO_3$	5.7	30.4	$4.34^2 \times 4.32$	0.579
La_2CuO_4	5.0	30	$3.8^2 \times 6^a$	0.647
$LaMnO_3$	3.9^b	16^b	3.86^3	0.884
$La_{2-2x}Sr_{1+2x}$-Mn_2O_7	4.9^b	38^b	$3.86^2 \times 3.9^c$	0.807
NiO	5.4	12	4.18^3	0.429
TiO_2	6–7.2	89–173	$4.59^2 \times 2.96$	0.643
MgO	2.964	9.816	4.2147^3	0.982
CdO	5.4	21.9	4.7^3	0.522
WO_3	5	100–300	$7.31 \times 7.54 \times 7.7$	0.445
NaCl	2.44	5.90	5.643^3	0.749
EuS	5.0	11.1	5.968^3	0.324
EuSe	5.0	9.4	6.1936^3	0.266

[a] Distance between CuO_2 planes.
[b] Ishikawa T private communication.
[c] Distance between MnO_2 planes.

table 5.1, the Fröhlich interaction alone provides the binding energy of two holes ($\approx 2E_p$) almost by one order of magnitude larger than the magnetic interaction $J \sim 0.1$ eV of the t–J model. The data in the Table represent lower boundaries for the polaron shift, since the deformation potential and/or the molecular-type (Jahn–Teller) e–ph interaction are not included in equation (5.4). There is virtually no screening of c-axis polarized optical phonons in cuprates because an upper limit for the out-of-plane plasmon frequency ($\lesssim 200$ cm^{-1} [188]) is well below the phonon frequency (section 4.4). Hence, equation (5.4) is perfectly applied at any doping.

The polaron shift about 1 eV results in the formation of small polarons no matter which criterion for their formation is applied. The bare half-bandwidth is about 1 eV or less in cuprates, so even a naive variational criterion of the small polaron formation ($E_p > D$) could be satisfied. Moreover, optical phonon frequencies are high in oxides, fullerenes and MgB_2, $\omega_0 \gtrsim 0.1$ eV. In this non-adabatic or intermediate regime, the variational criterion for small polaron formation fails. The correct criterion for small polaron formation is determined by the convergence of the $1/\lambda$ perturbation expansion as discussed in chapter 4.

Exact Monte Carlo calculations [94] show that the polaron theory based on this expansion describes quantitatively both strong-coupling (small polaron) and weak-coupling (large polaron) regimes without any restriction on the value of E_p for a long-range electron–phonon interaction with high-frequency phonons. The effective polaron–polaron attraction due to the overlap of the deformation fields is about $2E_p$ (section 4.4). It is sufficient to overcome the inter-site Coulomb repulsion at short distances, in particular if a weaker interaction with the deformation potential is taken into account. We show later that the adiabatic ratio in all novel superconductors is of the order or even larger than one for most essential optical phonons.

5.3 Low Fermi energy: pairing is individual in many cuprates

First-principle band-structure calculations show that copper, alkali metals and magnesium donate their outer electrons to oxygen, C_{60}, and boron in cuprates, doped fullerenes, and in MgB_2, respectively. In cuprates and MgB_2, the band structure is quasi-two-dimensional with a few degenerate hole pockets. Applying the parabolic approximation for the band dispersion we obtain the *renormalized* Fermi energy as

$$\epsilon_{\mathrm{F}} = \frac{\pi n_i d}{m_i^*} \tag{5.5}$$

where d is the interplane distance, and n_i, m_i^* are the density of holes and their effective mass in each of the hole sub-bands i renormalized by the electron–phonon (and by any other) interaction. One can express the renormalized band-structure parameters through the in-plane magnetic-field penetration depth at $T \simeq 0$, measured experimentally:

$$\lambda_{\mathrm{H}}^{-2} = 4\pi e^2 \sum_i \frac{n_i}{m_i^*}. \tag{5.6}$$

As a result, we obtain a *parameter-free* expression for the 'true' Fermi energy as

$$\epsilon_{\mathrm{F}} = \frac{d}{4ge^2\lambda_{\mathrm{H}}^2} \tag{5.7}$$

where g is the degeneracy of the spectrum, which is $g = 2$ in MgB_2. g may depend on doping in cuprates. One expects four hole pockets inside the Brillouin zone (BZ) due to the Mott–Hubbard gap in underdoped cuprates. If the hole band minima are shifted with doping to BZ boundaries, all their wavevectors would belong to the stars with two or more prongs. The groups of wavevectors for these stars have only one-dimensional representations. It means that the spectrum will be degenerate with respect to the number of prongs which the star has, i.e. $g \geqslant 2$. The only exception is the minimum at $\boldsymbol{k} = (\pi, \pi)$ with one prong and $g = 1$. Hence, in cuprates the degeneracy is $1 \leqslant g \leqslant 4$. Because equation (5.7) does not contain any other band-structure parameters, the estimate of ϵ_{F} using this equation does not depend very much on the parabolic approximation for the band dispersion.

Generally, the ratios n/m^* in equations (5.5) and (5.6) are not necessarily the same. The 'superfluid' density in equation (5.6) might be different from the total density of delocalized carriers in equation (5.5). However, in a translationally invariant system they must be the same [165]. This is also true even in the extreme case of a pure two-dimensional superfluid, where quantum fluctuations are important. One can, however, obtain a reduced value for the zero-temperature superfluid density in the dirty limit, $l \ll \xi(0)$, where $\xi(0)$ is the zero-temperature coherence length. The latter was measured directly in cuprates as the size of the vortex core. It is about 10 Å or even less. In contrast, the mean free path was found to be surprisingly large at low temperatures, $l \sim$ 100–1000 Å. Hence, it is rather probable that all novel superconductors, including MgB_2 are in the clean limit, $l \gg \xi(0)$, so that the parameter-free expression for ϵ_F, equation (5.7), is perfectly applicable.

A parameter-free estimate of the Fermi energy obtained by using equation (5.7) is presented in table 5.2. The renormalized Fermi energy in magnesium diboride and in more than 30 cuprates is less than 100 meV, if the degeneracy $g \geq 2$ is taken into account. This should be compared with the characteristic phonon frequency, which is estimated as the plasma frequency of boron or oxygen ions,

$$\omega_0 = \sqrt{\frac{4\pi Z^2 e^2 N}{M}}. \tag{5.8}$$

With $Z = 1$, $N = 2/V_{\text{cell}}$, $M = 10$ au, one obtains $\omega_0 \simeq 69$ meV for MgB_2, and $\omega_0 = 84$ meV with $Z = 2$, $N = 6/V_{\text{cell}}$, $M = 16$ au for $YBa_2Cu_3O_6$. Here V_{cell} is the volume of the (chemical) unit cell. The estimate agrees with the measured phonon spectra. As established experimentally in cuprates (section 5.1), the high-frequency phonons are strongly coupled with carriers. The parameter-free expression (5.7) does not apply to doped fullerenes with their three-dimensional energy spectrum. However, it is well established that they are also in the non-adiabatic regime [190], $\epsilon_F \lesssim \omega_0$.

A low Fermi energy is a serious problem for the Migdal–Eliashberg approach (chapter 3). Since the Fermi energy is small and phonon frequencies are high, the Coulomb pseudopotential μ_c^* is of the order of the bare Coulomb repulsion, $\mu_c^* \simeq \mu_c \simeq 1$ because the Tolmachev logarithm is ineffective. Hence, to get an experimental T_c, one has to have a strong coupling, $\lambda > \mu_c$. However, one cannot increase λ without accounting for the polaron collapse of the band (chapter 4). Even in the region of the applicability of the Eliashberg theory (i.e. at $\lambda \leq 0.5$), the non-crossing diagrams cannot be treated as vertex *corrections* as in [191], since they are comparable to the standard terms, when $\omega_0/\epsilon_F \gtrsim 1$. Because novel superconductors are in the non-adiabatic regime, interaction with phonons must be treated within the multi-polaron theory (chapter 4) at any value of λ.

In many cases (table 5.2), the renormalized Fermi energy is so small that

Table 5.2. The Fermi energy (multiplied by the degeneracy) of cuprates and MgB_2 [189].

Compound	T_c (K)	$\lambda_{H,ab}$ (Å)	d (Å)	$g\epsilon_F$ (meV)
$La_{1.8}Sr_{0.2}CuO_4$	36.2	2000	6.6	112
$La_{1.78}Sr_{0.22}CuO_4$	27.5	1980	6.6	114
$La_{1.76}Sr_{0.24}CuO_4$	20.0	2050	6.6	106
$La_{1.85}Sr_{0.15}CuO_4$	37.0	2400	6.6	77
$La_{1.9}Sr_{0.1}CuO_4$	30.0	3200	6.6	44
$La_{1.75}Sr_{0.25}CuO_4$	24.0	2800	6.6	57
$YBa_2Cu_3O_7$	92.5	1400	4.29	148
YBaCuO(2%Zn)	68.2	2600	4.29	43
YBaCuO(3%Zn)	55.0	3000	4.29	32
YBaCuO(5%Zn)	46.4	3700	4.29	21
$YBa_2Cu_3O_{6.7}$	66.0	2100	4.29	66
$YBa_2Cu_3O_{6.57}$	56.0	2900	4.29	34
$YBa_2Cu_3O_{6.92}$	91.5	1861	4.29	84
$YBa_2Cu_3O_{6.88}$	87.9	1864	4.29	84
$YBa_2Cu_3O_{6.84}$	83.7	1771	4.29	92
$YBa_2Cu_3O_{6.79}$	73.4	2156	4.29	62
$YBa_2Cu_3O_{6.77}$	67.9	2150	4.29	63
$YBa_2Cu_3O_{6.74}$	63.8	2022	4.29	71
$YBa_2Cu_3O_{6.7}$	60.0	2096	4.29	66
$YBa_2Cu_3O_{6.65}$	58.0	2035	4.29	70
$YBa_2Cu_3O_{6.6}$	56.0	2285	4.29	56
$HgBa_2CuO_{4.049}$	70.0	2160	9.5	138
$HgBa_2CuO_{4.055}$	78.2	1610	9.5	248
$HgBa_2CuO_{4.055}$	78.5	2000	9.5	161
$HgBa_2CuO_{4.066}$	88.5	1530	9.5	274
$HgBa_2CuO_{4.096}$	95.6	1450	9.5	305
$HgBa_2CuO_{4.097}$	95.3	1650	9.5	236
$HgBa_2CuO_{4.1}$	94.1	1580	9.5	257
$HgBa_2CuO_{4.101}$	93.4	1560	9.5	264
$HgBa_2CuO_{4.101}$	92.5	1390	9.5	332
$HgBa_2CuO_{4.105}$	90.9	1560	9.5	264
$HgBa_2CuO_{4.108}$	89.1	1770	9.5	205
MgB_2	39	1400	3.52	122

pairing is certainly individual, i.e. the bipolaron radius (equation (4.97)), is smaller than the inter-carrier distance. Indeed, this is the case, if

$$\epsilon_F \lesssim \pi \Delta. \tag{5.9}$$

The bipolaron binding energy is thought to be twice the so-called pseudogap experimentally measured in the normal state of many cuprates (section 6.5), $\Delta \gtrsim 100$ meV, so that equation (5.9) is well satisfied in underdoped and even in a few optimally and overdoped cuprates. One should note that the coherence length in the charged Bose gas has nothing to do with the size of the boson. It depends on the inter-particle distance and the mean-free path, equation (4.262), and might be as large as in the BCS superconductor. Hence, it would be incorrect to apply the ratio of the coherence length to the inter-carrier distance as a criterion of the BCS–Bose liquid crossover. The correct criterion is given by equation (5.9).

5.4 Bipolaron bands in high-T_c perovskites

Consideration of particular lattice structures shows that small inter-site bipolarons are perfectly mobile even when the electron–phonon coupling is strong and the bipolaron binding energy is large (section 4.6). Let us analyse the important case of copper-based high-T_c oxides. They are doped charged-transfer ionic insulators with narrow electron bands. Therefore, the interaction between holes can be analysed using computer simulation techniques based on a minimization of the ground-state energy of an ionic insulator with two holes, the lattice deformations and the Coulomb repulsion fully taken into account but neglecting the kinetic energy terms.

Using these techniques net inter-site interactions of an in-plane oxygen hole with an *apex* hole, figure 5.4, and of two in-plane oxygen holes were found to be atractive (the binding energies $\Delta = 119$ meV and $\Delta = 60$ meV, respectively), in La_2CuO_4 [192]. All other interactions were found to be repulsive. The Bloch bands of those bipolarons in the perovskites are obtained using the canonical transformations of chapter 4 [100, 108].

5.4.1 Apex bipolarons

An apex bipolaron can tunnel from one unit cell to another via a direct *single*-polaron tunnelling from one apex oxygen to its apex neighbour as in a ladder (section 4.6.3). Oxides are highly polarizable materials, so that we can apply the Fröhlich interaction with the coupling constant

$$\gamma(\boldsymbol{q}) \propto \frac{1}{q}$$

to calculate the bipolaron mass. Lattice polarization is coupled with the electron density, therefore the interaction is diagonal in the site representation and the coupling constant does not depend on a particular orbital state of the hole. The canonical displacement transformation (section 4.3.1) eliminates an essential part of the electron–phonon interaction. The transformed Hamiltonian is given by

$$\tilde{H} = \mathrm{e}^{S} H \mathrm{e}^{-S} = (T_p - E_p) \sum_{i(\mathrm{p})} n_{i(\mathrm{p})} + (T_d - E_d) \sum_{i(\mathrm{d})} n_{i(\mathrm{d})}$$
$$+ \sum_{i \neq j} \hat{\sigma}_{ij} c_i^{\dagger} c_j + \sum_{\boldsymbol{q}} \omega_{\boldsymbol{q}} (d_{\boldsymbol{q}}^{\dagger} d_{\boldsymbol{q}} + 1/2)$$

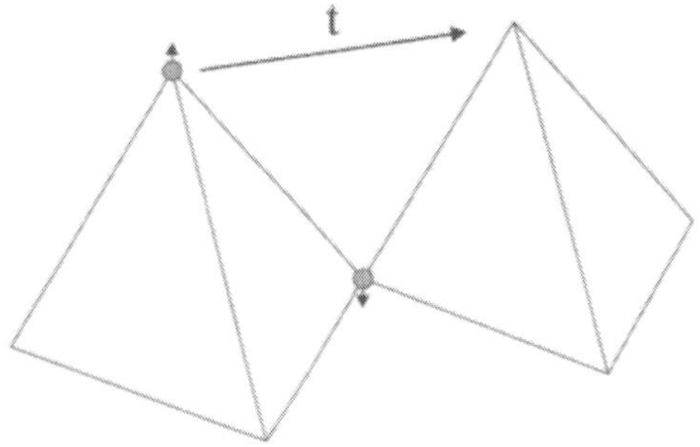

Figure 5.4. Apex bipolaron tunnelling in perovskites.

$$-\tfrac{1}{2}\sum_{q,i,j}(2\omega_q u_i(q)u_j^*(q) - V_{ij})\hat{n}_i\hat{n}_j. \qquad (5.10)$$

The oxygen (p) and the copper (d) diagonal terms include the polaron level shift, which is the same for oxygen and copper ions

$$E_p = E_d = \sum_q |u_j(q)|^2\omega_q. \qquad (5.11)$$

If the charge-transfer gap is large enough, $E_g \gg \omega_0$, the bandwidth narrowing factors are the same for the direct $\sigma_{pp'}$ and the second-order (via copper $\sigma_{pp'}^{(2)}$) oxygen–oxygen polaron hopping integrals ($T = 0$),

$$\sigma_{pp'} \equiv \langle 0|\hat{\sigma}_{pp'}|0\rangle = T_{pp'}\mathrm{e}^{-g_{pp'}^2} \qquad (5.12)$$

$$\sigma_{pp'}^{(2)} \equiv \sum_\nu \frac{\langle 0|\hat{\sigma}_{pd}|\nu\rangle\langle\nu|\hat{\sigma}_{dp'}|0\rangle}{E_0 - E_\nu} \approx -\frac{T_{pd}^2}{E_g}\mathrm{e}^{-g_{pp'}^2} \qquad (5.13)$$

where $|\nu\rangle$, E_ν are eigenstates and eigenvalues of the transformed Hamiltonian (5.10) without the third hopping term, $|0\rangle$ is the phonon vacuum and the reduction factor is

$$g_{pp'}^2 = \frac{1}{2N}\sum_q |\gamma(q)|^2(1 - \cos[q\cdot(m_p - m_{p'})]). \qquad (5.14)$$

These expressions are the result of straightforward calculations described later. Taking into account that $E_\nu - E_0 = E_g + \sum_q \omega_q n_q$, the second-order indirect hopping (equation (5.13)) is written as

$$\sigma_{pp'}^{(2)} = -\mathrm{i}\int_0^\infty \mathrm{d}t\, \mathrm{e}^{-\mathrm{i}E_g t}\langle 0|\hat{\sigma}_{pd}(t)\hat{\sigma}_{dp'}|0\rangle \qquad (5.15)$$

where

$$\hat{\sigma}_{pd}(t) = T_{pd}\exp\left(\sum_q u_d^*(q,t)d_q^\dagger - H.c.\right)\exp\left(\sum_q u_p(q,t)d_q - H.c.\right). \qquad (5.16)$$

Here $u_j(q,t) \equiv u_j(q)\exp(-i\omega_q t)$ and $n_q = 0, 1, 2, \ldots$ the phonon occupation numbers. Calculating the bracket in equation (5.15), one obtains

$$\langle \ldots \rangle = T_{pd}^2 e^{-g_{pd}^2} e^{-g_{dp'}^2} \exp\left(-\sum_q [u_d(\boldsymbol{q}) - u_p(\boldsymbol{q})][u_{p'}^*(\boldsymbol{q}) - u_d^*(\boldsymbol{q})] e^{-i\omega_q t} \right). \tag{5.17}$$

If ω_q is q-independent the integral in equation (5.15) is calculated by the expansion of the exponent in equation (5.17) as

$$\sigma_{pp'}^{(2)} = -\frac{T_{pd}^2}{E_g} e^{-g_{pd}^2} e^{-g_{dp'}^2} \sum_{k=0}^{\infty} \frac{(-1)^k (\sum_q [u_d(\boldsymbol{q}) - u_p(\boldsymbol{q})][u_{p'}^*(\boldsymbol{q}) - u_d^*(\boldsymbol{q})])^k}{k!(1 + k\omega_0/E_g)}. \tag{5.18}$$

Then equation (5.13) is obtained in the limit $E_g \gg \omega_0$. Equation (5.14) yields

$$g_{pp'}^2 = \frac{E_p}{\omega_0}\left(1 - \frac{\mathrm{Si}(q_d m)}{q_d m}\right), \tag{5.19}$$

if we approximate the Brillouin zone by a sphere of the radius q_d (the Debye approximation). Here

$$\mathrm{Si}(x) = \int_0^x \mathrm{d}t \, \frac{\sin t}{t}$$

$m = a/\sqrt{2}$ and $m = a$ for the in-plane oxygen–oxygen, and for apex–apex narrowing factors, respectively. In cuprates with $q_d \approx 0.7$ Å^{-1} and $a \simeq 3.8$ Å one obtains $g_{pp'}^2 \approx 0.2E_p/\omega_0$ for an in-plane hopping, and $g^2 \approx 0.3E_p/\omega_0$ for an apex–apex one. Hence, both band-narrowing factors are much less than one can expect in the Holstein model.

The *bipolaron* hopping integral, t, is obtained by projecting the Hamiltonian (5.10) onto a reduced Hilbert space containing only empty or doubly occupied elementary cells and averaging the result with respect to phonons. The wavefunction of an apex bipolaron localized, let us say, in the cell $\boldsymbol{m}$ is written as

$$|\boldsymbol{m}\rangle = \sum_{i=1}^{4} A_i c_i^\dagger c_{\mathrm{apex}}^\dagger |0\rangle \tag{5.20}$$

where i denotes the $p_{x,y}$ orbitals and spins of the four plane oxygen ions in the cell (figure 5.4) and $c_{\mathrm{apex}}^\dagger$ is the creation operator for the hole on one of the three apex oxygen orbitals with the spin, which is the same or opposite of the spin of the in-plane hole depending on the total spin of the bipolaron. The probability amplitudes A_i are normalized by the condition $|A_i| = 1/2$, if four plane orbitals p_{x1}, p_{y2}, p_{x3} and p_{y4} are involved or by $|A_i| = 1/\sqrt{2}$ if only two of them are relevant.

The matrix element of the Hamiltonian (5.10) of first order with respect to the transfer integral responsible for the bipolaron tunnelling to the nearest-neighbour cell $\boldsymbol{m} + \boldsymbol{a}$ is

$$t = \langle \boldsymbol{m}|\tilde{H}|\boldsymbol{m} + \boldsymbol{a}\rangle = |A_i|^2 T_{pp'}^{\mathrm{apex}} e^{-g^2} \tag{5.21}$$

where $T^{\text{apex}}_{pp'} e^{-g^2}$ is a single-polaron hopping integral between two apex ions. As a result, the hole bipolaron energy spectrum consists of two bands $E^{x,y}(\boldsymbol{K})$, formed by an overlap of p_x and p_y *apex polaron* orbitals, respectively, as in equation (4.269). In this equation t is the renormalized hopping integral between p orbitals of the same symmetry elongated in the direction of the hopping $(pp\sigma)$ and t' is the renormalized hopping integral in the perpendicular direction $(pp\pi)$. Their ratio $t/t' = T^{\text{apex}}_{pp'}/T'^{\text{apex}}_{pp'} = 4$ as follows from the tables of hopping integrals in solids. Two different bands are not mixed because $T^{\text{apex}}_{p_x,p'_y} = 0$ for the nearest neighbours. The random potential does not mix them either if it varies smoothly on the lattice scale. Hence, we can distinguish 'x' and 'y' bipolarons with a lighter effective mass in the x or y direction, respectively. The apex z bipolaron, if formed is approximately four times less mobile than x and y bipolarons. The bipolaron bandwidth is of the same order as the polaron one, which is a specific feature of the inter-site bipolarons (section 4.6). For a large part of the Brillouin zone near $(0, \pi)$ for 'x' and $(\pi, 0)$ for 'y' bipolarons, one can adopt the effective mass approximation

$$E^{x,y}_{\boldsymbol{K}} = \frac{K_x^2}{2m^{**}_{x,y}} + \frac{K_y^2}{2m^{**}_{y,x}} \tag{5.22}$$

with $K_{x,y}$ taken relative to the band bottom positions and $m^{**}_x = 1/t$, $m^{**}_y = 4m^{**}_x$.

5.4.2 In-plane bipolarons

Here we consider a two-dimensional lattice of ideal octahedra that can be regarded as a simplified model of the copper–oxygen perovskite layer as shown in figure 5.4. The lattice period is $a = 1$ and the distance between the apical sites and the central plane is $h = a/2 = 0.5$. For mathematical transparency, we assume that all in-plane atoms, both copper and oxygen, are static but apical oxygens are independent three-dimensional isotropic harmonic oscillators, so that the model Hamilonian of section 4.6.3 is applied. Due to poor screening, the hole–apical interaction is purely Coulombic:

$$g_\nu(\boldsymbol{m} - \boldsymbol{n}) = \frac{\kappa_\nu}{|\boldsymbol{m} - \boldsymbol{n}|^2}$$

where $\nu = x, y, z$. To account for the experimental fact that z-polarized phonons couple to holes stronger than others [172], we choose $\kappa_x = \kappa_y = \kappa_z/\sqrt{2}$. The direct hole–hole repulsion is

$$V_c(\boldsymbol{n} - \boldsymbol{n}') = \frac{V_c}{\sqrt{2}|\boldsymbol{n} - \boldsymbol{n}'|}$$

so that the repulsion between two holes in the nearest-neighbour (NN) configuration is V_c. We also include the bare NN hopping T_{NN}, the next nearest neighbour (NNN) hopping across copper $T_{\rm NNN}$ and the NNN hopping between the pyramids $T'_{\rm NNN}$. The polaron shift is given by the lattice sum (4.148) which, after summation over polarizations, yields

$$E_p = 2\kappa_x^2\omega_0 \sum_m \left(\frac{1}{|\boldsymbol{m}-\boldsymbol{n}|^4} + \frac{h^2}{|\boldsymbol{m}-\boldsymbol{n}|^6} \right) = 31.15\kappa_x^2\omega_0 \tag{5.23}$$

where the factor two accounts for two layers of apical sites. For reference, the Cartesian coordinates are $\boldsymbol{n} = (n_x + \frac{1}{2}, n_y + \frac{1}{2}, 0)$, $\boldsymbol{m} = (m_x, m_y, h)$ and n_x, n_y, m_x, m_y are integers. The polaron–polaron attraction is

$$V_{\rm ph}(\boldsymbol{n}-\boldsymbol{n}') = 4\omega_0\kappa_x^2 \sum_m \frac{h^2 + (\boldsymbol{m}-\boldsymbol{n}')\cdot(\boldsymbol{m}-\boldsymbol{n})}{|\boldsymbol{m}-\boldsymbol{n}'|^3|\boldsymbol{m}-\boldsymbol{n}|^3}. \tag{5.24}$$

Performing the lattice summations for the NN, NNN and NNN′ configurations one finds $V_{\rm ph} = 1.23E_p$, $0.80E_p$ and $0.82E_p$, respectively. As a result, we obtain a net inter-polaron interaction as $v_{\rm NN} = V_{\rm c} - 1.23E_p$, $v_{\rm NNN} = \frac{V_{\rm c}}{\sqrt{2}} - 0.80E_p$, $v'_{\rm NNN} = \frac{V_{\rm c}}{\sqrt{2}} - 0.82E_p$ and the mass renormalization exponents as $g^2_{\rm NN} = 0.38(E_p/\omega_0)$, $g^2_{\rm NNN} = 0.60(E_p/\omega_0)$ and $(g'_{\rm NNN})^2 = 0.59(E_p/\omega_0)$.

Let us now discuss different regimes of the model. At $V_{\rm c} > 1.23E_p$, no bipolarons are formed and the system is a polaronic Fermi liquid. The polarons tunnel in the *square* lattice with NN hopping $t = T_{\rm NN}\exp(-0.38E_p/\omega_0)$ and NNN hopping $t' = T_{\rm NNN}\exp(-0.60E_p/\omega_0)$. (Since $g^2_{\rm NNN} \approx (g'_{\rm NNN})^2$ one can neglect the difference between NNN hoppings within and between the octahedra.) A single-polaron spectrum is, therefore,

$$E_1(\boldsymbol{k}) = -E_p - 2t'[\cos k_x + \cos k_y] - 4t\cos(k_x/2)\cos(k_y/2). \tag{5.25}$$

The polaron mass is $m^* = 1/(t + 2t')$. Since in general $t > t'$, the mass is mostly determined by the NN hopping amplitude t.

If $V_{\rm c} < 1.23E_p$ then inter-site NN bipolarons form. The bipolarons tunnel in the plane via four resonating (degenerate) configurations A, B, C and D, as shown in figure 5.5.

In the first order in the renormalized hopping integral, one should retain only these lowest-energy configurations and discard all the processes that involve configurations with higher energies. The result of such a projection is a bipolaron Hamiltonian:

$$\begin{aligned} H_{\rm b} = {} & (V_{\rm c} - 3.23E_p)\sum_l [A_l^\dagger A_l + B_l^\dagger B_l + C_l^\dagger C_l + D_l^\dagger D_l] \\ & - t'\sum_l [A_l^\dagger B_l + B_l^\dagger C_l + C_l^\dagger D_l + D_l^\dagger A_l + H.c.] \\ & - t'\sum_n [A_{l-x}^\dagger B_l + B_{l+y}^\dagger C_l + C_{l+x}^\dagger D_l + D_{l-y}^\dagger A_l + H.c.] \end{aligned} \tag{5.26}$$

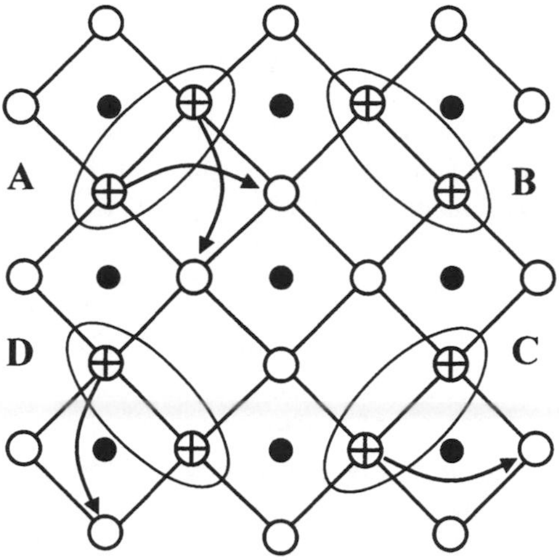

Figure 5.5. Four degenerate in-plane bipolaron configurations A, B, C and D. Some single-polaron hoppings are indicated by arrows.

where $\boldsymbol{l}$ numbers octahedra rather than individual sites, $\boldsymbol{x} = (1, 0)$ and $\boldsymbol{y} = (0, 1)$. A Fourier transformation and diagonalization of a 4×4 matrix yields the bipolaron spectrum:

$$E_2(\boldsymbol{k}) = V_c - 3.23E_p \pm 2t'[\cos(k_x/2) \pm \cos(k_y/2)]. \tag{5.27}$$

There are four bipolaronic sub-bands combined in the band of width $8t'$. The effective mass of the lowest band is $m^{**} = 2/t'$. The bipolaron binding energy is $\Delta \approx 1.23E_p - V_c$. The bipolaron already moves in the *first* order of polaron hopping. This remarkable property is entirely due to the strong on-site repulsion and long-range electron–phonon interaction that leads to a non-trivial connectivity of the lattice. This fact combines with a weak renormalization of t' yielding a *superlight* bipolaron with the mass $m^{**} \propto \exp(0.60E_p/\omega)$. We recall that in the Holstein model $m^{**} \propto \exp(2E_p/\omega)$ (section 4.6.1). Thus the mass of the Fröhlich bipolaron in the perovskite layer scales approximately as a *cubic root* of that of the Holstein one.

With an even stronger e–ph interaction, $V_c < 1.16E_p$, NNN bipolarons become stable. More importantly, holes can now form three- and four-particle clusters. The dominance of the potential energy over kinetic in the transformed Hamiltonian enables us to readily investigate these many-polaron cases. Three holes placed within one oxygen square have four degenerate states with the energy $2(V_c - 1.23E_p) + V_c/\sqrt{2} - 0.80E_p$. The first-order polaron hopping processes mix the states resulting in a ground-state linear combination with the energy $E_3 = 2.71V_c - 3.26E_p - \sqrt{4t^2 + t'^2}$. It is essential that between the squares such triads could move only in higher orders of polaron hopping. In the first order, they are immobile. A cluster of four holes has only one state within a square of oxygen atoms. Its energy is $E_4 = 4(V_c - 1.23E_p) + 2(V_c/\sqrt{2} - 0.80E_p) = 5.41V_c - 6.52E_p$. This cluster, as well as all bigger ones, are also immobile in first-order polaron hopping. We would like to stress that at distances much larger than the lattice constant the polaron–polaron interaction is always repulsive (section 4.4) and the formation of infinite clusters, stripes or strings is strictly prohibited (section 8.6). We conclude that at $V_c < 1.16E_p$, the system quickly becomes a charge segregated insulator.

The fact that within the window, $1.16E_p < V_c < 1.23E_p$, there are no three or more polaron bound states means that bipolarons repel each other. The system is effectively a charged Bose gas, which is a superconductor (section 4.7). The superconductivity window, that we have found, is quite narrow. This indicates that the superconducting state in cuprates requires a rather fine balance between electronic and ionic interactions.

5.5 Bipolaron model of cuprates

The considerations set out in sections 5.2, 5.3 and 5.4 leads us to a simple model of cuprates [179]. The main assumption is that *all electrons* are bound into small singlet and triplet *inter-site* bipolarons stabilized by e–ph and spin–fluctuation interactions. As the undoped plane has a half-filled $Cu3d^9$ band, there is no space for bipolarons to move if they are inter-site. Their Brillouin zone is half the original electron one and is completely filled with hard-core bosons. *Hole* pairs, which appear with doping, have enough space to move, and they are responsible for low-energy kinetics. Above T_c a material such as $YBa_2Cu_3O_{6+x}$ contains a non-degenerate gas of hole bipolarons in singlet and triplet states. Triplets are separated from singlets by a spin-gap J and have a lower mass due to a lower binding energy (figure 5.6). The main part of the electron–electron correlation energy (Hubbard U and inter-site Coulomb repulsion) and the electron–phonon interaction are taken into account in the binding energy of bipolarons Δ, and in their band-width renormalization as described in chapter 4. When the carrier density is small ($n_b \ll 1$ (as in cuprates)), bipolaronic operators are almost bosonic (section 4.7.1). The hard-core interaction does not play any role in this dilute limit, so only the Coulomb repulsion is relevant. This repulsion is significantly reduced due to a large static dielectric constant in oxides ($\epsilon_0 \gg 1$). Hence, carriers are almost-free charged bosons and thermally excited non-degenerate fermions, so that the canonical Boltzmann kinetics (section 1.1) and the Bogoliubov excitations of the charged Bose gas (section 4.7.3) are perfectly applied in the normal and superconducting states, respectively.

The population of singlet, n_s, triplet n_t and polaron, n_p bands is determined by the chemical potential $\mu \equiv T \ln y$, where y is found using the thermal equilibrium of singlet and triplet bipolarons and polarons:

$$2n_s + 2n_t + n_p = x. \tag{5.28}$$

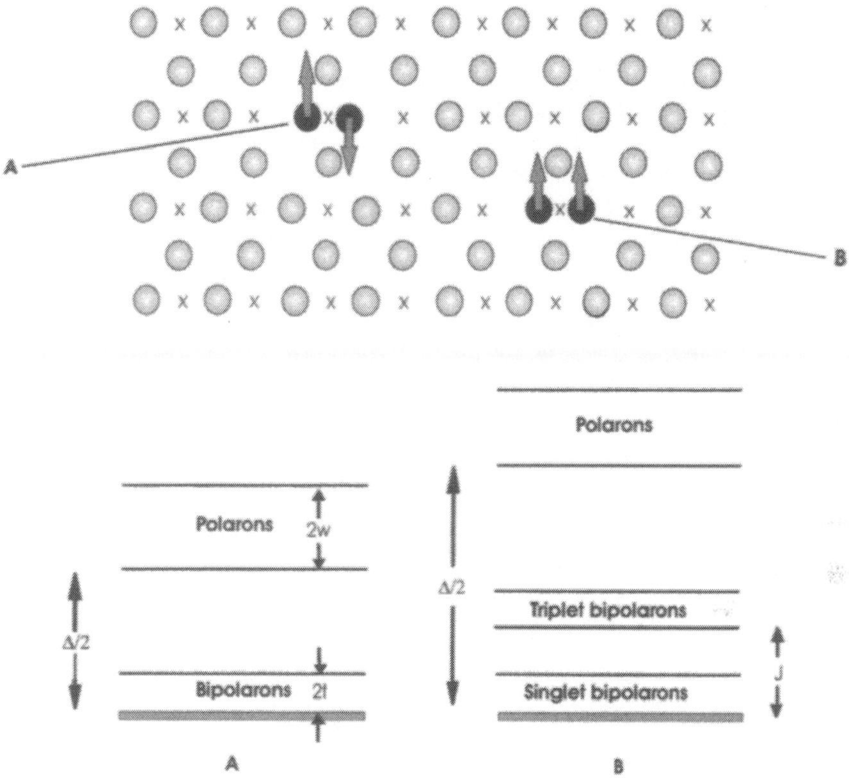

Figure 5.6. Bipolaron picture of high-temperature superconductors. A indicates degenerate singlet and triplet inter-site bipolarons. B shows non-degenerate singlet and triplet inter-site bipolaron, which naturally includes the addition of an extra excitation band. The crosses are copper sites and the circles are oxygen sites. w is the half-bandwidth of the polaron band, t is the half-bandwidth of the bipolaronic bands, $\Delta/2$ is the bipolaron binding energy per polaron and J is the exchange energy per bipolaron.

Applying the effective-mass approximation for quasi-two-dimensional energy spectra of all particles, we obtain for $0 \leqslant y < 1$:

$$-m_s^{**}\ln(1-y) - 3m_t^{**}\ln(1-y\mathrm{e}^{-J/T}) + m^*\ln(1+y^{1/2}\mathrm{e}^{-\Delta/(2T)}) = \frac{\pi x}{T} \tag{5.29}$$

in the normal state, and $y = 1$ in the superconducting state. Here x is the total number of holes per unit area. If the polaron energy spectrum is (quasi-)one-dimensional (as in section 6.5), an additional $T^{-1/2}$ appears in front of the logarithm in the third term on the left-hand side of equation (5.29).

We should also take into account localization of holes by a random potential, because doping inevitably introduces some disorder. The Coulomb repulsion restricts the number of charged bosons in each localized state, so that the distribution function will show a mobility edge E_c [194]. The number of bosons in a single potential well is determined by the competition between their long-range Coulomb repulsion ($\propto 4e^2/\xi$) and the binding energy, $E_c - \epsilon$. If the localization length diverges with the critical exponent $\nu < 1$ ($\xi \sim (E_c - \epsilon)^{-\nu}$), we can apply a 'single-well–single-particle' approximation assuming that there is only one boson in each potential well (see also section 8.1). Within this approximation *localized charged* bosons obey the Fermi–Dirac statistics, so that their density is given by

$$n_L(T) = \int_{-\infty}^{E_c} \frac{N_L(E)\,dE}{y^{-1}\exp(E/T)+1} \tag{5.30}$$

where $N_L(E)$ is the density of localized states. Near the mobility edge, it remains constant $N_L(E) \propto n_L/\gamma$, where γ is of the order of the binding energy in a single random potential well, and n_L is the number of localized states per unit area. The number of *empty* localized states turns out to be linear as a function of temperature in a wide temperature range $T \lesssim \gamma$ from equation (5.30). Then the conservation of the total number of carriers yields for the chemical potential:

$$\begin{aligned}\frac{\pi(x-2n_L)}{T} &= -m_s^{**}\ln(1-y) - 3m_t^{**}\ln(1-ye^{-J/T}) \\ &\quad + m^*\ln(1+y^{1/2}e^{-\Delta/(2T)}) - \frac{2\pi n_L}{\gamma}\ln(1+y^{-1}).\end{aligned} \tag{5.31}$$

If the number of localized states is about the same as the number of pairs ($n_L \approx x/2$), a solution of this equation does not depend on temperature in a wide temperature range above T_c. With y to be a constant ($y \approx 0.6$ in a wide range of parameters in equation (5.31)), the number of singlet bipolarons in the Bloch states is linear in temperature:

$$n_s(T) = (x/2 - n_L) + T\frac{n_L}{\gamma}\ln(1+y^{-1}). \tag{5.32}$$

The numbers of triplets and polarons are exponentially small at low temperatures, $T \ll J, \Delta$.

The model suggests a phase diagram of the cuprates as shown in figure 5.7. This phase diagram is based on the assumption that to account for the high values of T_c in cuprates, one has to consider electron–phonon interactions larger than those used in the intermediate-coupling theory of superconductivity (chapter 3). Regardless of the adiabatic ratio, the Migdal–Eliashberg theory of superconductivity and the Fermi-liquid theory break at $\lambda \approx 1$ (section 4.2). A many-electron system collapses into a small (bi)polaron regime at $\lambda \geq 1$ with well-separated vibration and charge-carrier degrees of freedom. Even though it seems that these carriers should have a mass too large to be mobile, the inclusion of the on-site Coulomb repulsion and a poor screening of the long-range electron–phonon interaction leads to *mobile* inter-site bipolarons (section 4.6).

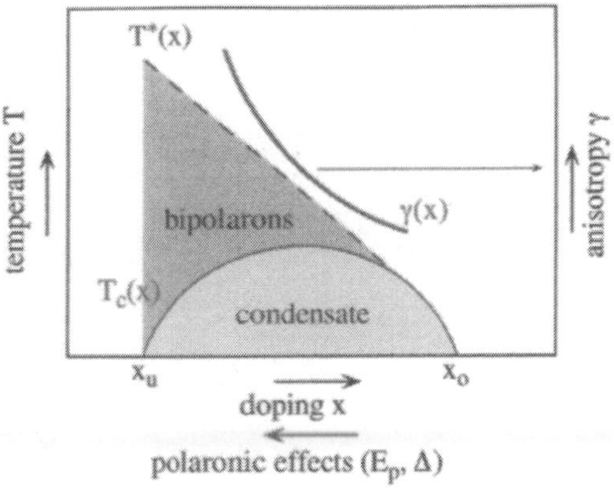

Figure 5.7. Phase diagram of superconducting cuprates in the bipolaron theory (courtesy of J Hofer).

Above T_c, the Bose gas of these bipolarons is non-degenerate and below T_c, their phase coherence sets in and hence superfluidity of the double-charged $2e$ bosons occurs. Of course, there are also thermally excited single polarons in the model (figure 5.6).

There is much evidence for the crossover regime at T^* and normal-state charge and spin gaps in cuprates (chapter 6). These energy gaps could be understood as half of the binding energy Δ and the singlet–triplet gap of preformed bipolarons, respectively [185] and T^* is a temperature, where the polaron density compares with the bipolaron one. Further evidence for bipolarons comes from a parameter-free estimate of the renormalized Fermi energy ϵ_F, which yields a very small value as discussed in section 5.3. There might be a crossover from the Bose–Einstein condensation to a BCS-like *polaronic* superconductivity (section 4.5) across the phase diagram [179]. If the Fermi liquid does exist at overdoping, then it is likely that the heavy doping causes an 'overcrowding effect' where polarons find it difficult to form bipolarons due to a larger number of competing holes. Many experimental observations can be explained using the bipolaron model (chapters 6, 7 and 8).

Chapter 6

Normal state of cuprates

6.1 In-plane resistivity and Hall ratio

Thermally excited phonons and (bi)polarons are well decoupled in the strong-coupling regime of the electron–phonon interaction (chapter 4), so that the standard Boltzmann equation of section 1.1 for renormalized carriers is applied. We make use of the τ approximation [193] in an electric field $\boldsymbol{E} = \nabla\phi$, a temperature gradient ∇T and in a magnetic field $\boldsymbol{B} \perp \boldsymbol{E}, \nabla T$. Bipolaron and single-polaron non-equilibrium distributions are found to be

$$f(\boldsymbol{k}) = f_0(E) + \tau \frac{\partial f_0}{\partial E} \boldsymbol{v} \cdot \{\boldsymbol{F} + \Theta \boldsymbol{n} \times \boldsymbol{F}\} \tag{6.1}$$

where

$$\boldsymbol{F} = (E - \mu)\nabla T/T + \nabla(\mu - 2e\phi)$$
$$f_0(E) = [y^{-1}\exp(E/T) - 1]^{-1}$$

for bipolarons with the energy $E = k^2/(2m^{**})$ and with the Hall angle $\Theta = \Theta_b = 2eB\tau_b/m^{**}$ and

$$\boldsymbol{F} = (E + \Delta/2 - \mu/2)\nabla T/T + \nabla(\mu/2 - e\phi)$$
$$f_0(E) = \{y^{-1/2}\exp[(E + \Delta/2)/T] + 1\}^{-1}$$

$E = k^2/(2m^*)$ and $\Theta = \Theta_p = eB\tau_p/m^*$ for thermally excited polarons. Here $\boldsymbol{n} = \boldsymbol{B}/B$ is a unit vector in the direction of the magnetic field. The Hall angles are assumed to be small ($\Theta \ll 1$), because the polarons and bipolarons are heavy. Equation (6.1) is used to calculate the electrical and thermal currents induced by the applied thermal and potential gradients:

$$j_\alpha = a_{\alpha\beta}\nabla_\beta(\mu - 2e\phi) + b_{\alpha\beta}\nabla_\beta T \tag{6.2}$$
$$w_\alpha = c_{\alpha\beta}\nabla_\beta(\mu - 2e\phi) + d_{\alpha\beta}\nabla_\beta T. \tag{6.3}$$

Equation (6.2) defines the current with the polaronic conductivity $\sigma_p = e^2\tau_p n_p/m^*$, where kinetic coeffficents are given by

$$a_{xx} = a_{yy} = \frac{1}{2e}\sigma_p(1+4A) \tag{6.4}$$
$$a_{yx} = -a_{xy} = \frac{1}{2e}\sigma_p(\Theta_p + 4A\Theta_\mathrm{b})$$
$$b_{xx} = b_{yy} = \frac{\sigma_p}{e}[\Gamma_p + \frac{\Delta-\mu}{2T} + 2A(\Gamma_\mathrm{b} - \mu/T)]$$
$$b_{yx} = -b_{xy} = \frac{\sigma_p}{e}\left[\Theta_p\left(\Gamma_p + \frac{\Delta-\mu}{2T}\right) + 2A\Theta_\mathrm{b}(\Gamma_\mathrm{b} - \mu/T)\right].$$

Equation (6.3) defines the heat flow with coefficients given by

$$c_{xx} = c_{yy} = \frac{\sigma_p}{2e^2}[T\Gamma_p + \Delta/2 + e\phi + 2A(T\Gamma_\mathrm{b} + 2e\phi)] \tag{6.5}$$
$$c_{yx} = -c_{xy} = \frac{\sigma_p}{2e^2}[\Theta_p(T\Gamma_p + \Delta/2 + e\phi) + 2A\Theta_\mathrm{b}(T\Gamma_\mathrm{b} + 2e\phi)]$$
$$d_{xx} = d_{yy} = \frac{\sigma_p}{e^2}\left\{T\gamma_p + \Gamma_p(\Delta - \mu/2 + e\phi) + (\Delta/2 + e\phi)\frac{\Delta-\mu}{2T}\right.$$
$$\left. + A[T\gamma_\mathrm{b} + \Gamma_\mathrm{b}(2e\phi - \mu) - 2e\phi\mu/T]\right\}$$
$$d_{yx} = -d_{xy} = \frac{\sigma_p}{e^2}\left\{\Theta_p\left[T\gamma_p + \Gamma_p(\Delta - \mu/2 + e\phi) + (\Delta/2 + e\phi)\frac{\Delta-\mu}{2T}\right]\right.$$
$$\left. + A\Theta_\mathrm{b}[T\gamma_\mathrm{b} + \Gamma_\mathrm{b}(2e\phi - \mu) - 2e\phi\mu/T)]\right\}.$$

Here

$$\Gamma = \frac{\int_0^\infty \mathrm{d}E\, E^2 \partial f_0/\partial E}{T\int_0^\infty \mathrm{d}E\, E\partial f_0/\partial E} = \frac{2\Phi(z,2,1)}{\Phi(z,1,1)} \tag{6.6}$$

and

$$\gamma = \frac{\int_0^\infty \mathrm{d}E\, E^3 \partial f_0/\partial E}{T^2\int_0^\infty \mathrm{d}E\, E\partial f_0/\partial E} = \frac{6\Phi(z,3,1)}{\Phi(z,1,1)}$$

are numerical coefficients, expressed in terms of the Lerch transcendent

$$\Phi(z,s,a) = \sum_{k=0}^{\infty} \frac{z^k}{(a+k)^s}$$

with $z = y$ in Γ_b, γ_b and $z = -y^{1/2}\exp[-\Delta/(2T)]$ in Γ_p, γ_p,

$$A = \frac{m^*\tau_\mathrm{b} n_\mathrm{b}}{m^{**}\tau_p n_p} \tag{6.7}$$

is the ratio of bipolaron and polaron contributions to the transport, which strongly depends on temperature. Here, for simplicity, we neglect the spin gap, which is small in optimally doped cuprates [195]. Then the bipolaron singlet and triplet states are nearly degenerate, so that bipolaron and polaron densities are expressed as

$$n_{\mathrm{b}} = \frac{2m^{**}T}{\pi}|\ln[1-y]| \tag{6.8}$$

$$n_p = \frac{m^*T}{\pi}\ln\left[1+y^{1/2}\exp\left(-\frac{\Delta}{2T}\right)\right]. \tag{6.9}$$

Using the kinetic coefficients (equations (6.4) and (6.5)), we obtain

$$\rho = \frac{1}{\sigma_p(1+4A)} \tag{6.10}$$

and

$$R_{\mathrm{H}} = \frac{1+4A\Theta_{\mathrm{b}}/\Theta_p}{en_p(1+4A)^2} \tag{6.11}$$

for the in-plane resistivity and the Hall ratio, respectively.

According to [194], the main scattering channel above T_{c} is due to particle–particle collisions with the relaxation time $\tau_{\mathrm{b,p}} \propto 1/T^2$. In case of the Bose–Einstein statistics, umklapp scattering can be neglected, so the scattering between bosons in extended states does not contribute to the resistivity in a clean sample. However, the inelastic scattering of an extended boson by bosons localized in the random potential and the scattering of extended bosons by each other make a contribution because the momentum is not conserved in two-particle collisions in the presence of the impurity potential. In the framework of the 'single-well–single-particle approximation', the repulsion between localized bosons plays the same role as the Pauli exclusion principle in fermionic systems. Then the relaxation rate is proportional to the temperature squared because only bosons within the energy shell of the order of T near the mobility edge contribute to the scattering and because the number of final states is proportional to T in a non-degenerate gas. The chemical potential is pinned near the mobility edge, so that $y \approx 0.6$ in a wide temperature range, if the number of localized states in the random potential is about the same as the number of bipolarons (section 5.5). This could be the case in $YBa_2Cu_3O_{6+x}$, where every excess oxygen ion x can localize the bipolaron. Hence, neglecting the polaron contribution to the in-plane transport which is exponentially small below T^*, we find that

$$\rho \propto T \tag{6.12}$$

and

$$R_{\mathrm{H}} \propto \frac{1}{T} \tag{6.13}$$

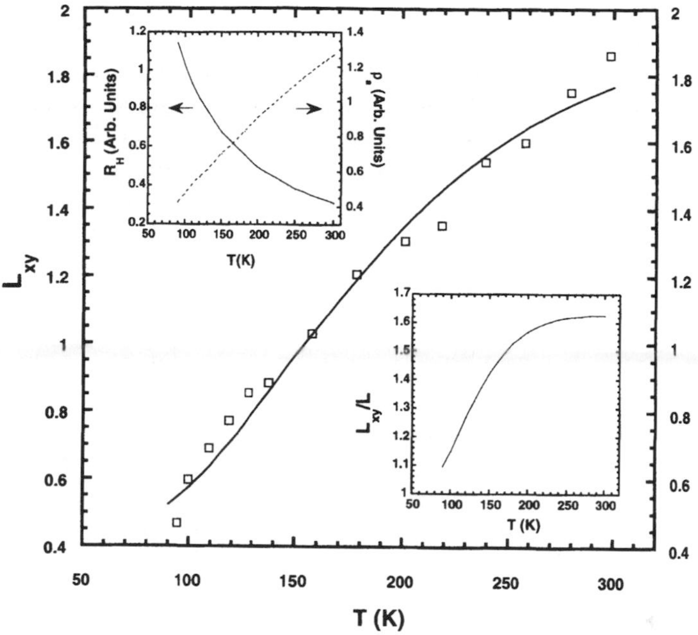

Figure 6.1. In-plane resistivity and the Hall ratio of bipolarons and thermally excited polarons for $\Delta/2 = 600$ K and $\Theta_b/\Theta_p = 2m^*\tau_b/(m^{**}\tau_p) = 0.44$ (inset), and the Hall Lorenz number as a function of temperature compared with the experimental data by Zhang *et al* in the optimally doped $YBa_2Cu_3O_{6.95}$ [196]. The second inset shows the ratio of the Hall Lorenz number to the Lorenz number.

as observed in many cuprates. Taking into account the polaron contribution in equations (6.10) and (6.11) does not change this result, if the binding energy is sufficiently large, as shown in figure 6.1. One can also take into account the transport relaxation rate due to a two-dimensional boson–acoustic phonon scattering,

$$\frac{1}{\tau_{b-ac}} \propto T \tag{6.14}$$

and the elastic relaxation rate due to the scattering by unoccupied impurity wells with their density proportional to temperature,

$$\frac{1}{\tau_{b-im}} \propto T. \tag{6.15}$$

Hence, neglecting the polaron contribution, we finally obtain [194]

$$R_H \approx \frac{1}{2e(x/2 - n_L + bn_L T)} \tag{6.16}$$

$$\rho \approx C \frac{T + \sigma_b T^2}{x/2 - n_L + b n_L T} \tag{6.17}$$

where σ_b is the relative boson–boson scattering cross sections, C is a temperature-independent constant, proportional to the carrier mass squared and b is about $1/\gamma$, where γ was introduced in section 5.5. These expressions contain important information about the number of bosons, localized states and the relative strength of the different scattering channels. They fit well the experimental data in a number of cuprates [194, 197]. Importantly, they predict the Hall ratio and resistivity scaling with doping as $1/x$ because $n_L \propto x$. This scaling was experimentally observed in underdoped and optimally doped cuprates [198]. There is a characteristic change in the linear slope of $\rho(T)$ (see, for example, [199]) below the temperature, where the spin gap is observed in NMR (section 6.4). This change was explained by Mott [200]: the spin gap is the singlet–triplet separation energy in the bipolaron model (figure 5.6). Because triplets are lighter than singlets, they lead to a smaller C in equation (6.17) and to a smaller temperature gradient of the in-plane resistivity at temperatures above J, where their density becomes comparable with the singlet density.

6.2 Normal-state resistivity below T_c

The scaling of the in-plane resistivity and of the Hall ratio with doping tells us that cuprates are doped semiconductors, which might be metallic Fermi liquids. But the linear (in temperature) resistivity and the temperature-dependent Hall ratio, discussed earlier, are only a tiny part of the non-Fermi-liquid characteristics of cuprates. Hence, one can ask a fundamental question: 'What are high-T_c cuprates in their normal state—metals or something else?'. The answer to this vexing question clearly depends on how precisely one can define a metal. To arrive at one single definition, one would require a measurement of the normal-state conductivity down to very low temperatures. Thus, metals show a non-zero low-temperature conductivity, while the electrical conductivity is zero in non-metals as $T \to 0$ K. On the basis of this simple—but potent—definition, of course, any superconductor must clearly be a metal, since the conductivity at $T = 0$ K is infinite. However, one can suppress the superconducting state by applying a large external magnetic field to measure a 'true' normal-state conductivity down to a very low temperature well below a zero field T_c. This important experiment has been carried out by Boebinger *et al* [201] by applying an ultra-high magnetic field (up to 60 T) to destroy the superconducting state. This then allows one to scrutinize the nature of the electronic structure of cuprates without the 'complication' of superconductivity at elevated temperatures. Remarkably, many high-T_c cuprates studied to date by this technique appear to be doped *three*-dimensional anisotropic insulators, with the anticipated divergent resistivity and temperature-independent anisotropy at low temperatures. These observations have been explained [202] as a result of the resonance scattering of bipolarons by lattice defects.

As we discussed in the previous section, there is no need to abandon the Boltzmann kinetics to explain the linear in-plane resistivity and the temperature-dependent Hall effect above T_c in cuprates in the framework of the bipolaron theory. A fraction of bipolarons is localized by disorder, so that the number of *delocalized* carriers is proportional to T at sufficiently high temperatures, while the boson–boson inelastic scattering rate is proportional to T^2. This allows us to explain that both the in-plane resistivity and the inverse Hall ratio are proportional to T. Let us now extend the theory towards low temperatures, where the transport relaxation time of bipolarons is determined by elastic boson–impurity scattering and single polarons are frozen out.

The number of extended bosons $n_b(T)$ above the mobility edge is determined in the 'single-well–single-particle' approximation (equation (5.30)) as

$$n_b(T) = \tfrac{1}{2}x - n_L(T) \tag{6.18}$$

where $\frac{1}{2}x$ is the total number of pairs and $n_L(T) \simeq n_L - N_L(0)T$ is the number of bosons localized by the random potential with $N_L(0)$ the density of localized states near the mobility edge. The Hall ratio, R_H, measures the inverse carrier density, so that

$$\frac{R_H(T)}{R_H(0)} = \frac{1}{1 + T/T_L} \tag{6.19}$$

where $T_L = (x - 2n_L)/2N_L(0)$. This simple expression fits the Hall ratio in $La_{2-x}Sr_xCuO_4$ at optimum doping ($x = 0.15$) (figure 6.2) with $T_L = 234$ K and the constant $R_H(0) = 2 \times 10^{-3}$ cm^3 C^{-1}. If the total number of carriers $\frac{1}{2}x$ is larger than the number of the potential wells n_L, the carrier density is practically temperature independent at low temperatures.

The characteristic kinetic energy of non-degenerate bipolarons in the normal state is of the order of the temperature rather than the Fermi energy in metals and in conventional semiconductors. The most effective scattering at low temperatures is then caused by attractive shallow potential wells. For slow particles, this is described by the familiar Wigner resonance cross section as

$$\sigma(E) = \frac{2\pi}{m^{**}} \frac{1}{E + |\epsilon|} \tag{6.20}$$

where

$$\epsilon = -\frac{\pi^2}{16} U_{\min} \left(\frac{U}{U_{\min}} - 1 \right)^2 \tag{6.21}$$

is the energy of a shallow virtual ($U < U_{\min}$) or real ($U > U_{\min}$) localized level, U is the well depth and $U_{\min} = \pi^2/8m^{**}a^2$ with the well size a. The transport relaxation rate is the sum of the scattering cross sections from different potential wells within the unit volume multiplied by the velocity $v = \sqrt{2E/m^{**}}$. There is a wide distribution of potential wells with respect to both U and a in cuprates. Therefore, one has to integrate the Wigner cross section (equation (6.20)) over U

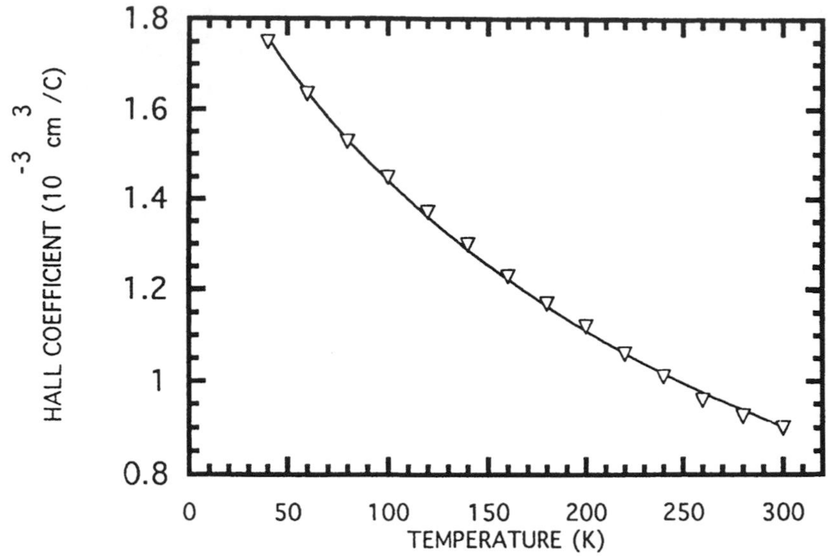

Figure 6.2. The Hall ratio in $La_{1.85}Sr_{0.15}CuO_4$ (triangles [198]) described by equation (6.19).

and over a. Performing the integration over U, we take into account only shallow wells with $U < U_{\min}$ because the deeper wells are occupied by localized carriers and cannot yield a resonant scattering. The result is:

$$\langle\sigma(E)\rangle \equiv \gamma^{-1}\int_0^{U_{\min}} \sigma(E)\,\mathrm{d}U = \frac{4\pi}{m^{**}\gamma a\sqrt{2m^{**}E}}\tan^{-1}\left(\frac{\pi^2}{8a\sqrt{2m^{**}E}}\right) \tag{6.22}$$

where the width of the U-distribution, γ, is supposed to be large compared with $U_{\min}$. Integrating over the size a, one has to take into account the fact that the Wigner formula (equation (6.20)) is applied only to slow particles with $a \leq \hbar/\sqrt{2m^{**}E}$. However, because the U-averaged cross section (equation (6.22)) behaves like $1/a^2$ at a large a, we can extend the integration over a up to infinity. As a result, we obtain the inverse mean free path:

$$l^{-1}(E) = \frac{n_{\mathrm{L}}}{A}\int_{a_{\min}}^{\infty} \langle\sigma(E)\rangle\,\mathrm{d}a \simeq \frac{\pi^2 N_{\mathrm{L}}(0)}{m^{**}A\sqrt{2m^{**}E}} \ln\frac{E_0}{E} \tag{6.23}$$

for $E \ll E_0$. Here A is the width of the size distribution of the random potential wells, $N_{\mathrm{L}}(0) \simeq n_{\mathrm{L}}/\gamma$, $E_0 = \pi^4/128m^{**}a_{\min}^2$ and $a_{\min}$ is the minimum size. We expect a very large value of A of the order of 10 Å or more due to the twin boundaries and impurity clusters, which are not screened. However, single impurities are screened. A simple estimate of the screening radius by the use

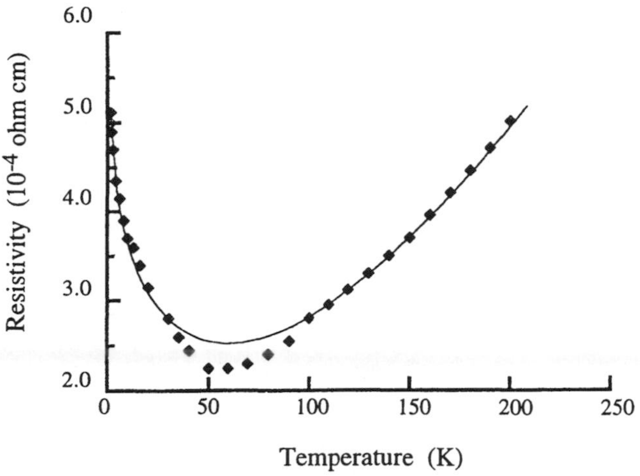

Figure 6.3. In-plane resistivity of $La_{1.87}Sr_{0.13}CuO_4$ (diamonds [201]) described by equation (6.26).

of the classical expression, $r_D = \sqrt{T\epsilon_0/(16\pi n_b e^2)}$ yields a value of a_{min} of the order of the interatomic spacing (~2 Å), which corresponds to quite a large $E_0 = 1500$ K if $m^{**} = 10m_e$. Hence, at low temperatures, we arrive at the logarithmic transport relaxation rate as

$$\frac{1}{\tau} \equiv vl^{-1}(E) = \frac{1}{\tau_0} \ln \frac{E_0}{E} \tag{6.24}$$

where $\tau_0 = m^{**2}A/\pi^2 N_L(0)$ is a constant.

The low-temperature resistivity is now derived by the use of the Boltzmann theory:

$$\rho(T) = \rho_0 \ln \frac{E_0}{T} \tag{6.25}$$

where $\rho_0^{-1} = 2(x - 2n_L)e^2\tau_0/m^{**}$. Combining both elastic (equation (6.25)) and inelastic (section 6.1) scattering rates and taking into account the temperature dependence of the extended boson density $n_b(T)$ (equation (6.18)), we find that

$$\frac{\rho(T)}{\rho_0} = \frac{\ln(E_0/T) + (T/T_b)^2}{1 + T/T_L} \tag{6.26}$$

with a constant T_b. The full curve in figure 6.3 is a fit to the experimental data with $\rho_0 = 7.2 \times 10^{-5}$ Ω cm, $E_0 = 1900$ K and $T_b = 62$ K which appears to be remarkably good.

The value of E_0 agrees well with the estimate. Because the bipolaron energy spectrum is three dimensional at low temperatures, there is no temperature dependence of the anisotropy ρ_c/ρ_{ab} at low T as observed.

6.3 Lorenz number: evidence for bipolarons

Kinetic evidence for $2e$ bipolarons in cuprates (sections 6.1 and 6.2) is strong but direct evidence that these materials contain a charged $2e$ Bose liquid in their normal state is highly desirable. Mott and the author [203] discussed the thermal conductivity κ. The contribution from the carriers, given by the Wiedemann–Franz ratio, depends strongly on the elementary charge as $\sim(e^*)^{-2}$ and should be significantly suppressed in the case of $e^* = 2e$ compared with the Fermi-liquid contribution. As a result, the Lorenz number L $(= (e/k_B)^2\kappa_e/(T\sigma))$ differs significantly from the Sommerfeld value L_e $(= \pi^2/3)$ of standard Fermi-liquid theory, if the carriers are bipolarons. Here κ_e, σ and e are the electronic thermal conductivity, the electrical conductivity, and the elementary charge, respectively. Reference [203] predicted a very low Lorenz number L_b of bipolarons—$L_b = 6L_e/(4\pi^2) \approx 0.15L_e$—due to the double charge of carriers and also due to their nearly classical distribution function above T_c.

Unfortunately, the extraction of the electron thermal conductivity has proven difficult since both the electron term, κ_e, and the phonon term, κ_{ph}, are comparable to each other in cuprates. Some experiments have attemped to get around this problem in a variety of ways [204,205]. In particular, Takenaka *et al* [204] found that κ_e is constant or weakly T-dependent in the normal state of $YBa_2Cu_3O_{6+x}$. This approximately T-independent κ_e, therefore, implies the violation of the Wiedemann–Franz law (since resistivity is found to be a nonlinear function of temperature) in the underdoped region. The breakdown of the Wiedemann–Franz law has also been seen in other cuprates [206,207].

A new way to determine the Lorenz number has been realized by Zhang *et al* [196] based on the thermal Hall conductivity. The thermal Hall effect allowed for an efficient way to separate the phonon heat current even when it is dominant. As a result, the 'Hall' Lorenz number $(L_H \equiv L_{xy} = (e/k_B)^2\kappa_{yx}/(T\sigma_{yx}))$ has been directly measured in $YBa_2Cu_3O_{6.95}$ because the transverse thermal κ_{xy} and the electrical σ_{xy} conductivities involve only electrons. Remarkably, the value of L_H just above T_c was found to be about the same as that predicted by the bipolaron model $(L_H \approx 0.15L_e)$. However, the experimental L_H showed a strong temperature dependence which violates the Wiedemann–Franz law. This experimental observation is hard to explain in the framework of any Fermi-liquid model. Here we demonstrate that the Wiedemann–Franz law breaks down because of the interference of polaron and bipolaron contributions to the heat transport [209]. When thermally excited polarons are included, the bipolaron model explains the violation of the Wiedemann–Franz law in cuprates and the Hall Lorenz number as seen in the experiment.

There is no electric current ($\boldsymbol{j} = 0$) in the thermal conductivity measurements. This constraint allows us to express the electric and chemical potential gradients $\nabla(\mu - 2e\phi)$ via the temperature gradient ∇T using equations (6.4) and (6.5). Then the thermal conductivity, κ, and the thermal Hall conductivity, κ_{xy}, are found to be

$$\kappa = d_{xx} - \frac{c_{xx}b_{xx}}{d_{xx}} \tag{6.27}$$

$$\kappa_{yx} = d_{yx} - \frac{c_{yx}b_{xx}}{a_{xx}} + \frac{c_{xx}(a_{xx}b_{yx} - a_{yx}b_{xx})}{a_{xx}^2}$$

and the Lorenz and Hall Lorenz numbers to be

$$L = \frac{L_p + 4AL_{\mathrm{b}}}{1+4A} + \frac{A[2\Gamma_p - \Gamma_{\mathrm{b}} + \Delta/T]^2}{(1+4A)^2} \tag{6.28}$$

$$L_{\mathrm{H}} = \frac{L_p + 4AL_{\mathrm{b}}\Theta_{\mathrm{b}}/\Theta_p}{1+4A\Theta_{\mathrm{b}}/\Theta_p} + \frac{A(4A+\Theta_{\mathrm{b}}/\Theta_p)[2\Gamma_p - \Gamma_{\mathrm{b}} + \Delta/T]^2}{(1+4A)^2(1+4A\Theta_{\mathrm{b}}/\Theta_p)}. \tag{6.29}$$

Here $L_p = (\gamma_p - \Gamma_p^2)$ and $L_{\mathrm{b}} = (\gamma_{\mathrm{b}} - \Gamma_{\mathrm{b}}^2)/4$ are the polaron and bipolaron Lorenz numbers. In the limit of a purely polaronic system (i.e. $A = 0$), the Lorenz numbers (equations (6.28) and (6.29)) are $L = L_{\mathrm{H}} = L_p$. In the opposite limit of a purely bipolaronic system (i.e. $A = \infty$), we obtain a reduced Lorenz number [203]: $L = L_{\mathrm{H}} = L_{\mathrm{b}}$. However, in general, equations (6.28) and (6.29) yield temperature-dependent Lorenz numbers that differ significantly from both limits. The main difference originates in the second terms in the right-hand side of equations (6.28) and (6.29), which describe an interference between the polaron and bipolaron contributions in the heat flow. In the low-temperature regime ($T \ll \Delta$), this contribution is exponentially small because the number of unpaired polarons is small. However, it is enhanced by the factor $(\Delta/T)^2$ and becomes important in the intermediate-temperature range $T_{\mathrm{c}} < T < T^*$. The contribution appears as a result of the recombination of a pair of polarons into a bipolaronic bound state at the cold end of the sample, which is reminiscent of the contribution of the electron-hole pairs to the heat flow in semiconductors [193]. These terms are mainly responsible for the breakdown of the Wiedemann–Franz law in the bipolaronic system.

The bipolaron model, which fits the in-plane resistivity and the Hall ratio, also fits the Hall Lorenz number measured by Zhang *et al* [196]. To reduce the number of fitting parameters, we take the charge pseudogap $\Delta/2 = 600$ K, as found by Mihailovic *et al* [195] for nearly optimally doped $YBa_2Cu_3O_{6+x}$ in their systematic analysis of charge and spin spectroscopies. As discussed in section 6.1, the main scattering channel above T_{c} is due to particle–particle collisions with a relaxation time $\tau_{\mathrm{b,p}} \propto 1/T^2$. The chemical potential is pinned near the mobility edge, so that $y \approx 0.6$ in a wide temperature range, if the number of localized states in the random potential is about the same as the number of bipolarons. As a result, there is only one fitting parameter in L_{H} (equation (6.29)) which is the ratio of the bipolaron and polaron Hall angles $\Theta_{\mathrm{b}}/\Theta_p$. The model gives quite a good fit (figure 6.1) with a reasonable value of $\Theta_b/\Theta_p = 0.44$. It also describes well the (quasi-)linear in-plane resistivity and the inverse Hall ratio, as observed in cuprates (inset in figure 6.1). The Hall Lorenz number appears to be slightly larger then the Lorenz number (figure 6.1, lower inset). Because the thermal Hall conductivity directly measures the Lorenz number, it can be used to measure the lattice contribution to the heat flow as well. When we subtract the electronic contribution determined by using the Lorenz number in figure 6.1, the lattice contribution to the diagonal heat flow appears to be much higher than anticipated in the framework of any Fermi-liquid model.

6.4 Spin pseudogap in NMR

Pairing of holes into singlets well above T_c should be seen as a drop of magnetic susceptibility. Indeed, a rapid decrease in the uniform magnetic susceptibility and of the nuclear magnetic relaxation rate $1/T_1$ with temperature lowering is a common feature of the normal state of novel superconductors. Let us see how the bipolaron model describes the temperature dependence of $1/T_1$ [185].

The conventional contact hyperfine coupling of nuclear spin on a site i with electron spins is described by the following Hamiltonian in the site representation:

$$H_i = \hat{A}_i \sum_j c_{j\uparrow}^\dagger c_{j\downarrow} + H.c. \tag{6.30}$$

where $\hat{A}_i$ is an operator acting on the nuclear spin and j is its nearest-neighbour site. Performing transformations to polarons and bipolarons as described in chapter 4, we obtain the effective spin–flip interaction of triplet bipolarons with the nuclear spin:

$$H_i \propto \sum_{j,l\neq l'} b_{j,l}^\dagger b_{j,l'} + H.c. \tag{6.31}$$

Here $l, l' = 0, \pm 1$ are the z-components of spin $S = 1$. The NMR width due to the spin-flip scattering of triplet bipolarons on nuclei is obtained using the Fermi–Dirac golden rule as in section 2.8,

$$\frac{1}{T_1} = -\frac{B}{t^2} \int_0^{2t} \mathrm{d}E \, \frac{\partial f(E)}{\partial E} \tag{6.32}$$

where $f(E) = [\exp(E + J - \mu)/T -]^{-1}$ is the distribution function and $2t$ is the bandwidth of triplet bipolarons. For simplicity, we take their DOS as a constant $(= 1/(2t))$. As a result, we obtain

$$\frac{1}{T_1} = \frac{BT \sinh(t/T)}{t^2[\cosh[(t + J)/T - \ln y] - \cosh(t/T)]} \tag{6.33}$$

where B is a temperature-independent hyperfine coupling constant.

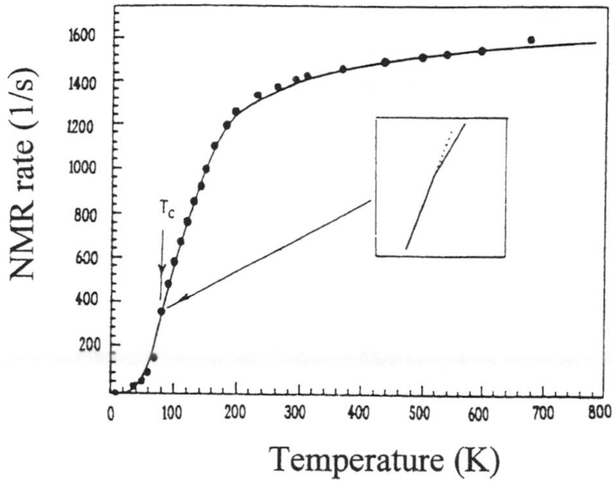

Figure 6.4. Temperature dependence of the nuclear spin relaxation rate $1/T_1$ [209] compared with the theory for $J = 150$ K and $t = 250$ K showing the absence of the Hebel–Slichter peak of Cu NMR in $YBa_2Cu_4O_8$.

Equation (6.33) describes all the essential features of the nuclear spin relaxation rate in copper-based oxides: the absence of the Hebel–Slichter coherent peak below T_c, the temperature-dependent Korringa ratio $(1/TT_1)$ above T_c and a large value of $1/T_1$ due to the small bandwidth $2t$. It fits the experimental data with reasonable values of the parameters, t and J (figure 6.4). NMR is almost unaffected by the superconducting transition in the bipolaron model, because there are no coherence factors (section 2.8) for triplets. There is only a tiny change in the slope in $1/T_1(T)$ near T_c due to a kink in the temperature dependence of the chemical potential $\mu(T)$ caused by the Bose–Einstein condensation of singlets (appendix B), as shown in the inset. A similar unusual behaviour of the NMR was found in underdoped $YBa_2Cu_3O_{6+x}$, and in many other cuprates. The Knight shift, which measures the spin susceptibility of carriers, also drops well above T_c in many copper oxides, in agreement with the bipolaron model. The value of the singlet–triplet bipolaron exchange energy J, determined from the fit to the experimental NMR width, is close to the 'spin' gap observed above and below T_c in $YBa_2Cu_3O_{6+x}$ with unpolarized [210], and polarized [211] neutron scattering.

6.5 *c*-axis transport and charge pseudogap

The bipolaron theory quantitatively explains the c-axis transport and anisotropy of cuprates [186, 197, 212, 213]. The crucial point is that polarons dominate in the c-axis transport at intermediate and high temperatures because they are much lighter than bipolarons in the c-direction. The latter can propagate across the planes only due to a simultaneous two-particle tunnelling, which is much less probable than a single-polaron tunnelling (chapter 4). Along the planes, polarons and inter-site bipolarons propagate with about the same effective mass. Hence, the polaron contribution to the in-plane transport is small at low temperatures due to their low density compared with the bipolaron density. As a result, we have a mixture of *non-degenerate* quasi-two-dimensional bosons and thermally excited fermions, which are capable of propagating along the c-axis. Polarons mainly contribute to c-axis transport, if the temperature is not very low, which leads to a fundamental relation between anisotropy and magnetic susceptibility [186].

To illustrate the point, let us consider a system containing only singlet pairs and thermally excited polarons. Quite generally the in-plane bipolaron and c-axis polaron, conductivities ($\sigma_{\rm ab}$ and $\sigma_{\rm c}$ respectively) and the uniform spin susceptibility (χ_s) are expressed as follows:

$$\sigma_{\rm ab}(T,x) = -4\int_0^\infty {\rm d}E\,\sigma_{\rm b}(E)\frac{\partial f_{\rm b}}{\partial E} \tag{6.34}$$

$$\sigma_{\rm c}(T,x) = -2\int_0^\infty {\rm d}E\,\sigma_{\rm pc}(E)\frac{\partial f_p}{\partial E} \tag{6.35}$$

$$\chi_s(T,x) = -2\mu_{\rm B}^2\int_0^\infty {\rm d}E\,N_p(E)\frac{\partial f_p}{\partial E} \tag{6.36}$$

where $f_{\rm b} = [y^{-1}\exp(E/T) - 1]^{-1}$ and $f_p = [y^{-1/2}\exp(E/T + \Delta/2T) + 1]^{-1}$ are the bipolaron and polaron distribution functions, respectively, and $\mu_{\rm B}$ is the Bohr magneton. Polarons are not degenerate at any temperature neither are bipolarons degenerate above $T_{\rm c}$, so that

$$f_{\rm b} \approx y\exp\left(-\frac{E}{T}\right) \tag{6.37}$$

and

$$f_p \approx y^{1/2}\exp\left(-\frac{E+\Delta/2}{T}\right). \tag{6.38}$$

If the scattering mechanism is the same for polarons and bipolarons, the ratio of the differential bipolaron and polaron conductivities ($\sigma_{\rm b}(E)$ and $\sigma_{\rm pc}(E)$, respectively) is independent of the energy and doping

$$\frac{\sigma_{\rm b}(E)}{\sigma_{\rm pc}(E)} \equiv A = \text{constant}. \tag{6.39}$$

There is a large difference in the values of the ppσ and ppπ hopping integrals between different oxygen sites (section 5.4). Therefore, we expect a highly anisotropic polaron energy spectrum with a quasi-one-dimensional polaron density of states as also observed in high-resolution ARPES experiments (section 8.4):

$$N_p(E) \simeq \frac{1}{4\pi da}\left(\frac{m^*}{2E}\right)^{1/2} \tag{6.40}$$

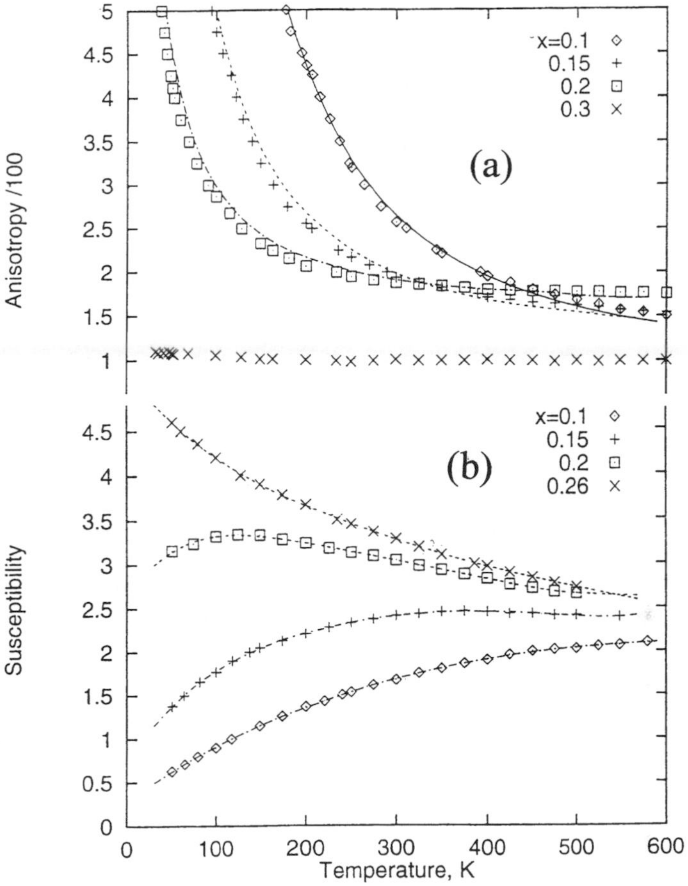

Figure 6.5. (*a*) Experimental anisotropy [214] compared with equation (6.41) and (*b*) using experimental magnetic susceptibilities of $La_{2-x}Sr_xCuO_4$ [215].

where a is the in-plane lattice constant. Then the anisotropy and the spin suceptibility are expressed as

$$\frac{\rho_c(T,x)}{\rho_{ab}(T,x)} = 2Ay^{1/2}\exp\left(\frac{\Delta}{2T}\right) \tag{6.41}$$

and

$$\chi_s(T,x) = \frac{\mu_B^2}{2da}\left(\frac{ym^*}{2\pi T}\right)^{1/2}\exp\left(-\frac{\Delta}{2T}\right). \tag{6.42}$$

The chemical potential, $y = 2\pi n_b(T,x)/(Tm_{ab}^{**})$, is calculated taking into account the localization of bipolarons in a random potential (section 5.5). The bipolaron density ($n_b(T,x)$ per cm^2) is linear in temperature and doping

($n_b(T, x) \approx xT/\gamma a^2$) in a wide temperature range above T_c which fits the temperature and doping dependence of the Hall ratio $R_H \simeq 1/(2en_b)$ (section 6.2). As a result, we find the temperature-independent $y \propto x$ and

$$\frac{\rho_c(T, x)}{\rho_{ab}(T, x)} \propto \frac{x}{T^{1/2}\chi_s(T, x)} \tag{6.43}$$

$$\chi_s(T, x) \propto \left(\frac{x}{T}\right)^{1/2} \exp\left(-\frac{\Delta}{2T}\right). \tag{6.44}$$

We expect some dependence of the binding energy on doping because of screening. Bipolarons are heavy non-degenerate particles, which well screen the electron–phonon interaction with low-frequency phonons. In a wide range of doping near the optimum value ($x = 0.15$), different experiments are consistent with

$$\Delta = \frac{\Delta_0}{x} \tag{6.45}$$

where Δ_0 is doping independent.

One can describe all qualitative features of c-axis resistivity and magnetic susceptibility of $La_{2-x}Sr_xCuO_4$ using equations (6.43), (6.44) and (6.45) *without any fitting parameters*. Anisotropy is *quantitatively* described by equation (6.43) using the experimental values of $\chi_s(T, x)$, as shown in figure 6.5. The temperature-independent anisotropy of a heavily doped sample, $x = 0.3$ in figure 6.5 is explained by the contribution of polarons to the in-plane transport. If the binding energy Δ is below 100 K, the polaron contribution flattens the temperature dependence of the anisotropy. An exponential temperature dependence for the c-axis resistivity and anisotropy was also interpreted within the framework of the bipolaron model in other cuprates, in particular in $Bi_2Sr_2CaCu_2O_{8+\delta}$ [197], $YBa_2Cu_3O_{6+x}$ [213] and $HgBa_2CuO_{4+\delta}$ [212]. Importantly, the uniform magnetic susceptibility above T_c *increases* with doping (figure 6.5(*b*)). It proves once again that cuprates are doped insulators, where low-energy charge and spin degrees of freedom are due to holes doped into a parent insulating matrix with no free carriers and no free spins. A rather low magnetic susceptibility of parent insulators in their paramagnetic phase is presumably due to a singlet pairing of copper spins as discussed in section 5.5.

6.6 Infrared conductivity

Studies of photoinduced carriers in dielectric 'parent' compounds like La_2CuO_4, $YBa_2Cu_3O_6$, and others demonstrated the formation of self-localized small polarons or bipolarons. In particular, Mihailovic *et al* [170] described the spectral shape of the photoconductivity with the small polaron transport theory. They also argued that a similar spectral shape and systematic trends in both the photoconductivity of optically doped dielectric samples and the infrared conductivity of chemically doped high-T_c oxides indicate that carriers in a

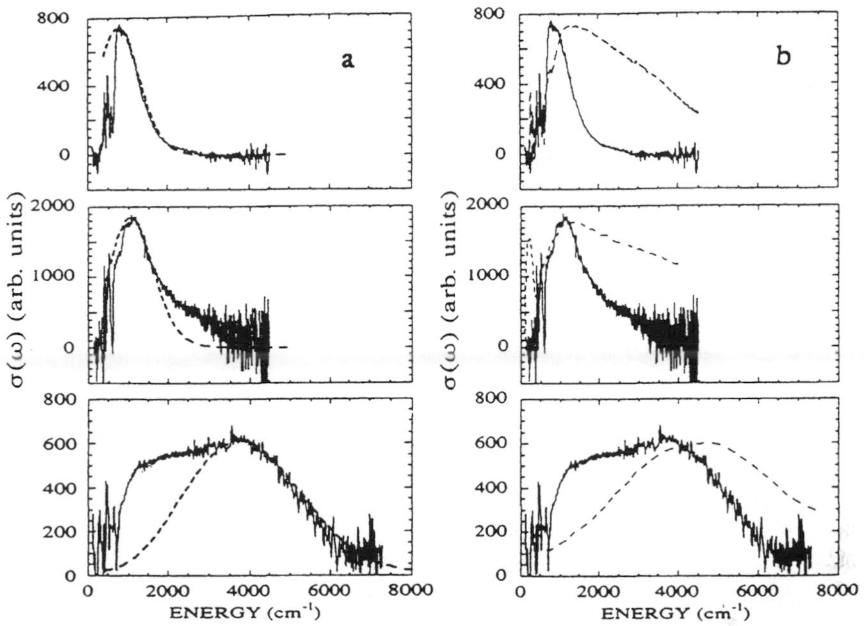

Figure 6.6. (*a*) Photoinduced infrared conductivity in the insulator precursors for $Tl_2Ba_2CaCu_2O_8$ (top), $YBa_2Cu_3O_7$ (middle) and $La_{2-x}Sr_xCuO_4$ (bottom) compared with the polaron infrared conductivity (broken curve); (*b*) the infrared conductivity for superconducting samples (broken curves) compared with (*a*).

concentrated (metallic) regime retain much of the character of carriers in a dilute (photoexcited) regime. The measured photoinduced infrared conductivity $\sigma(\nu)$ (full curves) in insulating parent compounds is shown in figure 6.6(*a*). Its characteristic maximum is due to the incoherent spectral weight of the polaron Green function (section 4.5) as follows from a comparison with a theoretical small-polaron infrared conductivity (broken curves) in figure 6.6(*a*). The infrared conductivity of high-T_c superconductors (broken curves) is compared with the photoinduced infrared conductivity (full curves) in respective insulator precursors in figure 6.6(*b*). One of the qualitative observations, which follow from a comparison of the infrared conductivities of insulating and superconducting materials is that in all perovskites exhibiting superconductivity, the peak frequency shifts toward lower values as the T_c of the material increases. The peak position in $\sigma(\nu)$ is related to the polaron level shift and, therefore, to the (bi)polaron mass. The critical temperature of the superconducting transition is proportional to $1/m^*$ in bipolaron theory (section 4.7.5). A lower peak frequency corresponds to a lower mass of bipolaronic carriers and, therefore, to a higher T_c.

Chapter 7

Superconducting transition

7.1 Parameter-free description of T_c

An ultimate goal of the theory of superconductivity is to provide an expression for T_c as a function of some well-defined parameters characterizing the material. In the framework of BCS theory, the Eliashberg equation (3.74) for the gap function properly takes into account a realistic phonon spectrum and retardation of the electron–phonon interaction. T_c is fairly approximated by McMillan's formula (3.96), which works well for simple metals and their alloys. But applying a theory of this kind to high-T_c cuprates is problematic. Since bare electron bands are narrow, strong correlations result in the Mott insulating state of undoped parent compounds. As a result, μ^* and λ are ill defined in doped cuprates and polaronic effects are important as in many doped semiconductors [13]. Hence, an estimate of T_c in cuprates within BCS theory appears to be an exercise in calculating μ^* rather than T_c itself. Also, one cannot increase λ without accounting for a polaron collapse of the band (chapter 4). This appears at $\lambda \simeq 1$ for uncorrelated electrons (holes) and even at a smaller value of bare electron–phonon coupling in strongly correlated models [216].

However, the bipolaron theory provides a parameter-free expression for T_c [217], which fits the experimentally measured T_c in many cuprates for any level of doping. T_c is calculated using the density sum rule as the Bose–Einstein condensation (BEC) temperature of $2e$ charged bosons on a lattice. Just before the discovery [1], we predicted T_c would be as high as $\sim$100 K using an estimate of the bipolaron effective mass [105]. Uemura [218] established a correlation of T_c with the in-plane magnetic field penetration depth measured by the μsR technique in many cuprates as $T_c \propto 1/\lambda_{ab}^2$. The technique is based on the implantation of spin-polarized muons. It monitors the time evolution of the muon spin polarization. He concluded that cuprates are neither BCS nor BEC superfluids but that they exist in the crossover region from one to the other, because the experimental T_c was to be found about three or more times lower than the BEC temperature.

Here we calculate the T_c of a bipolaronic superconductor taking properly into account the microscopic band structure of bipolarons in layered cuprates as derived in section 5.4. We arrive at a parameter-free expression for T_c, which in contrast to [218] involves not only the in-plane, λ_{ab}, but also the out-of-plane, λ_c, magnetic field penetration depth and a normal state Hall ratio R_H just above the transition. It describes the experimental data for a few dozen different samples clearly indicating that many cuprates are in the BEC rather than in the crossover regime.

The energy spectrum of bipolarons is twofold degenerate in cuprates (section 5.4). One can apply the effective mass approximation at $T \simeq T_c$, equation (5.22), because T_c should be less than the bipolaron bandwidth. Also a three-dimensional correction to the spectrum is important for BEC (appendix B) which is well described by the tight-binding approximation:

$$E_{\boldsymbol{K}}^{x,y} = \frac{\hbar^2 K_{x,y}^2}{2m_x^{**}} + \frac{\hbar^2 K_{y,x}^2}{2m_y^{**}} + 2t_{\perp}[1 - \cos(K_z d)] \tag{7.1}$$

where d is the interplane distance and $t_{\perp}$ is the inter-plane bipolaron hopping integral. Substituting the spectrum (equation (7.1)) into the density sum rule,

$$\sum_{\boldsymbol{K},i=(x,y)} [\exp(E_{\boldsymbol{K}}^i/T_c) - 1]^{-1} = n_b \tag{7.2}$$

one readily obtains T_c (in ordinary units):

$$k_B T_c = f\left(\frac{t_{\perp}}{k_B T_c}\right) \times \frac{3.31\hbar^2 (n_B/2)^{2/3}}{(m_x^{**} m_y^{**} m_c^{**})^{1/3}} \tag{7.3}$$

where the coefficient $f(x) \approx 1$ is shown in figure 7.1 as a function of the anisotropy, $t_{\perp}/(k_B T_c)$ and $m_c^{**} = \hbar^2/(2|t_{\perp}|d^2)$.

This expression is rather ambiguous because the effective mass tensor as well as the bipolaron density n_b are not well known. Fortunately, we can express the band-structure parameters via the in-plane magnetic field penetration depth

$$\lambda_{ab} = \left[\frac{m_x^{**} m_y^{**}}{8\pi n_B e^2 (m_x^{**} + m_y^{**})}\right]^{1/2}$$

and out-of-plane penetration depth,

$$\lambda_c = \left[\frac{m_c^{**}}{16\pi n_b e^2}\right]^{1/2}$$

(we use $c = 1$). The bipolaron density is expressed through the in-plane Hall ratio (above the transition) as

$$R_H = \frac{1}{2en_b} \times \frac{4m_x^{**} m_y^{**}}{(m_x^{**} + m_y^{**})^2} \tag{7.4}$$

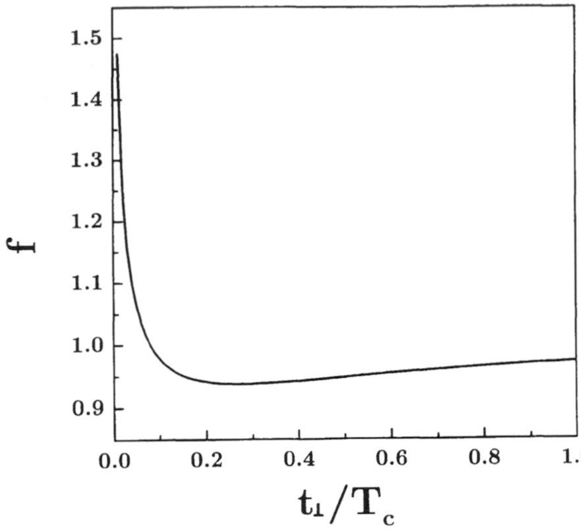

Figure 7.1. Correction coefficient to the three-dimensional Bose–Einstein condensation temperature as a function of anisotropy.

which leads to

$$T_c = 1.64 f\left(\frac{t_\perp}{k_B T_c}\right)\left(\frac{eR_H}{\lambda_{ab}^4\lambda_c^2}\right)^{1/3}. \tag{7.5}$$

Here T_c is measured in kelvin, eR_H in cm^3 and λ in cm. The coefficient f is about unity in a very wide range of $t_\perp/(k_B T_c) \geq 0.01$, figure 7.1. Hence, the bipolaron theory yields a parameter-free expression, which unambiguously tells us how near the cuprates are to the BEC regime:

$$T_c \approx T_c(3D) = 1.64\left(\frac{eR_H}{\lambda_{ab}^4\lambda_c^2}\right)^{1/3}. \tag{7.6}$$

We compare the last two expressions with the experimental T_c of more than 30 different cuprates, for which both λ_{ab} and λ_c have been measured along with $R_H(T_c + 0)$ in table 7.1 and figure 7.2. The Hall ratio has a strong temperature dependence above T_c (section 6.1). Therefore, we use the experimental Hall ratio just above the transition. In a few cases (mercury compounds), where $R_H(T_c + 0)$ is unknown, we take the inverse chemical density of carriers (divided by e) as R_H. For almost all samples, the theoretical T_c fits experimental values within an experimental error bar for the penetration depth (about ±10%). There are a few Zn-doped YBCO samples (figure 7.2) whose critical temperature is higher than the theoretical one. If we assume that the degeneracy of the bipolaron spectrum is removed by the random potential of Zn, then the theoretical T_c would be almost the same as the experimental values for these samples as well.

Table 7.1. Experimental data on T_c (K), *ab* and *c* penetration depth (nm), Hall coefficient (10^{-3} (cm^3 C^{-1})), and calculated values of T_c, respectively, for $La_{2-x}Sr_xCuO_4$ (La), YBaCuO (*x*%Zn) (Zn), $YBa_2Cu_3O_{7-x}$ (Y) and $HgBa_2CuO_{4+x}$ (Hg) compounds.

Compound	T_c^{exp}	λ_{ab}	λ_c	R_H	T_c (3D)	T_c	T_{KT}
La(0.2)	36.2	200	2 540	0.8	38	41	93
La(0.22)	27.5	198	2 620	0.62	35	36	95
La(0.24)	20.0	205	2 590	0.55	32	32	88
La(0.15)	37.0	240	3 220	1.7	33	39	65
La(0.1)	30.0	320	4 160	4.0	25	31	36
La(0.25)	24.0	280	3 640	0.52	17	19	47
Zn(0)	92.5	140	1 260	1.2	111	114	172
Zn(2)	68.2	260	1 420	1.2	45	46	50
Zn(3)	55.0	300	1 550	1.2	35	36	38
Zn(5)	46.4	370	1 640	1.2	26	26	30
Y(0.3)	66.0	210	4 530	1.75	31	51	77
Y(0.43)	56.0	290	7 170	1.45	14	28	40
Y(0.08)	91.5	186	1 240	1.7	87	88	98
Y(0.12)	87.9	186	1 565	1.8	75	82	97
Y(0.16)	83.7	177	1 557	1.9	83	89	108
Y(0.21)	73.4	216	2 559	2.1	47	59	73
Y(0.23)	67.9	215	2 630	2.3	46	58	73
Y(0.26)	63.8	202	2 740	2.0	48	60	83
Y(0.3)	60.0	210	2 880	1.75	43	54	77
Y(0.35)	58.0	204	3 890	1.6	35	50	82
Y(0.4)	56.0	229	4 320	1.5	28	42	65
Hg(0.049)	70.0	216	16 200	9.2	23	60	115
Hg(0.055)	78.2	161	10 300	8.2	43	92	206
Hg(0.055)	78.5	200	12 600	8.2	28	69	134
Hg(0.066)	88.5	153	7 040	6.85	56	105	229
Hg(0.096)	95.6	145	3 920	4.7	79	120	254
Hg(0.097)	95.3	165	4 390	4.66	61	99	197
Hg(0.1)	94.1	158	4 220	4.5	66	105	216
Hg(0.101)	93.4	156	3 980	4.48	70	107	220
Hg(0.101)	92.5	139	3 480	4.4	88	127	277
Hg(0.105)	90.9	156	3 920	4.3	69	106	220
Hg(0.108)	89.1	177	3 980	4.2	58	90	171

One can argue that, due to a large anisotropy, cuprates may belong to a two-dimensional '*XY*' universality class with the Kosterlitz–Thouless (KT) critical temperature T_{KT} of preformed bosons [219, 220] or Cooper pairs [164]. Should this be the case, one could hardly discriminate the Cooper pairs with respect to bipolarons. The KT critical temperature can be expressed through the in-plane

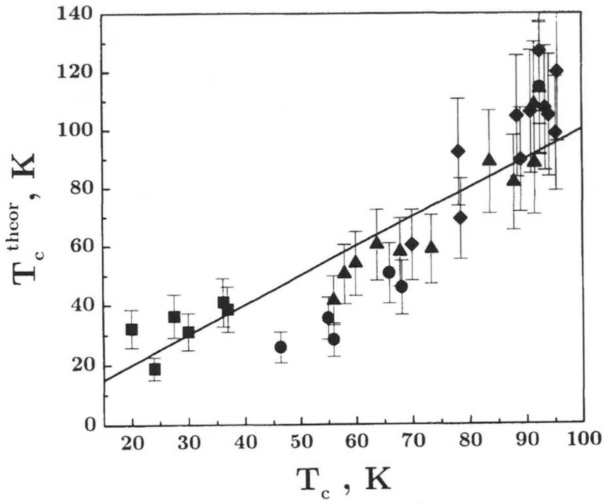

Figure 7.2. Theoretical critical temperature compared with the experiment (the theory is exact for samples on the straight line) for LaSrCuO compounds (squares), for Zn-substituted $YBa_2Cu_{1-x}Zn_xO_7$ (circles), for $YBa_2Cu_3O_{7-\delta}$ (triangles), and for $HgBa_2CuO_{4+\delta}$ (diamonds). Experimental data for the London penetration depth are taken from Xiang T *et al* 1998 *Int. J. Mod. Phys.* B **12** 1007 and Janossy B *et al* 1991 *Physica* C **181** 51 in $YBa_2Cu_3O_{7-\delta}$ and $YBa_2Cu_{1-x}Zn_xO_7$; from Grebennik V G *et al* 1990 *Hyperfine Interact.* **61** 1093 and Panagopoulos C Private communication in underdoped and overdoped $La_{2-x}Sr_xCuO_4$, respectively, and from Hofer J *et al* 1998 *Physica* C **297** 103 in $HgBa_2CuO_{4+\delta}$. The Hall coefficient above T_c is taken from Carrington A *et al* 1993 *Phys. Rev.* B **48** 13 051 and Cooper J R Private communication ($YBa_2Cu_3O_{7-\delta}$ and $YBa_2Cu_{1-x}Zn_xO_7$) and from Hwang H Y *et al* 1994 *Phys. Rev. Lett.* **72** 2636 ($La_{2-x}Sr_xCuO_4$).

penetration depth alone [164]

$$k_B T_{KT} \approx \frac{0.9 d\hbar^2}{16\pi e^2 \lambda_{ab}^2}. \tag{7.7}$$

It appears significantly higher than the experimental values in many cases (see table 7.1). There are also quite a few samples with about the same λ_{ab} and the same d but with very different values of T_c, which proves that the phase transition is not the KT transition. In contrast, our parameter-free fit of the experimental critical temperature and the critical behaviour (sections 7.3 and 7.4) favour 3D Bose–Einstein condensation of charged bosons as the mechanism for high T_c rather than any low-dimensional phase-fluctuation scenario.

The fluctuation theory [221] further confirms the three-dimensional character of the phase transition in cuprates. However, it does not mean that

Table 7.2. Mass enhancement in cuprates.

Compound	m_{ab}	m_c
La(0.2)	22.1	3558
La(0.15)	15.0	2698
La(0.1)	11.3	1909
Y(0.0)	7.2	584
Y(0.12)	8.3	600
Y(0.3)	10.6	1994

all cuprates are in the BEC regime with charged bosons as carriers. Some of them, in particular overdoped samples, might be in the BCS or intermediate regime, which makes the BCS–BEC crossover problem relevant. Starting from the pioneering works by Eagles [72] and Legget [222], this problem received particular attention in the framework of a *negative* Hubbard U model [223, 224]. Both analytical (diagrammatic [225], path integral [226]) and numerical [227] studies have addressed the intermediate-coupling regime beyond a variational approximation [223], including two-dimensional systems [227–229]. However, in using the negative Hubbard U model, we have to realize that this model, which predicts a smooth BCS-BEC crossover, cannot be applied to a strong electron–phonon interaction and polaron–bipolaron crossover (section 4.5). An essential effect of the polaron band-narrowing is missing in the Hubbard model. As discussed in section 4.7.5, the polaron collapse of the bandwidth is responsible for high T_c. It strongly affects the BCS–BEC crossover, significantly reducing the crossover region.

It is interesting to estimate the effective mass tensor using the penetration depth and the Hall ratio. These estimates for in-plane and out-of-plane boson masses are presented in table 7.2. They well argee with the inter-site bipolaron mass (sections 4.6 and 5.4). We note, however, that an absolute value of the effective mass in terms of the free electron mass does not describe the actual band mass renormalization if the bare (band) mass is unknown. Nevertheless, an assumption [141] that the number of carriers is determined by Luttinger's theorem (i.e. $n = 1 + x$) would lead to much heavier carriers with m^* about $100m_e$.

7.2 Isotope effect on T_c and on supercarrier mass

The advances in the fabrication of the isotope substituted samples made it possible to measure a sizeable isotope effect, $\alpha = -\mathrm{d}\ln T_c/\mathrm{d}\ln M$ in many high-T_c oxides. This led to a general conclusion that phonons are relevant for high T_c. Moreover, the isotope effect in cuprates was found to be quite different from the BCS prediction: $\alpha = 0.5$ (or less) (section 2.5). Several compounds showed

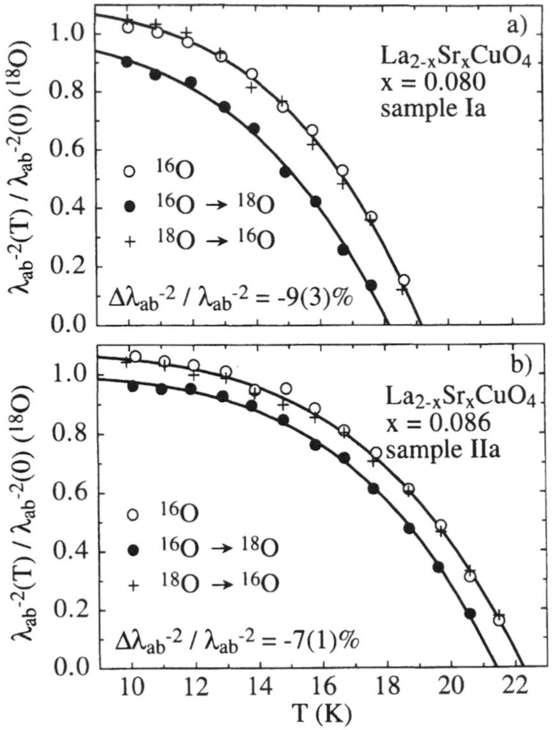

Figure 7.3. The isotope effect on the magnetic field penetration depth in two samples of $La_{2-x}Sr_xCuO_4$ [173] (courtesy of J Hofer).

$\alpha > 0.5$ [230], and a small negative value of α was found in Bi-2223 [231].

Essential features of the isotope effect, in particular large values in low-T_c cuprates, an overall trend to lower value as T_c increases [232] and a small or even negative α in some high-T_c cuprates were understood using equations (4.225) and (4.226) for the isotope exponents of (bi)polaronic superconductors [82]. With increasing ion mass, the bipolaron mass increases and the BEC temperature T_c decreases in the bipolaronic superconductor. In contrast, an increase in the ion mass leads to band narrowing and an enhancement of the polaron density of states and to an increase in T_c in polaronic superconductors. Hence, the isotope exponent in T_c can distinguish the BCS-like polaronic superconductivity with $\alpha < 0$ and the BEC of small bipolarons with $\alpha > 0$. Moreover, underdoped cuprates, which are definitely in the BEC regime (section 5.3), could have $\alpha > 0.5$ (equation (4.226)) as observed.

Another prediction of the bipolaron theory is an isotope effect on the carrier mass (equation (4.222)) which is linked with the isotope effect on T_c, according to equations (4.224) and (4.226).

Remarkably, this prediction was experimentally confirmed by Zhao *et al* [173] providing compelling evidence for polaronic carriers in doped cuprates. The effect was observed in the London penetration depth of isotope-substituted cuprates (figure 7.3). The carrier density is unchanged with the isotope substitution of O^{16} by O^{18}, so that the isotope effect on λ_{ab} directly measures the isotope's effect on the carrier mass. In particular, the carrier mass isotope exponent $\alpha_m = d\ln m^{**}/d\ln M$ was found to be as large as $\alpha_m = 0.8$ in $La_{1.895}Sr_{0.105}CuO_4$. Then, according to equation (4.222), the polaron mass enhancement should be $m^{**}/m \simeq 5$ in this material. Using equation (5.27), we obtain an in-plane bipolaron mass as large as $m^{**} \approx 10m_e$ with the bare hopping integral $T(\text{NNN}) = 0.2$ eV. The in-plane magnetic field penetration depth, calculated with this mass is $\lambda_{ab} = [m^{**}/8\pi ne^2]^{1/2} \approx 316$ nm, where n is the hole density. This agrees well with the experimental one, $\lambda_{ab} \simeq 320$ nm. Using the measured values of $\lambda_{ab} = 320$ nm, $\lambda_c = 4160$ nm and $R_H = 4 \times 10^{-3}\ \text{cm}^3\ \text{C}^{-1}$ (just above T_c), we obtain $T_c = 31$ K in astonishing agreement with the experimental value $T_c = 30$ K in this compound.

7.3 Specific heat anomaly

Bose liquids (or, more precisely, ^{4}He) show the characteristic λ-point singularity in their specific heat but superfluid Fermi liquids like BCS superconductors exhibit a sharp second-order phase transition accompanied by a finite jump in the specific heat (section 2.6). It was established beyond doubt [233–237] that the anomaly in high-T_c cuprates differs qualitatively from the BSC prediction. As was stressed by Salamon *et al* [238], the heat capacity is logarithmic near the transition and, consequently, cannot be adequately treated by mean-field BCS theory even if the Gaussian fluctuations are included. In particular, estimates using the Gaussian fluctuations yield an unusually small coherence volume (table 7.3), comparable with the unit cell volume [234] and *Gi* number (equation (1.53)) of the order of one.

Table 7.3. Coherence volume Ω in Å^3, the in-plane ξ_{ab} and out-of-plane ξ_c coherence lengths derived from a Ginzburg–Landau analysis of the specific heat [234].

Compound	Ω	ξ_{ab}^2 (Å^2)	ξ_c (Å)
$YBa_2Cu_3O_7$	400	125	3.2
$YBa_2Cu_3O_{7-0.025}$	309	119	2.6
$YBa_2Cu_3O_{7-0.05}$	250	119	2.1
$YBa_2Cu_3O_{7-0.1}$	143	119	1.2
$Ca_{0.8}Y_{0.2}Sr_2Tl_{0.5}Pb_{0.5}Cu_2O_7$	84	70	1.2
$Tl_{1.8}Ba_2Ca_{2.2}Cu_3O_{10}$		40	<0.9

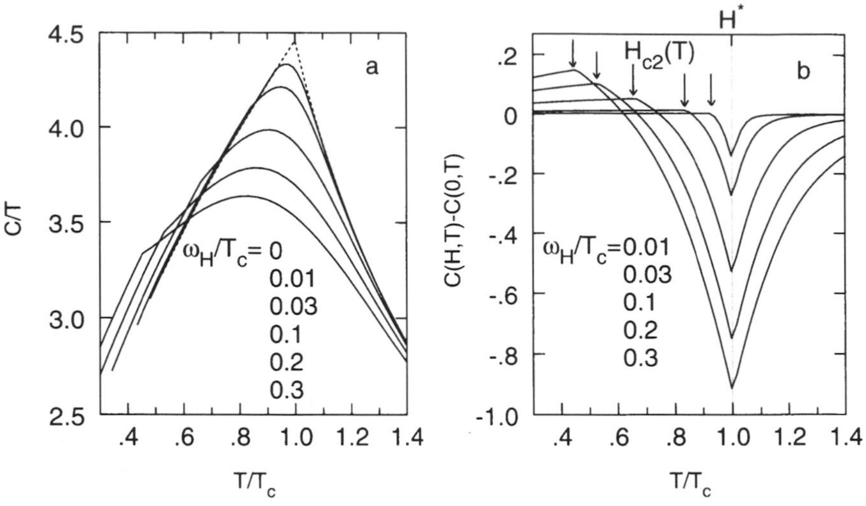

Figure 7.4. Temperature dependence of the specific heat divided by temperature (arbitrary units) of the charged Bose gas scattered off impurities for several fields ($\omega_{\rm H} = 2eB/m^{**}$). ($b$) shows two anomalies, the lowest one traces the resistive transition, while the highest anomaly is the normal-state feature.

The magnetic field dependence of the anomaly [239] is also unusual but it can be described by the bipolaron model [240, 241]. Calculations of the specific heat of charged bosons in a magnetic field require an analytical DOS, $N(\epsilon, B)$ of a particle, scattered by other particles and/or by a random potential of impurities. We can use the DOS in a magnetic field with an impurity scattering as in section 4.7.8, which allows for an analytical result (equation (4.266)). The specific heat coefficient

$$\frac{C(T,B)}{T} = \frac{\mathrm{d}}{T\,\mathrm{d}T}\int \mathrm{d}\epsilon\, \frac{N(\epsilon, B)\epsilon}{\exp[(\epsilon-\mu)/T]-1}$$

calculated with this DOS and with μ determined from $n_{\rm b} = \int \mathrm{d}\epsilon\, N(\epsilon, B) f(\epsilon)$ is shown in figure 7.4.

The broad maximum at $T \simeq T_{\rm c}$ is practically the same as in an ideal Bose gas without scattering [240] (appendix B.4.2). It barely shifts in the magnetic field. However, there is a second anomaly at lower temperatures, which is absent in the ideal gas. This shifts with the magnetic field, tracing precisely the resistive transition (section 7.4), as clearly seen from the difference between the specific heat in the field and zero-field curve, figure 7.4(b). The specific heat (figure 7.4) has a striking resemblance with the Geneva group's experiments on $DyBa_2Cu_3O_7$ and on $YBa_2Cu_3O_7$ [239], where both anomalies were observed. Within the bipolaron model, when the magnetic field is applied, it hardly changes the temperature dependence of the chemical potential near the zero-field T_c because the energy spectrum of thermally excited bosons is practically unchanged. This is because their characteristic energy (of the order of T_c) remains huge compared with the magnetic energy of the order of $2eB/m^{**}$. In contrast, the energy spectrum of low-energy bosons is strongly perturbed even by a weak magnetic field. As a result the chemical potential 'touches' the band edge at lower temperatures, while having almost the same 'kink'-like temperature dependence around T_c as in a zero field. While the lower anomaly corresponds to the true long-range order due to BEC, the higher one is just a 'memory' about the zero-field transition. This microscopic consideration shows that a genuine phase transition into a superconducting state is related to a resistive transition (section 7.4) and to the lower specific heat anomaly, while the broad higher anomaly is the normal-state feature of the bosonic system in an external magnetic field. Differing from the BCS superconductor, these two anomalies are well separated in the bosonic superconductor at any field but zero.

7.4 Universal upper critical field of unconventional superconductors

The upper critical field ($H_{c2}(T) = \Phi_0/2\pi\xi(T)^2$) is very different in a BCS superconductor (section 1.6.4) and in a charged Bose gas (section 4.7.8). While $H_{c2}(T)$ is linear in temperature near T_c in the Landau theory of second-order phase transitions, it has a positive curvature ($H_{c2}(T) \propto (T_c - T)^{3/2}$) in a CBG. Also at zero temperature, $H_{c2}(0)$ is normally below the Pauli pair-breaking limit given by $H_p \simeq 1.84T_c$ (in tesla) in the BCS theory but the limit can be exceeded by many times in CBG.

In cuprates [241–247], spin-ladders [248] and organic superconductors [249], high-magnetic field studies revealed a non-BCS upward curvature of resistive $H_{c2}(T)$. When measurements were performed on low-T_c unconventional superconductors [243, 244, 246, 248, 249], the Pauli limit was exceeded several times. A nonlinear temperature dependence in the vicinity of T_c was unambigously observed in a few samples [241, 245–247]. Importantly, a thermodynamically determined H_{c2} turned out to be much higher than the resistive H_{c2} [250] due to the contrasting magnetic field dependences of the specific heat anomaly and of resistive transition.

We believe that many unconventional superconductors are in the 'bosonic' limit of preformed real-space bipolarons, so their resistive H_{c2} is actually a critical field of the BEC of charged bosons [121]. Calculations carried out for the heat capacity of a CBG in section 7.3 led to the conclusion that the resistive H_{c2} and the thermodynamically determined H_{c2} are very different in bosonic superconductors. While the magnetic field destroys the condensate of ideal bosons, it hardly shifts the specific heat anomaly as observed.

A comprehensive scaling of resistive H_{c2} measurements in unconventional

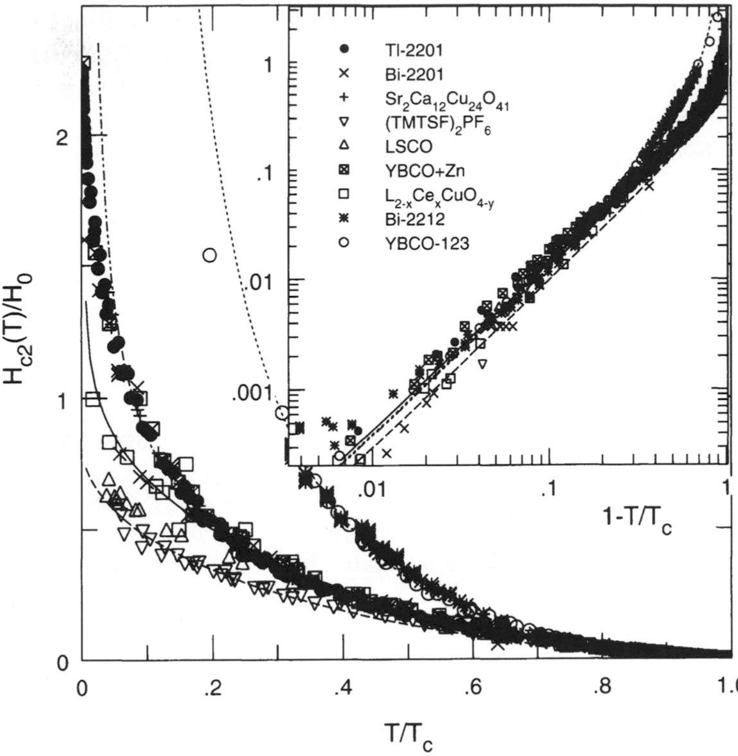

Figure 7.5. Resistive upper critical field (determined at 50% of the transition) of cuprates, spin-ladders and organic superconductors scaled according to equation (7.9). The parameter b is 1 (dots), 0.02 (chain), 0.0012 (full curve), and 0 (dashes). The inset shows a universal scaling of the same data near T_c on the logarithmic scale.

superconductors is shown in figure 7.5 [241] in the framework of the microscopic model of charged bosons scattered off impurities (section 4.7.8). The generalized equation (4.261), accounting for a temperature dependence of the number of delocalized bosons $n_b(T)$, can be written as follows:

$$H_{c2}(T) = H_0 \left[\frac{n_b(T)}{t n_b(T_c)} - t^{1/2} \right]^{3/2} \tag{7.8}$$

where T_c is the zero-field critical temperature and $t = T/T_c$. As shown in section 4.7.8, the scaling constant H_0 depends on the mean-free path l, $H_0 = \Phi_0/2\pi\xi_0^2$, with the characteristic (coherence) length $\xi_0 \simeq (l/n_b(T_c))^{1/4}$. In the vicinity of T_c, one obtains a parameter-free $H_{c2}(T) \propto (1-t)^{3/2}$ using equation (7.8) but the low-temperature behaviour depends on a particular scattering mechanism and a detailed structure for the density of localized states.

As suggested by the normal-state Hall measurements in cuprates (section 6.1), $n_b(T)$ can be parametrized as $n_b(T) = n_b(0) + \text{constant} \times T$, so that $H_{c2}(T)$ can be described by a single-parameter expression:

$$H_{c2}(T) = H_0 \left[\frac{b(1-t)}{t} + 1 - t^{1/2} \right]^{3/2}. \tag{7.9}$$

The parameter b is proportional to the number of delocalized bosons at zero temperature. We expect that this expression is applied in the whole temperature region except at ultra-low temperatures, where the Fermi–Dirac golden rule in the scaling fails. Exceeding the Pauli pair-breaking limit readily follows from the fact that the singlet-pair binding energy is related to the normal-state pseudogap temperature T^*, rather than to T_c. T^* is higher than T_c in bosonic superconductors and cuprates (figure 5.6).

The universal scaling of H_{c2} near T_c is confirmed by resistive measurements of the upper critical field of many cuprates, spin-ladders, and organic superconductors, as shown in figure 7.5. All measurements reveal a universal $(1-t)^{3/2}$ behaviour in a wide temperature region (inset), when they are fitted by equation (7.9). The low-temperature behaviour of $H_{c2}(T)/H_0$ is not universal but well described using the same equation with the single fitting parameter, b. The parameter is close to one in high-quality cuprates with a very narrow resistive transition [247]. It naturally becomes rather small in overdoped cuprates where randomness is more essential, so almost all bosons are localized (at least in one dimension) at zero temperature.

Chapter 8

Superconducting state of cuprates

Independent observations of normal-state pseudogaps in a number of magnetic and kinetic measurements (chapter 6) and unusual critical phenomena (chapter 7) tell us that many cuprates may not be BCS superconductors. Indeed their superconducting state is as anomalous as the normal one. In particular, there is strong evidence for a d-like order parameter (changing sign when the CuO_2 plane is rotated by $\pi/2$) in cuprates [251]. A number of phase-sensitive experiments [252] provide unambiguous evidence in this direction; furthermore, the low-temperature magnetic penetration depth [253, 254] was found to be linear in a few cuprates as expected for a d-wave BCS superconductor. However, superconductor–insulator–normal metal (SIN) and superconductor–insulator–superconductor (SIS) tunnelling studies (sections 8.2 and 8.3), the *c*-axis Josephson tunnelling [255] and some high-precision magnetic measurements [256] show a more usual s-like symmetry or even reveal an upturn in the temperature dependence of the penetration depth below some characteristic temperature [257]. Also both angle-resolved photoemission spectroscopy (ARPES) [264] and scanning tunnelling microscopy (STM) [265] have shown that the maximum energy gap is many times larger and the $2\Delta/T_c$ ratio is well above that expected in weak-coupling BCS theory ($\sim$3.5) or in its intermediate-coupling generalization (chapter 3).

Strong deviations from the Fermi/BCS-liquid behaviour are suggestive of a new electronic state in cuprates, which is a charged Bose liquid of bipolarons. In this chapter we discuss the low-temperature London penetration depth [258], tunnelling [259–261] and ARPES [262], the symmetry of the order parameter and superconducting stripes [263] in the framework of the bipolaron theory.

8.1 Low-temperature penetration depth

If the total number of bipolarons in one unit cell is $x/2$ of which n_L are in localized states and n_b are in delocalized states, then the number in the condensate n_c is

$$n_c = \tfrac{1}{2}x - n_L - n_b \tag{8.1}$$

and the London penetration depth $\lambda \propto 1/\sqrt{n_c}$. Taking the delocalized bipolarons to be a three-dimensional gas, we have $n_b \propto T^{3/2}$. Thus in the low-temperature limit, we can neglect n_b and make the approximation

$$n_c \approx x/2 - n_L. \tag{8.2}$$

In this limit, $\lambda(T) - \lambda(0)$ is small and so

$$\lambda(T) - \lambda(0) \propto n_L(T) - n_L(0) \tag{8.3}$$

i.e. λ has the same temperature dependence as n_L. For small amounts of disorder delocalized bipolarons may contribute to the low-temperature dependence of $\lambda(T)$ as well; for non-interacting bipolarons moving in d dimensions, this would give $\lambda \propto T^{d/2}$.

In our picture, interacting bosons fill up all the localized single-particle states in a random potential and Bose-condense into the first extended state at the mobility edge, E_c. For convenience we choose $E_c = 0$. When two or more charged bosons are in a single localized state of energy E, there may be significant Coulomb energy and we take this into account as follows. The localization length ξ is thought to depend on E via

$$\xi \propto \frac{1}{(-E)^{\nu}} \tag{8.4}$$

where $\nu > 0$. The Coulomb potential energy of p charged bosons confined within a radius ξ can be expected to be

$$\text{potential energy} \simeq \frac{4p(p-1)e^2}{\epsilon_0 \xi}. \tag{8.5}$$

Thus, the total energy of p bosons in a localized state of energy E is taken to be

$$w(E) = pE + p(p-1)\kappa(-E)^{\nu} \tag{8.6}$$

where $\kappa > 0$. We can thus define an energy scale E_1:

$$E_1 = \kappa^{1/(1-\nu)}. \tag{8.7}$$

From here on *we choose our units of energy such that* $E_1 = 1$. We take the total energy of charged bosons in localized states to be the sum of the energies of the bosons in individual potential wells. The partition function Z for such a system is then the product of the partition functions $z(E)$ for each of the wells,

$$z(E) = \mathrm{e}^{\alpha p_0^2} \sum_{p=0}^{\infty} \mathrm{e}^{-\alpha(p-p_0)^2} \tag{8.8}$$

where

$$p_0 = \tfrac{1}{2}\{1 + (-E)^{1-\nu}\} \tag{8.9}$$

$$\alpha = \frac{(-E)^{\nu}}{\theta} \tag{8.10}$$

and

$$\theta = \frac{T}{E_1}. \tag{8.11}$$

The average number of bosons in localized states n_L is

$$n_L = \int_{-\infty}^{0} dE\, \langle p \rangle N_L(E) \tag{8.12}$$

where the mean occupancy $\langle p \rangle$ of a single localized state is taken to be

$$\langle p \rangle = \frac{\sum_{p=0}^{\infty} p e^{-\alpha(p-p_0)^2}}{\sum_{p=0}^{\infty} e^{-\alpha(p-p_0)^2}} \tag{8.13}$$

and $N_L(E)$ is the one-particle density of localized states per unit cell below the mobility edge (section 5.5).

We now focus on the temperature dependence of n_L at low temperature ($\theta \ll 1$) for the case where the (dimensionless) width of the impurity tail γ is large ($\gamma > 1$). We note that the parameter γ is the ratio of the width of the tail to the characteristic Coulomb repulsion, E_1. In the following, we consider the case $\nu > 1$ first and then $\nu < 1$. If $\nu > 1$, we can approximate n_L as

$$\frac{n_L}{N_L} \approx 1 + \frac{\nu - 1}{2(2-\nu)\gamma} + \frac{2\theta}{(2-\nu)\gamma} \ln 2. \tag{8.14}$$

So we expect n_L to be close to the total number of wells N_L and to increase linearly with temperature. Figure 8.1(*a*) compares this analytical formula with an accurate numerical calculation for the case $\nu = 1.5$, $\gamma = 20$. We also note that even when $\gamma < 1$, $n_L(\theta)$ will still be linear with the same slope provided $\theta \ll \gamma$. If $\nu < 1$ we obtain, keeping only the lowest power of θ (valid provided $\theta^{1/\nu} \ll \theta$)

$$\frac{n_L}{N_L} = \frac{1}{2} + \frac{\Gamma(2-\nu)\gamma^{1-\nu}}{2} + \frac{1-\nu}{2(2-\nu)\gamma} - \frac{\theta}{\gamma} \ln 2. \tag{8.15}$$

Hence, in this case, n_L decreases linearly with increasing temperature (in the low-temperature limit). Figure 8.1(*b*) compares this analytical formula with the numerical calculation for the case $\nu = 0.65$, $\gamma = 20$. We note that such a value of ν is typical for amorphous semiconductors [266]. A large value of γ is also expected in disordered cuprates with their large static dielectric constant.

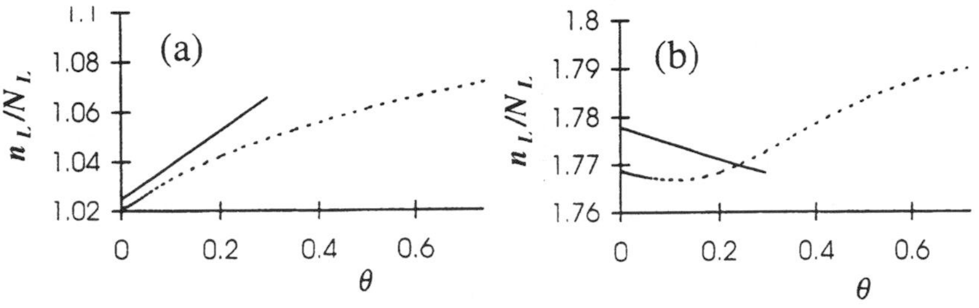

Figure 8.1. Dependence of the density of localized bosons n_L on temperature θ for $\gamma = 20$: (*a*) $\nu = 1.5$, (*b*) $\nu = 0.65$. The full lines correspond to the low-temperature predictions from equations (8.14) and (8.15) while the broken lines are derived from an accurate numerical calculation.

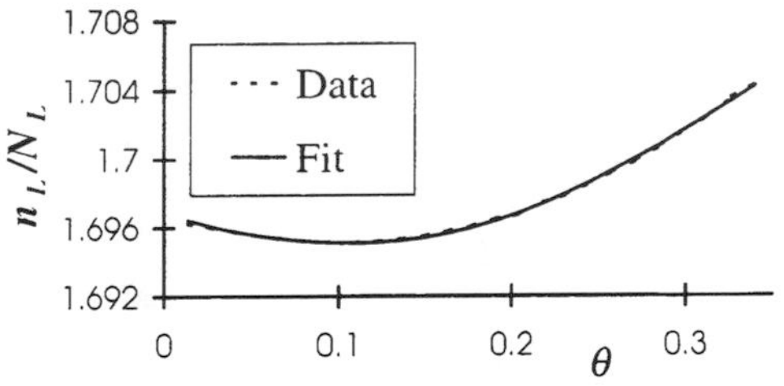

Figure 8.2. Fit to the London penetration depth measured by Walter *et al* [257] for a YBCO film. The parameter values from the fit were $E_1 = 74$ K, $\gamma = 20$ and $\nu = 0.67$.

Figure 8.2 shows that the low-temperature experimental data [257] on the London penetration depth λ of YBCO films can be fitted very well by this theory with $\nu < 1$. It is more usual to see λ increase linearly with temperature [253, 254] and this would correspond to $\nu > 1$ (or to the predominance of the effect of delocalized bipolarons moving in two dimensions). We believe that $\nu < 1$ is more probable for a rapidly varying random potential while $\nu \geq 1$ is more likely for a slowly varying one. Both $\nu < 1$ and $\nu \geq 1$ are observed in doped semiconductors [266]. Hence, it is not surprising that a drastically different low-temperature dependence of the London penetration depth is observed in different samples of doped cuprates. In the framework of this approach, $\lambda(T)$ is related to the localization of carriers at low temperatures rather than to any energy scale characteristic of the condensate or to its symmetry. The excitation spectrum of the charged Bose liquid determines, however, the temperature dependence of $\lambda(T)$ at higher temperatures.

The key parameter of the temperature dependence of the London penetration depth is the exponent ν of the localization length of bosons. Experimentally, the more usual linear increase of λ with temperature is observed in samples which either have no disorder or a shallow and smooth random potential profile. Here we expect $\nu \geq 1$ and a substantial contribution from delocalized bipolarons moving principally in two dimensions and thus, that λ increases linearly with temperature, as observed [254]. However, heavy-ion bombardment [257] introduces rather deep and narrow potential wells for which we might expect $\nu < 1$. This would explain the upturn in the temperature dependence of λ in the disordered films [257].

8.2 SIN tunnelling and Andreev reflection

There is compelling experimental evidence that the pairing of carriers takes place well above T_c in cuprates (chapter 6), the clearest being uniform susceptibility [186, 267] and tunnelling [265]. The gap in tunnelling and photoemission is almost temperature independent below T_c [265, 268] and exists above T_c [265, 269–271]. Kinetic [198] and thermodynamic [272] data suggest that the gap opens in both charge and spin channels at any relevant temperature in a wide range of doping. At the same time, reflection experiments, in which an incoming electron from the normal side of a normal/superconducting contact is reflected as a hole along the same trajectory (section 2.13), revealed a much smaller gap edge than the bias at the tunnelling conductance maxima in a few underdoped cuprates [273]. Other tunnelling measurements [274, 275] also showed distinctly different superconducting- and normal-state gaps.

In the framework of bipolaron theory, we can consider a simplified model, which describes the temperature dependence of the gap and tunnelling spectra in cuprates and accounts for two different energy scales in the electron-hole reflection [260]. The assumption is that the attraction potential in cuprates is large compared with the (renormalized) Fermi energy of polarons. The model is a generic one-dimensional Hamiltonian including the kinetic energy of carriers in the effective mass (m^*) approximation and a local attraction potential, $V(x - x') = -U\delta(x - x')$, as

$$H = \sum_s \int \mathrm{d}x\, \psi_s^\dagger(x) \left(-\frac{1}{2m^*}\frac{\mathrm{d}^2}{\mathrm{d}x^2} - \mu \right) \psi_s(x) - U \int \mathrm{d}x\, \psi_\uparrow^\dagger(x)\psi_\downarrow^\dagger(x)\psi_\downarrow(x)\psi_\uparrow(x) \tag{8.16}$$

where $s = \uparrow, \downarrow$ is the spin. The first band in cuprates to be doped is the oxygen band inside the Hubbard gap (section 5.2). This band is quasi-one-dimensional as discussed in section 5.4, so that a one-dimensional approximation (equation (8.16)) is a realistic starting point. Solving a two-particle problem with the δ-function potential, one obtains a bound state with the binding energy $2\Delta_p = m^*U^2/4$ and with the radius of the bound state $r = 2/(m^*U)$. We assume that this radius is less than the inter-carrier distance in cuprates. It is then that real-space bipolarons are formed. If three-dimensional corrections to the energy spectrum of pairs are taken into account, the ground state of the system is the Bose–Einstein condensate (BEC). The chemical potential is pinned below the band edge by about Δ_p both in the superconducting and normal states, so that the normal state single-particle gap is Δ_p.

Now we take into account that in the superconducting state ($T < T_c$) single-particle excitations interact with the condensate via the same potential U. Applying the Bogoliubov approximation (section 2.2) we reduce the Hamiltonian (8.16) to a quadratic form:

$$H = \sum_s \int \mathrm{d}x\, \psi_s^\dagger(x)\left(-\frac{1}{2m^*}\frac{\mathrm{d}^2}{\mathrm{d}x^2} - \mu\right)\psi_s(x) + \int \mathrm{d}x\,[\Delta_c \psi_\uparrow^\dagger(x)\psi_\downarrow^\dagger(x) + H.c.] \tag{8.17}$$

where the coherent pairing potential, $\Delta_c = -U\langle\psi_\downarrow(x)\psi_\uparrow(x)\rangle$, is proportional to the square root of the condensate density, $\Delta_c = \text{constant} \times n_c^{1/2}(T)$. The single-particle excitation energy spectrum, $E(k)$, is found using the Bogoliubov transformation as

$$E(k) = [(k^2/2m^* + \Delta_p)^2 + \Delta_c^2]^{1/2} \tag{8.18}$$

if one assumes that the condensate density does not depend on position. This spectrum is quite different from the BCS quasi-particles because the chemical potential is negative with respect to the bottom of the single-particle band, $\mu = -\Delta_p$. The single-particle gap, $\Delta/2$, defined as the minimum of $E(k)$, is given by

$$\Delta/2 = [\Delta_p^2 + \Delta_c^2(T)]^{1/2}. \tag{8.19}$$

It varies with temperature from $\Delta(0)/2 = [\Delta_p^2 + \Delta_c^2(0)]^{1/2}$ at zero temperature down to the temperature-independent Δ_p above T_c. The theoretical temperature dependence, equation (8.19), describes well the pioneering experimental observation of the anomalous gap in $YBa_2Cu_3O_{7-\delta}$ in the electron-energy-loss spectra by Demuth *et al* [277] (figure 8.3) with $\Delta_c^2(T) = \Delta_c^2(0) \times [1 - (T/T_c)^n]$ below T_c and zero above T_c, and $n = 4$. We note that $n = 4$ is an exponent which is expected in the two-fluid model of any superfluid [276].

A normal metal–superconductor (SIN) tunnelling conductance via a dielectric contact, $\mathrm{d}I/\mathrm{d}V$ is proportional to the density of states, $\rho(E)$, of the spectrum (equation (8.18)). Taking into account the scattering of single-particle excitations by a random potential, as well as thermal lattice and spin fluctuations (see section 8.4.3), one finds, at $T = 0$

$$\mathrm{d}I/\mathrm{d}V = \text{constant} \times \left[\rho\left(\frac{2eV - \Delta}{\Gamma_0}\right) + A\rho\left(\frac{-2eV - \Delta}{\Gamma_0}\right)\right] \tag{8.20}$$

with

$$\rho(\xi) = \frac{4}{\pi^2} \times \frac{\mathrm{Ai}(-2\xi)\mathrm{Ai}'(-2\xi) + \mathrm{Bi}(-2\xi)\mathrm{Bi}'(-2\xi)}{[\mathrm{Ai}^2(-2\xi) + \mathrm{Bi}^2(-2\xi)]^2} \tag{8.21}$$

A is an asymmetry coefficient, explained in [259], $\mathrm{Ai}(x)$ and $\mathrm{Bi}(x)$ the Airy functions and Γ_0 is the scattering rate.

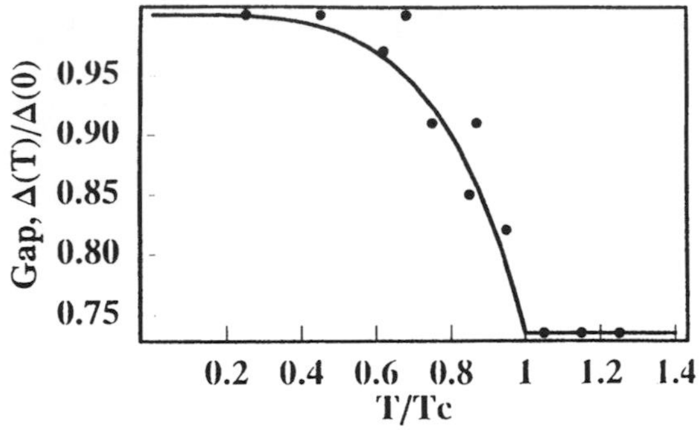

Figure 8.3. Temperature dependence of the gap (equation (8.19) (line)) compared with the experiment [277] (dots) for $\Delta_p = 0.7\Delta(0)$.

We compare the conductance (equation (8.20)) with one of the best STM spectra measured in Ni-substituted $Bi_2Sr_2CaCu_2O_{8+x}$ single crystals by Hancottee *et al* [268] in figure 8.4(*a*). This experiment shows anomalously large $\Delta/T_c > 12$ with the temperature dependence of the gap similar to that in figure 8.3 below T_c. The theoretical conductance (equation (8.20)) describes well the anomalous gap/T_c ratio, injection/emission asymmetry, zero-bias conductance at zero temperature and the spectral shape inside and outside the gap region. There is no doubt that the gap (figure 8.4(*a*)) is s-like. The conductance (equation (8.20)) also fits well the conductance curve obtained on 'pure' Bi-2212 single crystals, as shown in figure 8.4(*b*), while a simple d-wave BCS density of states cannot describe the excess spectral weight in the peaks and the shape of the conductance outside the gap region. We note that the scattering rate, Γ_0, is apparently smaller in the 'pure' sample than in the Ni-substituted sample, as expected.

A simple theory of tunnelling into a bosonic (bipolaronic) superconductor in a metallic (no-barrier) regime follows from this model. As in the canonical BCS approach applied to normal metal–superconductor tunnelling by Blonder *et al* [47], the incoming electron produces only outgoing particles in the superconductor ($x > l$), allowing for a reflected electron and (Andreev) hole in the normal metal ($x < 0$) (section 2.13). There is also a buffer layer of the thickness l at the normal metal–superconductor boundary ($x = 0$), where the chemical potential with respect to the bottom of the conduction band changes gradually from a positive large value μ in metal to a negative value $-\Delta_p$ in a bosonic superconductor. We approximate this buffer layer by a layer with a constant chemical potential μ_b ($-\Delta_p < \mu_b < \mu$) and with the same strength of the pairing potential Δ_c as in a bulk superconductor. The Bogoliubov equations may be written as usual (section 2.13), with the only difference being that the chemical potential with respect to the bottom of the band is a function of the

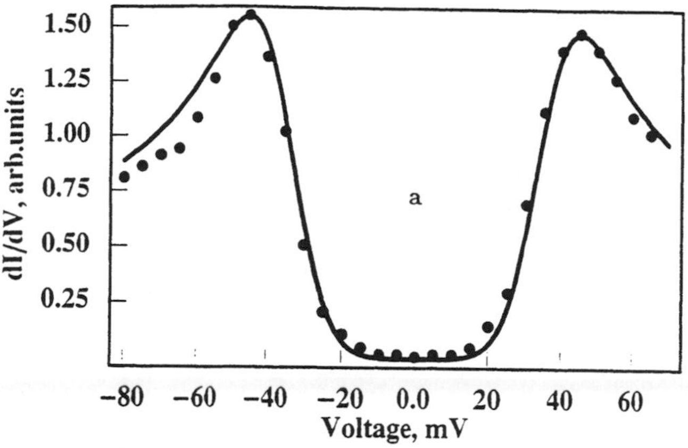

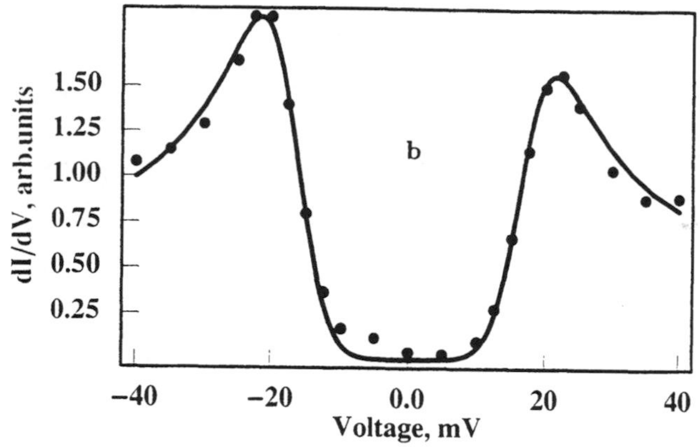

Figure 8.4. Theoretical tunnelling conductance (equation (8.20) (line)) compared with the experimental STM conductance (dots) in (*a*) Ni-substituted $Bi_2Sr_2CaCu_2O_{8+x}$ [268] with $\Delta = 90$ meV, $A = 1.05$, $\Gamma_0 = 40$ meV and (*b*) in 'pure' Bi-2212 [268] with $\Delta = 43$ meV, $A = 1.2$ and $\Gamma_0 = 18$ meV.

coordinate x:

$$E\psi(x) = \begin{pmatrix} -(1/2m)\,\mathrm{d}^2/\mathrm{d}x^2 - \mu(x) & \Delta_c \\ \Delta_c & (1/2m)\,\mathrm{d}^2/\mathrm{d}x^2 + \mu(x) \end{pmatrix} \psi(x). \quad (8.22)$$

Thus, the two-component wavefunction in normal metal is given by

$$\psi_n(x < 0) = \begin{pmatrix} 1 \\ 0 \end{pmatrix} \mathrm{e}^{\mathrm{i}q^+x} + b \begin{pmatrix} 1 \\ 0 \end{pmatrix} \mathrm{e}^{-\mathrm{i}q^+x} + a \begin{pmatrix} 0 \\ 1 \end{pmatrix} \mathrm{e}^{-\mathrm{i}q^-x} \quad (8.23)$$

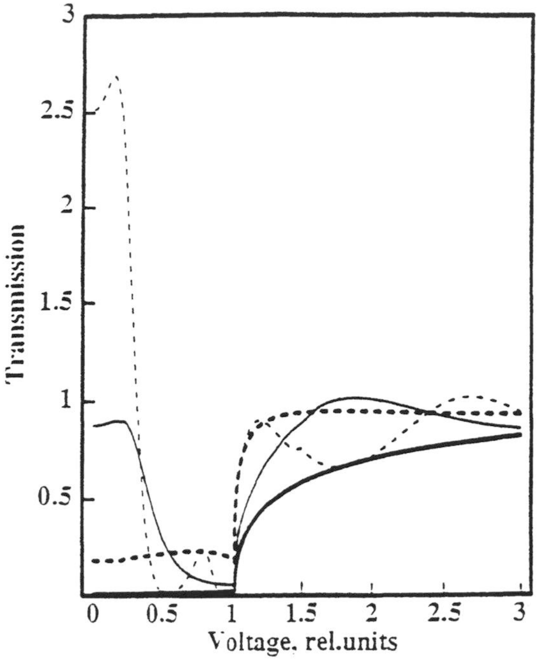

Figure 8.5. Transmission *versus* voltage (measured in units of Δ_p/e) for $\Delta_c = 0.2\Delta_p$, $\mu = 10\Delta_p$, $\mu_b = 2\Delta_p$ and $l = 0$ (bold line), $l = 1$ (bold dashes), $l = 4$ (thin line), and $l = 8$ (thin dashes) (in units of $1/(2m\Delta_p)^{1/2}$).

while in the buffer layer, it has the form

$$\psi_b(0 < x < l) = \alpha \begin{pmatrix} 1 \\ \frac{\Delta_c}{E+\xi} \end{pmatrix} e^{ip^+x} + \beta \begin{pmatrix} 1 \\ \frac{\Delta_c}{E-\xi} \end{pmatrix} e^{-ip^-x} + \gamma \begin{pmatrix} 1 \\ \frac{\Delta_c}{E+\xi} \end{pmatrix} e^{-ip^+x} + \delta \begin{pmatrix} 1 \\ \frac{\Delta_c}{E-\xi} \end{pmatrix} e^{ip^-x} \tag{8.24}$$

where the momenta associated with the energy E are $q^\pm = [2m(\mu \pm E)]^{1/2}$ and $p^\pm = [2m(\mu_b \pm \xi)]^{1/2}$ with $\xi = (E^2 - \Delta_c^2)^{1/2}$. A well-behaved solution in the superconductor with a negative chemical potential is given by

$$\psi_s(x > l) = c \begin{pmatrix} 1 \\ \frac{\Delta_c}{E+\xi} \end{pmatrix} e^{ik^+x} + d \begin{pmatrix} 1 \\ \frac{\Delta_c}{E-\xi} \end{pmatrix} e^{ik^-x} \tag{8.25}$$

where the momenta associated with the energy E are $k^\pm = [2m(-\Delta_p \pm \xi)]^{1/2}$.

The coefficients $a, b, c, d, \alpha, \beta, \gamma, \delta$ are determined from the boundary conditions, which are the continuity of $\psi(x)$ and its derivatives at $x = 0$ and $x = l$. Applying the boundary conditions and carrying out an algebraic reduction,

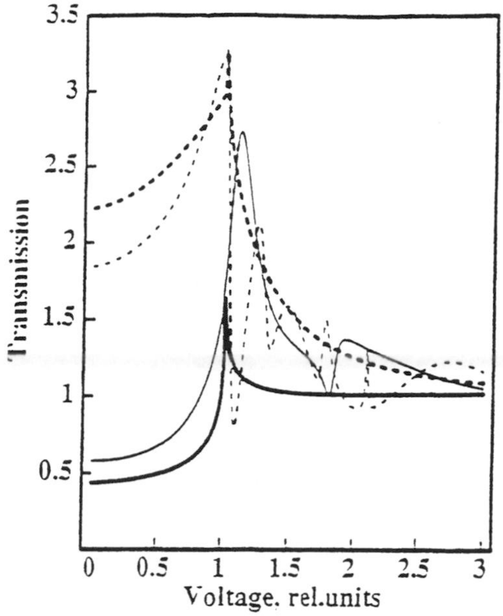

Figure 8.6. Transmission *versus* voltage (measured in units of Δ_c/e) for $\Delta_p = 0.2\Delta_c$, $\mu = 10\Delta_c$, $\mu_b = 2\Delta_c$ and $l = 0$ (bold line), $l = 1$ (bold dashes), $l = 4$ (thin line), and $l = 8$ (thin dashes) (in units of $1/(2m\Delta_c)^{1/2}$).

we find

$$a = 2\Delta_c q^+(p^+ f^- g^+ - p^- f^+ g^-)/D \tag{8.26}$$

$$b = -1 + \frac{2q^+}{D}[(E+\xi)f^+(q^- f^- - p^- g^-) - (E-\xi)f^-(q^- f^+ - p^+ g^+)] \tag{8.27}$$

with

$$\begin{aligned} D = (E+\xi)&(q^+ f^+ + p^+ g^+)(q^- f^- - p^- g^-) \\ &- (E-\xi)(q^+ f^- + p^- g^-)(q^- f^+ - p^+ g^+) \end{aligned} \tag{8.28}$$

and

$$\begin{aligned} f^\pm &= p^\pm \cos(p^\pm l) - \mathrm{i}k^\pm \sin(p^\pm l) \\ g^\pm &= k^\pm \cos(p^\pm l) - \mathrm{i}p^\pm \sin(p^\pm l). \end{aligned} \tag{8.29}$$

The transmission coefficient in the electrical current, $1 + |a|^2 - |b|^2$ is shown in figure 8.5 for different values of l, when the coherent gap Δ_c is smaller than the pair-breaking gap Δ_p and in figure 8.6 in the opposite case, $\Delta_p < \Delta_c$. In the first case, figure 8.5, we find two distinct energy scales, one is Δ_c in the subgap region due to the electron-hole reflection and the other one is $\Delta/2$, which is a single-particle band edge. However, there is only one gap, Δ_c, which can be seen in the second case, figure 8.6. We note that the transmission has no subgap structure if the buffer layer is absent ($l = 0$) in both cases. In the extreme case of a wide buffer layer, $l \gg (2m\Delta_p)^{-1/2}$ (figure 8.5) or $l \gg (2m\Delta_c)^{-1/2}$ (figure 8.6), there are some oscillations of the transmission due to the bound states inside the buffer layer.

We expect that $\Delta_p \gg \Delta_c$ in underdoped cuprates (figure 8.5) while $\Delta_p \leq \Delta_c$ in optimally doped cuprates (figure 8.6). Thus, the model accounts for two different gaps experimentally observed in Giaver tunnelling and electron-hole reflection in underdoped cuprates and for a single gap in optimally doped samples [273]. An oscillating structure, observed in underdoped $YBa_2Cu_3O_{7-\delta}$ [273], is also found in the theoretical conductance at finite l (figure 8.5). The transmissions (figures 8.5 and 8.6) are due to a coherent tunnelling into the condensate and into a single-particle band of the bosonic superconductor. There is an incoherent transmission into localized single-particle impurity states and into incoherent ('supracondensate') bound pair states as well, which might explain a significant featureless background in the subgap region [273].

8.3 SIS tunnelling

Within the standard approximation [76], the tunnelling current, $I(V)$, between two parts of a superconductor separated by an insulating barrier is proportional to a convolution of the Fourier component of the single-hole retarded Green's function, $G^{\mathrm{R}}(\boldsymbol{k}, \omega)$, with itself as

$$I(V) \propto \sum_{\boldsymbol{k},\boldsymbol{p}} \int_{-\infty}^{\infty} \mathrm{d}\omega \, \mathrm{Im}\, G^{\mathrm{R}}(\boldsymbol{k}, \omega) \, \mathrm{Im}\, G^{\mathrm{R}}(\boldsymbol{p}, e|V| - \omega) \tag{8.30}$$

where V is the voltage in the junction.

The problem of a hole on a lattice coupled with the bosonic field of lattice vibrations has a solution in terms of the coherent (Glauber) states in the strong-coupling limit, $\lambda > 1$, where the Migdal–Eliashberg theory cannot be applied (section 4.3.4),

$$G^{\mathrm{R}}(\boldsymbol{k}, \omega) = Z \sum_{l=0}^{\infty} \sum_{\boldsymbol{q}_1,\ldots,\boldsymbol{q}_l} \frac{\prod_{r=1}^{l} |\gamma(\boldsymbol{q}_r)|^2}{(2N)^l l!(\omega - \sum_{r=1}^{l} \omega_{\boldsymbol{q}_r} - \epsilon(\boldsymbol{k} + \sum_{r=1}^{l} \boldsymbol{q}_r) + \mathrm{i}\delta)}. \tag{8.31}$$

The hole energy spectrum, $\epsilon(\boldsymbol{k})$, is renormalized due to the polaron narrowing of the band and (in the superconducting state) also due to interaction with the BEC of bipolarons as discussed in section 8.2,

$$\epsilon(\boldsymbol{k}) = [\xi^2(\boldsymbol{k}) + \Delta_c^2]^{1/2}. \tag{8.32}$$

Here $\xi(\mathbf{k}) = Z'E(\mathbf{k}) - \mu$ is the renormalized polaron band dispersion with the chemical potential μ and $E(\mathbf{k}) = \sum_{\mathbf{m}} T(\mathbf{m}) \exp(-\mathrm{i}\mathbf{k} \cdot \mathbf{m})$ is the bare band dispersion in a rigid lattice.

Quite differently from the BCS superconductor, the chemical potential μ is negative in the bipolaronic system, so that the edge of the single-hole band is found above the chemical potential at $-\mu = \Delta_p$. Near the edge the parabolic one-dimensional approximation for the oxygen hole is applied, compatible with the ARPES data [262],

$$\epsilon_{\mathbf{k}} \simeq \frac{k_x^2}{2m^*} + \Delta/2. \tag{8.33}$$

Differently from the canonical Migdal–Eliashberg Green function (chapter 3), there is no damping ('dephasing') of low-energy polaronic excitations in equation (8.31) due to the electron–phonon coupling alone. This coupling leads to a coherent dressing of low-energy carriers by phonons, which is seen in the Green fucntion as phonon sidebands with $l \geq 1$.

However, elastic scattering by impurities yields a finite life-time for the Bloch polaronic states. For the sake of analytical transparency, here we model this scattering as a constant imaginary self-energy, replacing iδ in equation (8.31) by a finite i$\Gamma/2$. In fact, the 'elastic' self-energy can be found explicitly as a function of energy and momentum (see section 8.4.3). Its particular energy/momentum dependence is essential in the subgap region of tunnelling, where it determines the value of zero-bias conductance. However, it hardly plays any role in the peak region and higher voltages.

Substiting equation (8.33) into equations (8.31) and (8.30), and performing the intergration with respect to frequency and both momenta, we obtain for the tunnelling conductance, $\sigma(V) = \mathrm{d}I/\mathrm{d}V$,

$$\sigma(V) \propto \sum_{l,l'=0}^{\infty} \sum_{\mathbf{q},\mathbf{q}'} \frac{\prod_{r=1}^{l} \prod_{r'=1}^{l'} |\gamma(\mathbf{q}_r)|^2 |\gamma(\mathbf{q}'_{r'})|^2}{(2N)^{l+l'} l! l'!} \times L\left[e|V| - \Delta - \sum_{r=1}^{l} \omega_{\mathbf{q}_r} - \sum_{r'=1}^{l'} \omega(\mathbf{q}'_{r'}), \Gamma\right] \tag{8.34}$$

where $L[x, \Gamma] = \Gamma/(x^2 + \Gamma^2)$. To perform the remaining integrations and summations, we introduce a model analogue of the Eliashberg spectral function ($\alpha^2 F(\omega)$ (section 3.4)) by replacing the $\mathbf{q}$-sums in equation (8.34) by

$$\frac{1}{2N} \sum_{\mathbf{q}} |\gamma(\mathbf{q})|^2 A(\omega_{\mathbf{q}}) = \frac{g^2}{\pi} \int \mathrm{d}\omega \, L[\omega - \omega_0, \delta\omega] A(\omega) \tag{8.35}$$

for any arbitrary function of the phonon frequency $A(\omega_{\mathbf{q}})$. In this way we introduce the characteristic frequency, ω_0, of phonons strongly coupled with holes, their average number g^2 in the polaronic cloud and their dispersion $\delta\omega$.

As soon as $\delta\omega$ is less than ω_0, we can extend the integration over phonon frequencies from $-\infty$ to ∞ and obtain

$$\sigma(V) \propto \sum_{l,l'=0}^{\infty} \frac{g^{2(l+l')}}{l!l'!} L[e|V| - \Delta - (l+l')\omega_0, \Gamma + \delta\omega(l+l')]. \tag{8.36}$$

By replacing the Lorentzian in equation (8.36) by the Fourier integral, we perform the summation over l and l' with the final result for the conductance as follows:

$$\sigma(V) \propto \int_0^\infty \mathrm{d}t \, \exp[2g^2 e^{-\delta\omega t} \cos(\omega_0 t) - \Gamma t] \times \cos[2g^2 e^{-\delta\omega t} \sin(\omega_0 t) - (e|V| - \Delta)t]. \tag{8.37}$$

From the isotope effect on the carrier mass, phonon densities of states, experimental values of the normal-state pseudogap and the residual resistivity (chapters 6 and 7), one estimates the coupling strength g^2 to be of the order of one, the characteristic phonon frequency about 30–100 meV, the phonon frequency dispersion about a few tens meV, the gap $\Delta/2$ about 30–50 meV and the impurity scattering rate of the order of 10 meV.

The SIS conductance (equation (8.37)) calculated with parameters in this range is shown in figure 8.7 for three different values of coupling. The conductance shape is remarkably different from the BCS density of states, both s-wave and d-wave. There is no Ohm's law in the normal region, $e|V| > \Delta$, the dip/hump features (due to phonon sidebands) are clearly seen already in the first derivative of the current, there is a substantial incoherent spectral weight beyond the quasi-particle peak for the strong coupling, $g^2 \geq 1$, and there is an unusual shape for the quasi-particle peaks. All these features as well as the temperature dependence of the gap are beyond BSC theory no matter what the symmetry of the gap is. However, they agree with the experimental SIS tunnelling spectra in cuprates [265,268,269,274,275]. In particular, the theory quantitatively describes one of the best tunnelling spectra obtained on $Bi_2Sr_2CaCu_2O_{8+\delta}$ single crystals by the break-junction technique [268] (figure 8.8). Some excess zero-bias conductance compared with the experiment is due to our approximation of the 'elastic' self-energy. The exact (energy-dependent) self-energy provides an agreement in this subgap region as well, as shown in figure 8.9. The dynamic conductance of Bi-2212 mesas [274] at low temperatures is almost identical to the theoretical one in figure 8.7 for $g^2 = 0.5625$.

The unusual shape of the main peaks (figure 8.8) is a simple consequence of the (quasi-)one-dimensional hole density of states near the edge of the oxygen band. The coherent ($l = l' = 0$) contribution to the current with no elastic scattering ($\Gamma = 0$) is given by

$$I_0 \propto \int_\Delta^\infty \frac{\mathrm{d}\epsilon}{(\epsilon - \Delta/2)^{1/2}} \int_\Delta^\infty \frac{\mathrm{d}\epsilon'}{(\epsilon' - \Delta/2)^{1/2}} \delta(\epsilon + \epsilon' - e|V|) \tag{8.38}$$

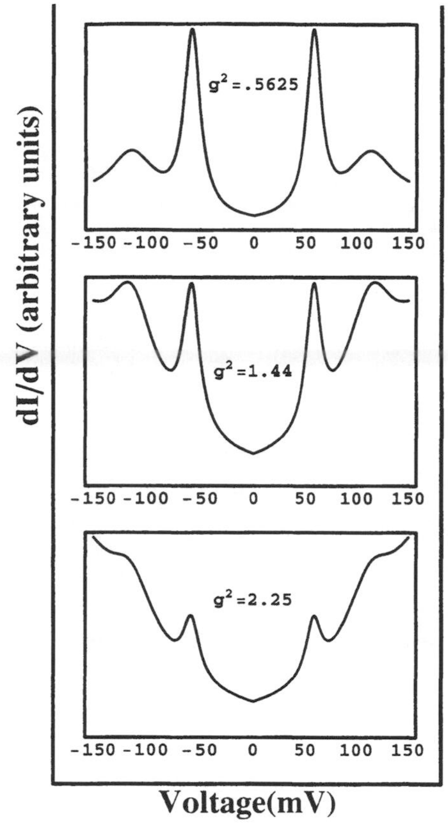

Figure 8.7. SIS tunnelling conductance in a bipolaronic superconductor for different values of the electron–phonon coupling, g^2, and $\Delta/2 = 29$ meV, $\omega_0 = 55$ meV, $\delta\omega = 20$ meV, $\Gamma = 8.5$ meV.

so that the conductance is a δ function:

$$\sigma_0(V) \propto \delta(e|V| - \Delta). \tag{8.39}$$

Hence, the width of the main peaks in the SIS tunnelling (figure 8.8) measures the elastic scattering rate directly.

The disappearance of the quasi-particle sharp peaks above T_c in Bi-cuprates can also be explained in the framework of bipolaron theory. Below T_c, bipolaronic condensate provides an effective screening of the long-range (Coulomb) potential of impurities, while above T_c the scattering rate might increase by many times (section 8.5). This sudden increase in Γ in the normal state washes out sharp peaks from tunnelling and ARPES spectra. The temperature-dependent part, $\Delta_c(T)$, of the total gap (equation (8.19)) is responsible for the temperature shift of the

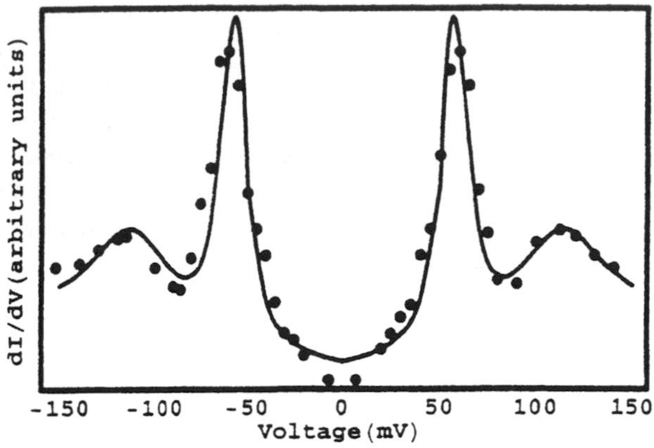

Figure 8.8. Theoretical conductance of figure 8.7 for $g^2 = 0.5625$ (full line) compared with the tunnelling spectrum obtained on $Bi_2Sr_2CaCu_2O_{8+\delta}$ single crystals by the break-junction technique [268] (dots).

main peaks in the superconducting state of an SIS junction, while the incoherent temperature-independent part Δ_p survives well above T_c as observed [274, 275].

8.4 ARPES

Let us discuss ARPES in doped charge-transfer Mott insulators in the framework of bipolaron theory [262] to describe some unusual ARPES features of high-T_c $YBa_2Cu_3O_{7-\delta}$ (Y123), $YBa_2Cu_4O_8$ (Y124) and several other materials.

Perhaps the most intriguing feature of ARPES in cuprates is an extremely narrow and intense peak lying below the Fermi energy, which is most clearly seen near the Y and X points in Y124 [278], and Y123 [279]. Its angular dependence and spectral shape as well as the origin of the featureless (but dispersive) background remain unclear. Some authors [278] refer to the peak as an extended van Hove singularity (evHs) arising from a plane (CuO_2) strongly correlated band. They also implicate the resulting (quasi-)one-dimensional DOS singularity as a possible origin for the high transition temperature. However, polarized ARPES studies of untwinned Y123 crystals of exceptional quality [279] unambiguously show that the peak is a narrow resonance arising primarily from the quasi-one-dimensional CuO_3 chains in the buffer layers rather than from the planes. Interestingly, a very similar narrow peak was observed by Park *et al* [280] in high-resolution ARPES near the gap edge of the cubic *semiconductor* FeSi with no Fermi surface at all.

As discussed in section 5.2, cuprates and many other transition metal compounds are charge transfer Mott–Hubbard insulators, where the first band to be doped is the oxygen band lying within the Hubbard gap, as shown in figure 5.3.

A single photoexited oxygen hole is described by the polaron spectral function of section 4.3.4. Its low-energy part is affected by low-frequency thermal lattice, spin and random fluctuations. The latter can be described as a 'Gaussian white noise' potential. Also p-hole polarons in oxides are almost one-dimensional due to a large difference in the ppσ and ppπ hopping integrals. This allows us to explain the shape of the narrow peaks in ARPES using the spectral density $A(k, E)$ of a one-dimensional particle in a Gaussian white noise potential [281].

8.4.1 Photocurrent

The interaction of the crystal with the electromagnetic field of frequency ν is described by the following Hamiltonian (in the dipole approximation):

$$H_{\text{int}} = (8\pi I)^{1/2} \sin(\nu t) \sum_{\boldsymbol{k},\boldsymbol{k}'} (\boldsymbol{e} \cdot \boldsymbol{d}_{\boldsymbol{k}\boldsymbol{k}'}) c_{\boldsymbol{k}}^{\dagger} h_{\boldsymbol{k}'}^{\dagger} + H.c. \tag{8.40}$$

where I is the intensity of the radiation with polarization $\boldsymbol{e}$, $\boldsymbol{k}$ is the momentum of the final state (i.e. of the photoelectron registered by the detector), $\boldsymbol{k}'$ is the (quasi-)momentum of the hole remaining in the sample after the emission and $c_{\boldsymbol{k}}^{\dagger}$ and $h_{\boldsymbol{k}'}^{\dagger}$ are their creation operators, respectively. For simplicity, we suppress the band index in $h_{\boldsymbol{k}'}^{\dagger}$. Due to the translational symmetry of the Bloch states, $|\boldsymbol{k}'\rangle \equiv u_{-\boldsymbol{k}'}(\boldsymbol{r}) \exp(-\mathrm{i}\boldsymbol{k}' \cdot \boldsymbol{r})$ (appendix A), there is a momentum conservation in the dipole matrix element,

$$\boldsymbol{d}_{\boldsymbol{k}\boldsymbol{k}'} = \boldsymbol{d}(\boldsymbol{k}) \delta_{\boldsymbol{k}+\boldsymbol{k}',\boldsymbol{G}} \tag{8.41}$$

with

$$\boldsymbol{d}(\boldsymbol{k}) = \mathrm{i}e(N/v_0)^{1/2} \nabla_{\boldsymbol{k}} \int_{v_0} \mathrm{e}^{-\mathrm{i}\boldsymbol{G}\cdot\boldsymbol{r}} u_{\boldsymbol{k}-\boldsymbol{G}}(\boldsymbol{r})\, \mathrm{d}\boldsymbol{r} \tag{8.42}$$

and v_0 is the unit cell volume ($\boldsymbol{G}$ is a reciprocal-lattice vector). The Fermi–Dirac golden rule gives the photocurrent to be

$$I(\boldsymbol{k}, E) = 4\pi^2 I |\boldsymbol{e} \cdot \boldsymbol{d}(\boldsymbol{k})|^2 \sum_{i,f} \mathrm{e}^{\Omega + \mu N_i - E_i} |\langle f | h_{\boldsymbol{k}-\boldsymbol{G}}^{\dagger} | i \rangle f|^2 \delta(E + E_{\mathrm{f}} - E_i) \tag{8.43}$$

where E is the binding energy, $E_{i,f}$ is the energy of the initial and final states and Ω, μ, N_i are the thermodynamic and chemical potentials and number of holes, respectively. By definition, the sum in equation (8.43) is $n(E)A(\boldsymbol{k} - \boldsymbol{G}, -E)$, where the spectral function

$$A(\boldsymbol{k} - \boldsymbol{G}, E) = -\frac{1}{\pi} \operatorname{Im} G^{\mathrm{R}}(\boldsymbol{k} - \boldsymbol{G}, E) \tag{8.44}$$

is proportional to the imaginary part of the retarded GF (appendix D) and $n(E) = [\exp(E/T)+1]^{-1}$ is the Fermi distribution. In the following, we consider temperatures well below the experimental energy resolution, so that $n(E) = 1$, if E is negative, and zero otherwise, and we put $\boldsymbol{G} = 0$.

The spectral function depends on essential interactions of a single hole with the rest of the system. The most important interaction in oxides is the Fröhlich electron–phonon interaction with c-axis polarized high-frequency phonons (section 5.2), which leads to the polaron spectral function (4.79). As a result, we obtain

$$I(\boldsymbol{k}, E) \sim |d(\boldsymbol{k})|^2 n(E) Z \delta(E + \xi_{\boldsymbol{k}}) + I_{\text{incoh}}(\boldsymbol{k}, E) \tag{8.45}$$

where $I_{\text{incoh}}(\boldsymbol{k}, E)$ is an incoherent part, which spreads from about $-\omega_0$ down to $-2E_p$. There might be some multi-phonon structure in $I_{\text{incoh}}(\boldsymbol{k}, E)$ as observed in tunnelling (section 8.3).

Here we concentrate on the angular, spectral and polarization dependence of the first *coherent* term in equation (8.45). The present experimental resolution [264] allows the intrinsic damping of the coherent quasi-particle excitations to be probed. The damping appears due to a random field and low-frequency lattice and spin fluctuations described by the polaron self-energy, $\Sigma(\boldsymbol{k}, E)$, so that the coherent part of the spectral function is given by

$$A_{\text{c}}(\boldsymbol{k}, E) = -\frac{Z}{\pi} \frac{\operatorname{Im} \Sigma(\boldsymbol{k}, E)}{[E + \operatorname{Re} \Sigma(\boldsymbol{k}, E) - \xi_{\boldsymbol{k}})]^2 + [\operatorname{Im} \Sigma(\boldsymbol{k}, E)]^2}. \tag{8.46}$$

Hence, the theory of narrow ARPES peaks is reduced to determining the self-energy of a hole.

8.4.2 Self-energy of one-dimensional hole in a non-crossing approximation

Due to energy conservation, small polarons exist in the Bloch states at temperatures below the optical phonon frequency $T < \omega_0/2$ (section 4.3.2). A finite polaron self-energy appears due to (quasi-)elastic scattering off impurities, a low-frequency deformation potential and spin fluctuations. First we apply the simplest non-crossing (ladder) approximation (chapter 3) to define an analytical $\Sigma(\boldsymbol{k}, E)$. Within this approximation the self-energy is $\boldsymbol{k}$-independent for a short-range scattering potential like a deformation or a screened impurity potential, so that

$$\Sigma(E) \propto \sum_{\boldsymbol{k}} G^{\text{R}}(\boldsymbol{k}, E) \tag{8.47}$$

where $G^{\text{R}}(\boldsymbol{k}, E) = [E - \xi_{\boldsymbol{k}} - \Sigma(E)]^{-1}$.

The hole energy spectrum is parametrized in a tight-binding model as

$$\xi_{\boldsymbol{k}}^{x,y} = 2t \cos(k_{x,y}a) - 2t' \cos(k_{y,x}a) - \mu. \tag{8.48}$$

We assume that the minima of two polaron bands (equation (8.48)), are found at the Brillouin zone boundary in X $(\pi, 0)$ and Y $(0, \pi)$.

As previously mentioned, the oxygen hole is (quasi-)one-dimensional due to a large difference between the oxygen hopping integrals for the orbitals elongated parallel to and perpendicular to the oxygen–oxygen hopping $t' \ll t$. This allows us to apply a one-dimensional approximation, reducing equation (8.48) to two one-dimensional parabolic bands near the X and Y points, $\xi_k^{x,y} = k^2/2m^* - \mu$ with $m^* = 1/2ta^2$ and k taking relative to $(\pi, 0)$ and $(0, \pi)$, respectively. Then, equation (8.47) for the self-energy in the non-crossing approximation takes the following form:

$$\Sigma(\epsilon) = -2^{-3/2}[\Sigma(\epsilon) - \epsilon]^{-1/2} \tag{8.49}$$

for each doublet component. Here we introduce a dimensionless energy (and self-energy), $\epsilon \equiv (E + \mu)/\Gamma_0$ using $\Gamma_0 = (D^2 m^*)^{1/3}$ as the energy unit. The constant D is the second moment of the Gaussian white noise potential, comprising thermal and random fluctuations as $D = 2(V_0^2 T/M + n_{\mathrm{im}}\alpha^2)$, where V_0 is the amplitude of the deformation potential, M is the elastic modulus, n_{im} is the impurity density and α is the coefficient of the δ-function impurity potential (i.e. the strength of the scattering potential). The solution is

$$\begin{aligned}\Sigma(\epsilon) = \frac{\epsilon}{3} &- \left(\frac{1 + \mathrm{i}3^{1/2}}{2}\right)\left[\frac{1}{16} + \frac{\epsilon^3}{27} + \left(\frac{1}{256} + \frac{\epsilon^3}{216}\right)^{1/2}\right]^{1/3} \\ &- \left(\frac{1 - \mathrm{i}3^{1/2}}{2}\right)\left[\frac{1}{16} + \frac{\epsilon^3}{27} - \left(\frac{1}{256} + \frac{\epsilon^3}{216}\right)^{1/2}\right]^{1/3} .\end{aligned} \tag{8.50}$$

While the energy resolution in the present ARPES studies is almost perfect [264], the momentum resolution remains finite in most experiments, $\delta > 0.1\pi/a$. Hence, we have to integrate the spectral function (equation (8.46)) with a Gaussian momentum resolution to obtain the experimental photocurrent:

$$I(\boldsymbol{k}, E) \propto \int_{-\infty}^{\infty} \mathrm{d}k'\, A_c(k', -E) \exp\left[-\frac{(k - k')^2}{\delta^2}\right]. \tag{8.51}$$

The integral is expressed in terms of $\Sigma(\epsilon)$ and the error function $w(z)$ as follows:

$$I(\boldsymbol{k}, E) \propto -\frac{2Z}{\delta} \operatorname{Im}\{\Sigma(\epsilon)[w(z_1) + w(z_2)]\} \tag{8.52}$$

where $z_{1,2} = [\pm k - \mathrm{i}/2\Sigma(\epsilon)]/\delta$, $w(z) = \mathrm{e}^{-z^2} \operatorname{erfc}(-\mathrm{i}z)$ and $\epsilon = (-E + \mu)/\Gamma_0$. This photocurrent is plotted as broken lines in figure 8.9 for two momenta, $k = 0.04\pi/a$ (almost Y or X points of the Brillouin zone) and $k = 0.3\pi/a$. The chemical potential is placed in the charge transfer gap below the bottom of the hole band, $\mu = -20$ meV, the momentum resolution is taken as $\delta = 0.28\pi/a$ and the damping $\Gamma_0 = 19$ meV. The imaginary part of the self-energy (equation (8.50)) disappears below $\epsilon = -3/2^{5/3} \simeq -0.9449$. Hence, this

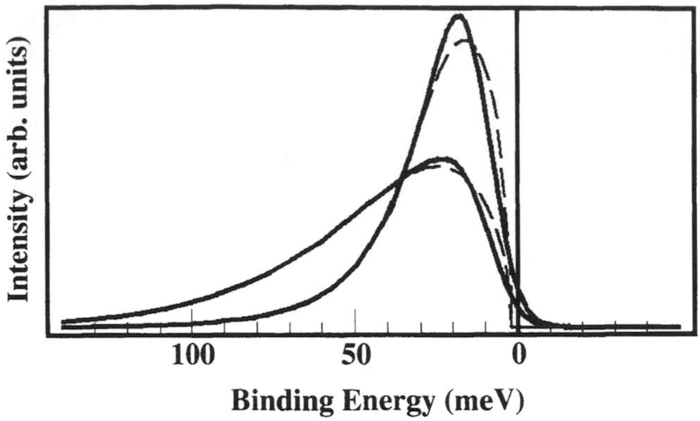

Figure 8.9. The polaron spectral function, integrated with the momentum resolution function for two angles, $k = 0.04\pi/a$ (upper curves) and $k = 0.30\pi/a$ with the damping $\Gamma_0 = 19$ meV, the momentum resolution $\delta = 0.28\pi/a$ and the polaron mass $m^* = 9.9m_e$. The bipolaron binding energy $2|\mu| = 40$ meV. The broken curves are the spectral density integrated with the momentum resolution in the non-crossing approximation.

approximation gives a well-defined gap rather than a pseudogap. Actually, the non-crossing approximation fails to describe the localized states inside the gap (i.e. a tail of the density of states). We have to go beyond a simple ladder to describe the single-electron tunnelling inside the gap and the ARPES spectra at small binding energies.

8.4.3 Exact spectral function of a one-dimensional hole

The exact spectral function for a one-dimensional particle in a random Gaussian white noise potential was calculated by Halperin [281] and the density of states by Frisch and Lloyd [282]. Halperin derived two pairs of differential equations from whose solutions the spectral function may be calculated:

$$A_c(k, \epsilon) = 4 \int_{-\infty}^{\infty} p_0(-z)\, \mathrm{Re}\, p_1(z)\, \mathrm{d}z. \tag{8.53}$$

Here $p_{0,1}(z)$ obeys two differential equations:

$$\left[\frac{\mathrm{d}^2}{\mathrm{d}z^2} + \frac{\mathrm{d}}{\mathrm{d}z}(z^2 + 2\epsilon)\right] p_0 = 0 \tag{8.54}$$

and

$$\left[\frac{\mathrm{d}^2}{\mathrm{d}z^2} + \frac{\mathrm{d}}{\mathrm{d}z}(z^2 + 2\epsilon) - z - \mathrm{i}k\right] p_1 + p_0 = 0 \tag{8.55}$$

with the boundary conditions

$$\lim_{z\to\infty} z^{2-n} p_n(z) = \lim_{z\to-\infty} z^{2-n} p_n(z) \tag{8.56}$$

where k is measured in units of $k_0 = (D^{1/2}m^*)^{2/3}$. The first equation may be integrated to give

$$p_0(z) = \frac{\exp(-z^3/3 - 2z\epsilon)\int_{-\infty}^{z}\exp(u^3/3 + 2u\epsilon)\,\mathrm{d}u}{\pi^{1/2}\int_0^{\infty} u^{-1/2}\exp(-u^3/12 - 2u\epsilon)\,\mathrm{d}u}. \tag{8.57}$$

The equation for $p_1(z)$ has no known analytic solution and, hence, must be solved numerically. There is, however, an asymptotic expression for $A_c(k,\epsilon)$ in the tail where $|\epsilon| \gg 1$:

$$A_c(k,\epsilon) \propto 2\pi(-2\epsilon)^{1/2}\exp\left[-\frac{4}{3}(-2\epsilon)^{3/2}\right]\cosh^2\left[\frac{\pi k}{(-8\epsilon)^{1/2}}\right]. \tag{8.58}$$

The result for $A_c(k, -E)$ integrated with the Gaussian momentum resolution is shown in figure 8.9 for two values of the momentum (full lines). In contrast to the non-crossing approximation, the exact spectral function (averaged with the momentum resolution function) has a tail due to the states localized by disorder within the normal-state gap. However, besides this tail, the non-crossing approximation gives very good agreement and for a binding energy greater than about 30 meV, it is practically exact.

The cumulative DOS (appendix B)

$$K_0(\epsilon) = (2\pi)^{-1}\int_{-\infty}^{\epsilon}\mathrm{d}\epsilon'\int_{-\infty}^{\infty}\mathrm{d}k\,A_p(k,\epsilon') \tag{8.59}$$

is expressed analytically [282] in terms of the tabulated Airy functions $\mathrm{Ai}(x)$ and $\mathrm{Bi}(x)$ as

$$K_0(\epsilon) = \pi^{-2}[\mathrm{Ai}^2(-2\epsilon) + \mathrm{Bi}^2(-2\epsilon)]^{-1}. \tag{8.60}$$

DOS $\rho(\epsilon) = \mathrm{d}K_0(\epsilon)/\mathrm{d}\epsilon$ fits well the voltage–current tunnelling characteristics of cuprates, as discussed in section 8.2.

8.4.4 ARPES in Y124 and Y123

With the polaronic doublet (equation (8.48)) placed above the chemical potential, we can quantitatively describe high-resolution ARPES in Y123 [279] and Y124 [278]. The exact one-dimensional polaron spectral function, integrated with the experimental momentum resolution (shown in figure 8.10), provides a quantitative fit to the ARPES spectra in Y124 along the Y–Γ direction. The angular dispersion is described with the polaron mass $m^* = 9.9m_e$. The spectral shape is reproduced well with $\Gamma_0 = 19$ meV. There is also quantitative agreement

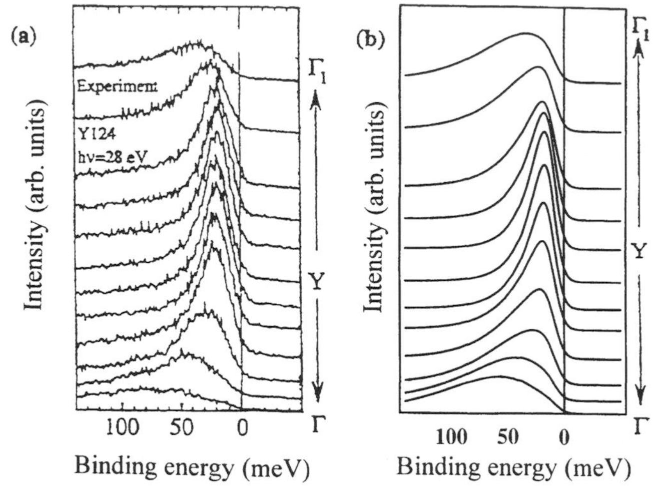

Figure 8.10. Theoretical ARPES spectra for the Y–Γ direction (*b*). Parameters are those of figure 8.9. The theory provides a quantitative fit to experiment (*a*) [278] in this scanning direction.

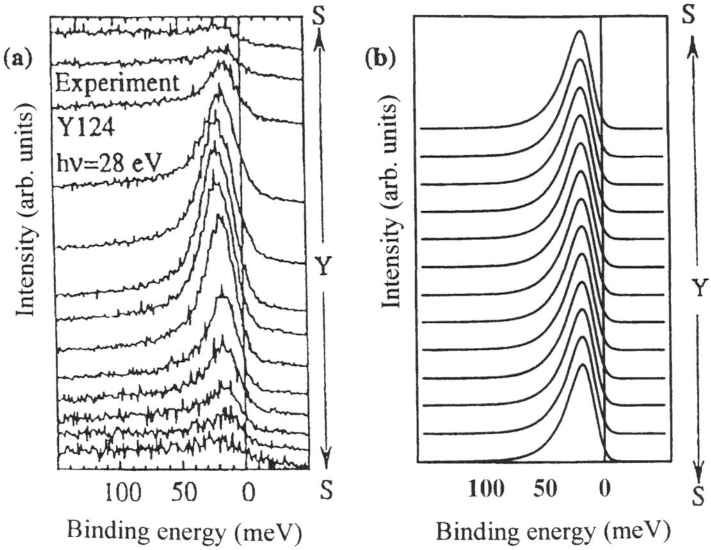

Figure 8.11. Theoretical ARPES spectra in Y124 for the Y–S direction (*b*). The theoretical fit agrees well with experiment (*a*) [278] in a restricted range of k_x near the Y-point and, outside this range, the theoretical peaks are somewhat higher than in experiment.

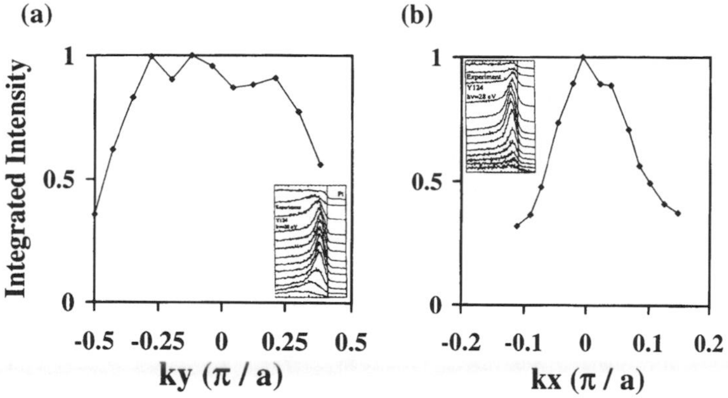

Figure 8.12. Energy-integrated ARPES intensity in Y124 in the Y–Γ (*a*) and Y–S (*b*) directions. Momenta are measured relative to the Y-point of the Brillouin zone.

between theory and experiment in the perpendicular direction Y–S, in a restricted region of small k_x, where almost no dispersion is observed around Y (figure 8.11).

However, there is a significant loss of the energy-integrated intensity along both directions (figure 8.12) which the theoretical spectral function alone cannot account for. The energy-integrated ARPES spectra obey the sum rule,

$$\int_{-\infty}^{\infty} \mathrm{d}E\, I(\boldsymbol{k}, E) \propto |d(\boldsymbol{k})|^2 \tag{8.61}$$

if the chemical potential is pinned inside the charge-transfer gap. Therefore, we have to conclude that the dipole matrix element depends on $\boldsymbol{k}$. The rapid loss of the integrated intensity in the Y–S direction was interpreted by some authors [283] as a Fermi-surface crossing. While the Fermi-surface crossing might be compatible with our scenario (see figure 5.3(*b*)), it is hard to believe that it has been really observed in Y124. Indeed, the peaks in the Y–S direction are all 15 meV or more below the Fermi level—at a temperature of 1 meV. If the loss of spectral weight were due to a Fermi-surface crossing, one would expect the peaks to approach much closer to the Fermi level.

Also the experimental spectral shape of the intensity at $\boldsymbol{k} = \boldsymbol{k}_\mathrm{F}$ is incompatible with any theoretical scenario, including different marginal Fermi-liquid models, as shown in figure 8.13. The spectral function on the Fermi surface should be close to a simple Lorentzian:

$$A_\mathrm{c}(\boldsymbol{k}_\mathrm{F}, E) \sim \frac{|E|^\beta}{E^2 + \text{constant} \times E^{2\beta}} \tag{8.62}$$

because the imaginary part of the self-energy behaves as $|E|^\beta$ with $0 \leq \beta \leq 2$ in a Fermi or in a marginal Fermi liquid. In contrast, the experimental intensity

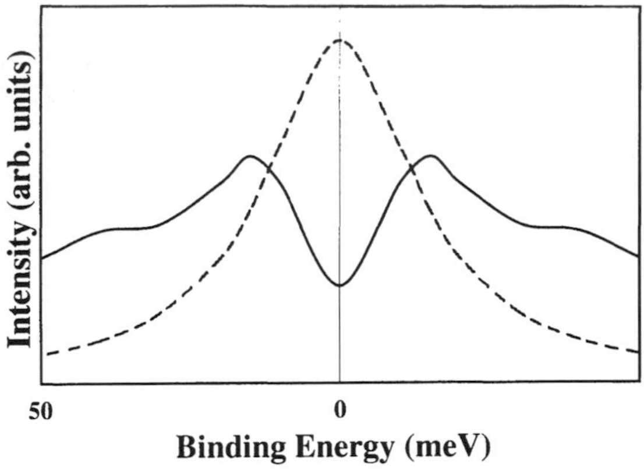

Figure 8.13. The experimental ARPES signal (full line) on the alleged Fermi surface does not correspond to a Fermi-liquid spectral function (broken line). We assume particle–hole symmetry to obtain the spectral function for negative binding energy.

shows a pronounced minimum at the alleged Fermi surface (figure 8.13). If there is indeed no Fermi-surface crossing, why then do some determinations point to a large Fermi surface in cuprates, which is drastically incompatible with their kinetic and thermodynamic properties? It might be due to the fact that the oxygen hole band has its minima at large $\boldsymbol{k}$ inside or even on the boundary of the Brillouin zone. Then ARPES show intense peaks near large $\boldsymbol{k}$ imitating a large Fermi-surface.

8.5 Sharp increase of the quasi-particle lifetime below T_c

It has been observed at certain points in the Brillouin zone that the ARPES peak in bismuth cuprates is relatively sharp at low temperatures in the superconduncting state but that it almost disappears into the background above the transition [284, 285]. A very sharp increase in the quasi-particle lifetime in the superconducting state has been observed in the thermal Hall conductivity measurements [286]. The abrupt appearence of the quasi-particle state below T_c is also implied in the tunnelling I–V characteristics [274, 275].

Here we show that a large increase in the quasi-particle lifetime below T_c is due to the screening of scatterers by the Bose–Einstein condensate of charged bipolarons [203, 287]. To illustrate the point, let us calculate the scattering cross section of a charged particle (mass m, charge e) scattered by a static Coulomb potential $V(r)$ screened by a charged Bose gas. The general theory of potential scattering in terms of phase shifts was developed in the earliest days of quantum mechanics [288]. While, in principle, this allows scattering cross sections to be calculated for an arbitary potential, in practice the equations for the radial part of the wavefunction may only be solved analytically for a few potentials and, in the standard formulation, are not in a suitable form for numerical computation. The 'variable phase' approach [289] solves this problem by taking the phase-shift functions of the radial coordinate. The Schrödinger equation for each radial component of the wavefunction then reduces to a first-order differential equation for the corresponding phase shift.

In dimensionless units ($\hbar = 2m = 1$), the Schrödinger equation for the radial part of the angular momentum component (l) of the wavefunction of a particle with wavevector k undergoing potential scattering is

$$u_l''(r) + [k^2 - l(l+1)/r^2 - V(r)]u_l(r) = 0. \tag{8.63}$$

The scattering phase shift, δ_l, is obtained by comparison with the asymptotic relation

$$u_l(r) \xrightarrow{r\to\infty} \sin(kr - l\pi/2 + \delta_l) \tag{8.64}$$

and the scattering cross section is then

$$\sigma = \frac{4\pi}{k^2}\sum_{l=0}^{\infty} \sin^2 \delta_l. \tag{8.65}$$

In the variable phase method [289], we must satisfy the condition that

$$V(r) \xrightarrow{r\to 0} V_0 r^{-n} \tag{8.66}$$

with $n < 2$. The angular momentum phase shift is then

$$\delta_l = \lim_{r\to\infty} \delta_l(r) \tag{8.67}$$

where the phase function, $\delta_l(r)$, satisfies the phase equation

$$\delta_l'(r) = -k^{-1}V(r)[\cos\delta_l(r) j_l(kr) - \sin\delta_l(r) n_l(kr)]^2 \tag{8.68}$$

with

$$\delta_l(r) \xrightarrow{r\to 0} -\frac{V_0 r^{-n}}{k^2}\frac{(kr)^{2l+3}}{(2l+3-n)[(2l+1)!!]^2} \tag{8.69}$$

and $j_l(x)$ and $n_l(x)$ are the Riccati–Bessel functions. In the $l = 0$ case, the phase equation reduces to

$$\delta_0'(r) = -k^{-1}V(r)\sin^2[kr + \delta_0(r)]. \tag{8.70}$$

In the slow-particle limit, we may neglect higher-order contributions to the scattering cross section, so that

$$\sigma = \frac{4\pi}{k^2}\sin^2\delta_0. \tag{8.71}$$

The effective potential about a point charge in the charged Bose gas (CBG) was calculated by Hore and Frankel [290]. Its static dielectric function is:

$$\epsilon(q,0) = 1 + \frac{4\pi(e^*)^2}{q^2\epsilon_0\Omega}\sum_{\boldsymbol{k}}\left(\frac{f_{\boldsymbol{k}} - f_{\boldsymbol{k}-\boldsymbol{q}}}{q^2/2m^{**} - \boldsymbol{k}\cdot\boldsymbol{q}/m^{**}}\right) \tag{8.72}$$

in which $e^* = 2e$ the boson charge, Ω is the volume of the system and

$$f_{\boldsymbol{k}} = \frac{1}{\exp\left(\frac{k^2/2m^{**}-\mu}{T}\right) - 1}.$$

Eliminating the chemical potential, for small q the dielectric function for $T < T_c$ is

$$\epsilon(q,0) = 1 + \frac{4(m^{**}\omega_p)^2}{q^4}\left[1 - \left(\frac{T}{T_c}\right)^{3/2}\right] + O\left(\frac{1}{q^3}\right) \tag{8.73}$$

and, for $T \to \infty$,

$$\epsilon(q,0) = 1 + \frac{1}{q^2}\frac{m^{**}\omega_p^2}{T}\left[1 + \frac{\zeta(3/2)}{2^{3/2}}\left(\frac{T_c}{T}\right)^{3/2} + \cdots\right] + O(q^0) \tag{8.74}$$

with $\omega_p^2 = 4\pi(e^*)^2 n_b/(\epsilon_0 m^{**})$ and n_b the boson density. If the unscreened scattering potential is the Coulomb potential ($V(r) = V_0/r$), then performing the inverse Fourier transforms, one finds that, for $T < T_c$ [290],

$$\lim_{r\to\infty} V(r) = \frac{V_0}{r}\exp[-K_s r]\cos[K_s r] \equiv V_s(r) \tag{8.75}$$

with

$$K_s = (m^{**}\omega_p)^{1/2}\left[1 - \left(\frac{T}{T_c}\right)^{3/2}\right]^{1/4} \tag{8.76}$$

and for $T \to \infty$,

$$\lim_{r\to\infty} V(r) = \frac{V_0}{r}\exp[-K_n r] \equiv V_n(r) \tag{8.77}$$

with

$$K_n = \left(\frac{m^{**}\omega_p^2}{T}\right)^{1/2}\left[1 + \frac{\zeta(3/2)}{2^{3/2}}\left(\frac{T_c}{T}\right)^{3/2} + \cdots\right]^{1/2}. \tag{8.78}$$

The $T < T_c$ result is exact for all r at $T = 0$.

There are two further important analytical results; the first (Levinson's theorem [289]) states that for 'regular' potentials (which include all those with which we shall be concerned), the zero-energy phase shift is equal to π multiplied by the number of bound states of the potential. The second is the well-known

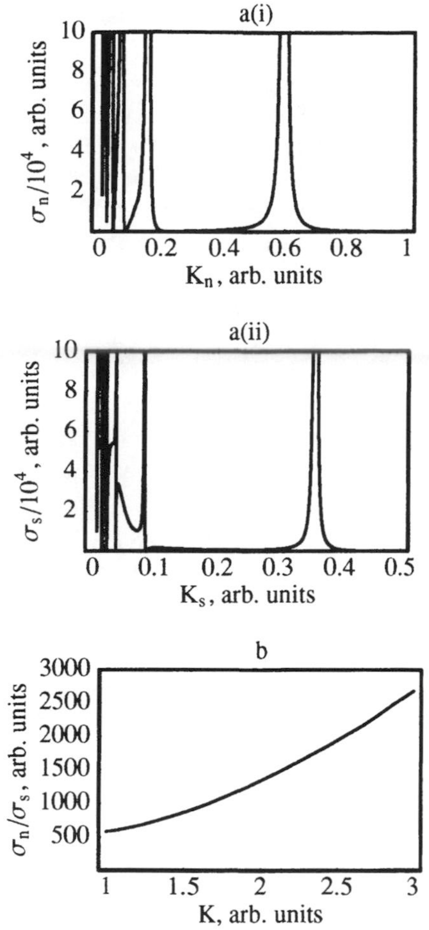

Figure 8.14. (*a*) Plots of zero-energy scattering cross sections (i) σ_n and (ii) σ_s against screening wavevector K for the potentials (i) $V_n(r) = -(1/r)e^{-K_n r}$ and (ii) $V_s(r) = -(1/r)\mathrm{e}^{-K_s r}\cos(K_s r)$. (*b*) Plot of σ_n/σ_s for a range of $K = K_n = K_s$ in which both potentials have no bound states. In each case the units are those used to derive the phase equation.

Wigner resonance scattering formula, which states that for slow-particle scattering of a particle with energy E off a potential with a shallow bound state of the binding energy $\epsilon \lesssim E$, the total scattering cross section is

$$\sigma = \frac{2\pi}{m}\frac{1}{E + |\epsilon|}. \tag{8.79}$$

The zero-energy scattering cross sections for the potentials $V_n(r) =$

$-(V_0/r)e^{-Kr}$ and $V_s(r) = -(V_0/r)e^{-Kr}\cos(Kr)$ are shown in figure 8.14(*a*). These graphs are plotted for $V_0 = 1$; in each case, the equivalent graph for arbitary V_0 may be found by rescaling σ and K. According to the Wigner formula (equation (8.79)) as K decreases, when a new bound state appears there should be a peak in the cross section, as there will then be a minimum in the binding energy of the shallowest bound state. This is the origin of the peaks in figure 8.14(*a*), which may be checked using Levinson's theorem. It can also be seen that as K decreases, the first few bound states appear at higher K in the ordinary Yukawa potential; this agrees with the intuitive conclusion that the bound states should, in general, be deeper in a non-oscillatory potential. Another intuitive expectation which is also borne out is that for a given V_0 and K, the non-oscillatory potential should be the stronger scatterer. In figure 8.14(*b*) it may be seen that this is the case when K is large enough, so that both potentials have no bound states (the difference in cross sections is then, in fact, about three orders of magnitude).

At zero temperature, the screening wavevector is $K_0 = (m^{**}\omega_p)^{1/2}$, and at a temperature $T = T_c + 0$ just above the transition, $K_{T_c} = (m^{**}\omega_p^2/T_c)^{1/2}$. Substituting ω_p and $T_c \simeq 3.3n_b^{2/3}/m^{**}$, we obtain

$$\frac{K_{T_c}}{K_0} = \left(\frac{2.1e\sqrt{m^{**}}}{\epsilon_0^{1/2} n_b^{1/6}}\right)^{1/2}. \tag{8.80}$$

From this, we see that the ratio is only marginally dependent on the boson density, so substituting for $n_b = 10^{21}$ cm^{-3}, e and m_e, we obtain

$$\frac{K_{\alpha T_c}}{K_0} = 3.0\left(\frac{m^{**}}{m_e\epsilon_0}\right)^{1/4}. \tag{8.81}$$

With a realistic boson mass $m^{**} = 10m_e$ and dielectic constant $\epsilon_0 = 100$, K_{T_c} and K_0, while different, are of the same order of magnitude. If the screening wavevectors are such that neither the normal state nor condensate impurity potentials have bound states, with these parameters it would then follow that the quasi-particle lifetime is much greater in the superconducting state, figure 8.14(*b*). This effect could then explain the appearance of sharp quasi-particle peaks in ARPES and tunnelling, and the enhancement of the thermal conductivity in cuprates below T_c [203] due to a many-fold increase in the quasi-particle lifetime.

8.6 Symmetry of the order parameter and stripes

In bipolaron theory, the symmetry of the BEC on a lattice should be distinguished from the 'internal' symmetry of a single-bipolaron wavefunction and from the symmetry of a single-particle excitation gap. As described in section 4.7.9, a Bose condensate of bipolarons might be d-wave, if bipolaron bands have their minima at finite $\boldsymbol{K}$ in the centre-of-mass Brillouin zone. The d-wave condensate reveals itself as a *checkerboard* modulation of the hole density and of the global gap (equation (8.19)) in the real (Wannier) space (figure 4.14). At the same time, the single-particle excitation spectrum might be an anysotropic s-wave providing an explanation of conflicting experimental observations.

The two-dimensional pattern (figure 4.14) is oriented along the diagonals, i.e. the d-wave bipolaron condensate is 'striped'. Hence, there is a fundamental connection between stripes detected by different techniques [292,293] in cuprates and the symmetry of the order parameter [123]. Originally antiferromagnetic interactions were thought to give rise to spin and charge segregation (stripes) [291]. However, the role of long-range Coulomb and Fröhlich interactions has not been properly addressed. Here we show that the Fröhlich electron–phonon interaction combined with the direct Coulomb repulsion does not lead to charge segregation like strings or stripes in doped insulators and the antiferromagnetic exchange interaction is not sufficient to produce long stripes either [294]. However, the Fröhlich interaction significantly reduces the Coulomb repulsion, and allows much weaker short-range electron–phonon and antiferromagnetic interactions to bound carriers into small bipolarons. Then the d-wave Bose condensate of bipolarons naturally explains superstripes in cuprates.

As discussed in section 4.4. the extention of the deformation surrounding (Fröhlich) polarons is large, so their deformation fields overlap at a finite density. However, taking into account both the long-range attraction of polarons due to the lattice deformations *and* the direct Coulomb repulsion, the net long-range interaction is repulsive. At distances larger than the lattice constant ($|\boldsymbol{m}-\boldsymbol{n}| \geq a \equiv 1$), this interaction is significantly reduced to

$$v_{ij} = \frac{e^2}{\epsilon_0 |\boldsymbol{m}-\boldsymbol{n}|}. \tag{8.82}$$

Optical phonons reduce the bare Coulomb repulsion at large distances in ionic solids if $\epsilon_0 \gg 1$, which is the case in oxides.

Let us first consider a non-adibatic and intermediate regime when the characteristic phonon energy is comparable with the kinetic energy of holes. In this case the problem is reduced to narrow-band fermions with a repulsive interaction (equation (8.82)) at large distances and a short-range attraction at atomic distancies. Because the net long-range repulsion is relatively weak, the relevant dimensionless parameter r_s ($= m^* e^2/\epsilon_0(4\pi n/3)^{1/3}$) is not very large in doped cuprates and the Wigner crystallization does not appear at any physically interesting density. In contrast, polarons could be bound into small bipolarons and/or into small clusters as discussed in sections 4.6.3 and 5.4.2 but, in any case, the long-range repulsion would prevent any clustering in infinitely charged domains.

In the opposite adiabatic limit, one can apply a discrete version [80] of the continuous nonlinear Pekar equation [61], taking into account the Coulomb repulsion and lattice deformation for a single-polaron wavefunction, $\psi_{\boldsymbol{n}}$ (the amplitude of the Wannier state $|\mathbf{n}\rangle$):

$$-\sum_{\mathbf{m}\neq 0} t(\mathbf{m})[\psi_{\mathbf{n}} - \psi_{\mathbf{n}+\mathbf{m}}] - e\phi_{\mathbf{n}}\psi_{\mathbf{n}} = E\psi_{\mathbf{n}}. \tag{8.83}$$

The potential $\phi_{\mathbf{n},k}$ acting on a fermion, k, at the site $\mathbf{n}$ is created by the polarization of the lattice, $\phi^{l}_{\mathbf{n},k}$, and by the Coulomb repulsion with the other $M-1$ fermions, $\phi^{\mathrm{c}}_{\mathbf{n},k}$:

$$\phi_{\mathbf{n},k} = \phi^{l}_{\mathbf{n},k} + \phi^{\mathrm{c}}_{\mathbf{n},k}. \tag{8.84}$$

Both potentials satisfy the discrete Poisson equation:

$$\kappa\Delta\phi^{\mathrm{l}}_{\mathbf{n},k} = 4\pi e\sum_{p=1}^{M}|\psi_{\mathbf{n},p}|^2 \tag{8.85}$$

and

$$\epsilon_\infty\Delta\phi^{\mathrm{c}}_{\mathbf{n},k} = -4\pi e\sum_{p=1,p\neq k}^{M}|\psi_{\mathbf{n},p}|^2 \tag{8.86}$$

with $\Delta\phi_{\mathbf{n}} = \sum_{\mathbf{m}}(\phi_{\mathbf{n}} - \phi_{\mathbf{n}+\mathbf{m}})$. Then the functional J [61], describing the total energy in this self-consistent Hartree approximation, is given by

$$J = -\sum_{\mathbf{n},p,\mathbf{m}\neq 0}\psi^{*}_{\mathbf{n},p}t(\mathbf{m})[\psi_{\mathbf{n},p} - \psi_{\mathbf{n}+\mathbf{m},p}] - \frac{2\pi e^2}{\kappa}\sum_{\mathbf{n},p,\mathbf{m},q}|\psi_{\mathbf{n},p}|^2\Delta^{-1}|\psi_{\mathbf{m},q}|^2$$
$$+\frac{2\pi e^2}{\epsilon_\infty}\sum_{\mathbf{n},p,\mathbf{m},q\neq p}|\psi_{\mathbf{n},p}|^2\Delta^{-1}|\psi_{\mathbf{m},q}|^2.$$

The single-particle function of a hole trapped in a string of the length N is $\psi_n = N^{-1/2}\exp(ikn)$ with periodic boundary conditions, so that the functional J is expressed as $J = T + U$, where $T = -2t(N-1)\sin(\pi M/N)/[N\sin(\pi/N)]$ is the kinetic energy, proportional to t, and

$$U = -\frac{e^2}{\kappa}M^2 I_N + \frac{e^2}{\epsilon_\infty}M(M-1)I_N \tag{8.87}$$

corresponds to the polarization and Coulomb energies. Here the integral I_N is given by

$$I_N = \frac{\pi}{(2\pi)^3}\int_{-\pi}^{\pi}\mathrm{d}x\int_{-\pi}^{\pi}\mathrm{d}y\int_{-\pi}^{\pi}\mathrm{d}z\,\frac{\sin(Nx/2)^2}{N^2\sin(x/2)^2}(3-\cos x-\cos y-\cos z)^{-1}.$$

It has the following asymptotics at $N \gg 1$:

$$I_N = \frac{1.31 + \ln N}{N}. \tag{8.88}$$

If we split the first (attractive) term in equation (8.87) into two parts by replacing M^2 for $M + M(M - 1)$, then it becomes clear that the net interaction between polarons remains repulsive in the adiabatic regime as well because $\kappa > \epsilon_\infty$. And this shows the absence of strings or stripes. The energy of M well-separated polarons is lower than the energy of polarons trapped in a string.

When a short-range e–ph (or exchange) interaction is taken into account, a string of finite length can appear (sections 4.6.3 and 5.4.2). We can readily estimate its length by the use of equation (8.87) for any type of short-range interaction. For example, dispersive phonons, $\omega_{\boldsymbol{q}} = \omega + \delta\omega(\cos q_x + \cos q_y + \cos q_z)$ with a q-independent matrix element $\gamma(\boldsymbol{q}) = \gamma$, yield a short-range polaron–polaron attraction:

$$v_{\text{att}}(\boldsymbol{n} - \boldsymbol{m}) = -E_{\text{att}}(\delta\omega/\omega)\delta_{|\boldsymbol{n}-\boldsymbol{m}|,1} \tag{8.89}$$

where $E_{\text{att}} = \gamma^2\omega/2$. Taking into account the long-range repulsion as well, the potential energy of the string with $M = N$ polarons is now

$$U = \frac{e^2}{\epsilon_0}N^2 I_N - \frac{N E_{\text{att}}\delta\omega}{\omega}. \tag{8.90}$$

Minimizing this energy yields the length of the string as

$$N = \exp\left(\frac{\epsilon_0 E_{\text{att}}\delta\omega}{e^2\omega} - 2.31\right). \tag{8.91}$$

Actually, this expression provides a fair estimate of the string length for any kind of attraction (not only generated by phonon dispersion) but also by antiferromagnetic exchange and/or by Jahn–Teller type of e–ph interactions. Due to the numerical coefficient in the exponent in equation (8.91), one can expect only short strings (if any) with realistic values of E_{att} (about 0.1 eV) and with a static dielectric constant $\epsilon_0 \leq 100$.

We conclude that there are no strings in ionic narrow-band doped insulators with only the Fröhlich interaction. Short-range electron–phonon and/or antiferromagnetic interactions could give rise to a bipolaronic liquid and/or short strings *only* because the long-range Fröhlich interaction significantly reduces the Coulomb repulsion in highly polarizable ionic insulators. However, with the typical values of $\epsilon_0 = 30$, $a = 3.8$ Å, one obtains $N \leqslant 2$ in equation (8.91) even with E_{att} as high as 0.3 eV. Hence, the short-range attractive forces are not strong enough to segregate charges into strings of any length, at least not in all high-T_c cuprates.

If (bi)polaronic carriers in many cuprates are in a liquid state, one can pose the key question of how one can see stripes at all. In fact, the bipolaron condensate might be striped owing to the bipolaron energy band dispersion, as previously discussed. In this scenario, the hole density, which is about twice that of the condensate density at low temperatures, is striped, with the characteristic period of stripes determined by inverse wavevectors corresponding to bipolaron band minima. Such an interpretation of stripes is consistent with the inelastic neutron scattering in $YBa_2Cu_3O_{7-\delta}$, where the incommensurate peaks were observed *only* in the *superconducting* state [154]. The vanishing at T_c of the incommensurate peaks is inconsistent with any other stripe picture, where a characteristic distance needs to be observed in the normal state as well. In contrast, with the d-wave *striped* Bose–Einstein condensate, the incommensurate neutron peaks should disappear above T_c, as observed. Importantly, a checkerboard modulation (figure 4.14) has been observed in STM experiments with [295] and without [296] an applied magnetic field in Bi cuprates.

Chapter 9

Conclusion

The main purpose of this book was to lead the reader from the basic principles through detailed derivations to a description of the many facinating phenomena in conventional and unconventional (high T_c) superconductors. The seminal work by Bardeen, Cooper and Schrieffer taken further by Eliashberg to intermediate coupling solved the major scientific problem of Condensed Matter Physics in the first half of the 20th century. High-temperature superconductors present a challenge to the conventional theory. While the BCS theory gives a qualitatively correct description of some novel superconductors like magnesium diborade and doped fullerenes, if the phonon dressing of carriers (i.e. polaron formation) is properly taken into account, cuprates remain a real problem. Here strong antiferromagnetic and charge fluctuations and the Fröhlich and Jahn–Teller electron–phonon interactions have been identified as an essential piece of physics. We have discussed the multi-polaron approach to the problem based on our recent extension of BCS theory to the strong-coupling regime. The low-energy physics in this regime is that of small bipolarons, which are real-space electron (hole) pairs dressed by phonons. They are itinerant quasi-particles existing in the Bloch states at temperatures below the characteristic phonon frequency. We have discussed a few applications of the bipolaron theory to cuprates, in particular the bipolaron theory of the normal state, of the superconducting critical temperature and the upper crtitical field, the isotope effect, normal and superconducting gaps, the magnetic-field penetration depth, tunnelling and the Andreev reflection, angle-resolved photoemission, stripes and the symmetry of the order parameter. These and some other experimental observations have been satisfactorily explained using this particular approach, which provides evidence for a novel state of electronic matter in layered cuprates. This is the charged Bose liquid of bipolarons. A direct measurement of a double elementary charge $2e$ on carriers in the normal state could be decisive.

Appendix A

Bloch states

A.1 Bloch theorem

If we neglect the interaction of carriers with impurities and ion vibrations, and take the Coulomb interaction between carriers in the mean-field (Hartree–Fock) approximation (similar to the central-field approximation for atoms) into account, we arrive at a one-electron Hamiltonian with a periodic (crystal field) potential $V(\boldsymbol{r})$. The wavefunction obeys the Schrödinger equation

$$\left[-\frac{\nabla^2}{2m_\mathrm{e}} + V(\boldsymbol{r})\right]\psi(\boldsymbol{r}) = E\psi(\boldsymbol{r}). \tag{A.1}$$

The Bloch theorem states that one-particle eigenstates in the periodic potential are sorted by the wavevector $\boldsymbol{k}$ in the first Brillouin zone and by the band index n and have the form

$$\psi_{n\boldsymbol{k}}(\boldsymbol{r}) = u_{n\boldsymbol{k}}(\boldsymbol{r})\mathrm{e}^{\mathrm{i}\boldsymbol{k}\cdot\boldsymbol{r}}. \tag{A.2}$$

To prove the theorem, let us define the translation operator $T_{\boldsymbol{l}}$, which shifts the argument by the lattice vector $\boldsymbol{l}$ while acting upon any function $F(\boldsymbol{r})$:

$$T_{\boldsymbol{l}}F(\boldsymbol{r}) = F(\boldsymbol{r}+\boldsymbol{l}). \tag{A.3}$$

Since the Hamiltonian has the translation symmetry, $H(\boldsymbol{r}+\boldsymbol{l}) = H(\boldsymbol{r})$, it commutes with $T_{\boldsymbol{l}}$. Hence, the eigenstates of H can be chosen to be simultaneously eigenstates of all the $T_{\boldsymbol{l}}$:

$$T_{\boldsymbol{l}}\psi(\boldsymbol{r}) = c(\boldsymbol{l})\psi(\boldsymbol{r}) \tag{A.4}$$

where $c(\boldsymbol{l})$ is a number depending on $\boldsymbol{l}$. The eigenvalues of the translation operators are related as follows:

$$c(\boldsymbol{l}+\boldsymbol{l}') = c(\boldsymbol{l})c(\boldsymbol{l}') \tag{A.5}$$

since shifting the argument by $\boldsymbol{l} + \boldsymbol{l}'$ leads to the same function as two successive translations, $T_{l+l'} = T_l T_{l'}$. This relation is satisfied if

$$c(\boldsymbol{l}) = \exp(\mathrm{i}\boldsymbol{k} \cdot \boldsymbol{l}) \tag{A.6}$$

with any $\boldsymbol{k}$. Imposing an appropriate boundary condition on the wavefunction makes the allowed wavevectors $\boldsymbol{k}$ real and confined to the first Brillouin zone. The most convenient condition is the Born–von Karman periodic boundary condition

$$\psi(\boldsymbol{r} + N_j \boldsymbol{a}_j) = \psi(\boldsymbol{r}) \tag{A.7}$$

where $j = 1, 2, 3$, the $\boldsymbol{a}_j$ are three primitive vectors of the crystal lattice and N_j are large integers, such that $N = N_1 N_2 N_3$ is the total number of primitive cells in the crystal ($N \simeq 10^{23}\ \mathrm{cm}^{-3}$). This requires that

$$\exp(\mathrm{i}N_j \boldsymbol{k} \cdot \boldsymbol{a}_j) = 1. \tag{A.8}$$

Therefore, the general form for allowed wavevectors is

$$\boldsymbol{k} = \frac{n_1}{N_1}\boldsymbol{b}_1 + \frac{n_2}{N_2}\boldsymbol{b}_2 + \frac{n_3}{N_3}\boldsymbol{b}_3. \tag{A.9}$$

Here, n_i are integers and $\boldsymbol{b}_i$ are primitive vectors for a reciprocal lattice, which satisfy

$$\boldsymbol{b}_i \boldsymbol{a}_j = 2\pi \delta_{ij} \tag{A.10}$$

where δ_{ij} is the Kronecker delta symbol ($\delta_{ij} = 1$, if $i = j$, and zero otherwise). Replacing $\boldsymbol{k}$ by $\boldsymbol{k} + \boldsymbol{G}$ does not change any of the $c(\boldsymbol{l})$ if $\boldsymbol{G}$ is a reciprocal lattice vector (i.e. a linear combination of the $\boldsymbol{b}_i$ with integer coefficients). Hence, all eigenstates of the periodic Hamiltonian can be sorted by the wavevectors confined to the primitive cell of the reciprocal lattice, for example to the region

$$-\pi < \boldsymbol{k}\boldsymbol{a}_j \leqslant \pi \tag{A.11}$$

which is the first Brillouin zone. In particular for a simple cubic lattice with period a,

$$\boldsymbol{k} = \frac{2\pi}{a} \sum_{i=1,2,3} \frac{n_i}{N_i}\boldsymbol{e}_i \tag{A.12}$$

where $-N_i/2 < n_i \leqslant N_i/2$ and $\boldsymbol{e}_i$ are unit vectors parallel to the primitive lattice vectors. The number of allowed wavevectors in the first Brillouin zone is equal to the number of cells, N, in the crystal. The volume, $\Delta\boldsymbol{k}$, of the reciprocal $\boldsymbol{k}$-space per allowed value of $\boldsymbol{k}$ is just the volume of a little parallelepiped with edges $\boldsymbol{b}_i/N_i$:

$$\Delta\boldsymbol{k} = \frac{1}{N}\boldsymbol{b}_1 \cdot \boldsymbol{b}_2 \times \boldsymbol{b}_3. \tag{A.13}$$

Since the volume of the reciprocal lattice primitive cell is $(2\pi)^3 N/V$, where V is the volume of the crystal, $\Delta \boldsymbol{k}$ can be written in the alternative form:

$$\Delta \boldsymbol{k} = \frac{(2\pi)^3}{V}. \tag{A.14}$$

This allows us to replace the sum over all allowed $\boldsymbol{k}$ by an integral over the first Brillouin zone,

$$\sum_{\boldsymbol{k}} F(\boldsymbol{k}) = \frac{V}{(2\pi)^3} \sum_{\boldsymbol{k}} F(\boldsymbol{k}) \Delta \boldsymbol{k} = \frac{V}{(2\pi)^3} \int_{\mathrm{BZ}} F(\boldsymbol{k})\, \mathrm{d}\boldsymbol{k} \tag{A.15}$$

for any analytical function $F(\boldsymbol{k})$. Different eigenstates of the periodic Hamiltonian are also the eigenstates of the translation operator. Hence, they transform under translations as

$$\psi_{\boldsymbol{k}}(\boldsymbol{r} + \boldsymbol{l}) = \mathrm{e}^{\mathrm{i}\boldsymbol{k}\cdot\boldsymbol{l}} \psi_{\boldsymbol{k}}(\boldsymbol{r}). \tag{A.16}$$

If we write $\psi(\boldsymbol{r})$ in the form

$$\psi_{\boldsymbol{k}}(\boldsymbol{r}) = u_{\boldsymbol{k}}(\boldsymbol{r}) \mathrm{e}^{\mathrm{i}\boldsymbol{k}\cdot\boldsymbol{r}} \tag{A.17}$$

then $u_{\boldsymbol{k}}(\boldsymbol{r})$ should be periodic ($u_{\boldsymbol{k}}(\boldsymbol{r} + \boldsymbol{l}) = u_{\boldsymbol{k}}(\boldsymbol{r})$) as follows from equation (A.16). Substituting $\psi_{\boldsymbol{k}}(\boldsymbol{r})$ into the Schrödinger equation yields the following equation for the periodic part of the wavefunction:

$$\left(-\frac{1}{2m_{\mathrm{e}}}(\nabla^2 + 2\mathrm{i}\boldsymbol{k} \cdot \nabla) + V(\boldsymbol{r})\right) u_{\boldsymbol{k}}(\boldsymbol{r}) = \left(E_{\boldsymbol{k}} - \frac{k^2}{2m_{\mathrm{e}}}\right) u_{\boldsymbol{k}}(\boldsymbol{r}). \tag{A.18}$$

This equation should be solved within the primitive unit cell and extended periodically for the rest of the crystal. Just like for particles in a box, the eigenstates of the equation are each identified by a discrete quantum number $n = 0, 1, 2, \ldots$. Hence, the eigenstates of an electron in a crystal are described by almost continuous wavevectors $\boldsymbol{k}$ in the first Brillouin zone, and by a discrete band index n. The wavevector, $\boldsymbol{k}$, plays a fundamental role in the electron energy band structure and dynamics, similar to that of the free electron momentum $\boldsymbol{p}$. However, although the free-electron wavevector and the momentum are the same (if $\hbar = 1$), the Bloch $\boldsymbol{k}$ is *not* proportional to the electron momentum on a lattice. When acting on $\psi_{\boldsymbol{k}}(\boldsymbol{r})$, the momentum operator $\boldsymbol{p} = -\mathrm{i}\nabla$ yields

$$-\mathrm{i}\nabla \psi_{\boldsymbol{k}}(\boldsymbol{r}) = \boldsymbol{k}\psi_{\boldsymbol{k}}(\boldsymbol{r}) - \mathrm{i}\mathrm{e}^{\mathrm{i}\boldsymbol{k}\cdot\boldsymbol{r}} \nabla u_{\boldsymbol{k}}(\boldsymbol{r}) \tag{A.19}$$

which is not a constant multiplied by $\psi_{\boldsymbol{k}}(\boldsymbol{r})$ because of the second term. Hence, $\psi_{\boldsymbol{k}}(\boldsymbol{r})$ is not a momentum eigenstate.

A.2 Effective mass approximation

Electron wavefunctions in semiconductors and narrow-band metals differ significantly from plane waves because the periodic part of the Bloch function is strongly modulated with a characteristic length of the order of the lattice constant. In general, the band structure should be calculated numerically. However, in doped semiconductors, states in the vicinity of special points of the Brillouin zone matter only at low temperatures and doping. These are the points where the energy dispersion, $E_{n\boldsymbol{k}}$, of the conduction (empty) band has a minimum and the valence band dispersion has a maximum. While their positions and band edges are determined experimentally or numerically, the energy dispersion nearby can be calculated analytically by applying the so-called '$\boldsymbol{k} \cdot \boldsymbol{p}$' perturbation theory. To illustrate the theory, we consider a simple cubic semiconductor with the gap E_{g}, which has a non-degenerate conduction band of s or d-like symmetry at $\boldsymbol{k} = 0$ (Γ point) and three p-like valence bands degenerate at $\boldsymbol{k} = 0$ and transforming at this point like x, y and z under the rotation of the crystal space group. There is no spin–orbit interaction in this example, so that spin is irrelevant. According to the Bloch theorem,

$$\psi_{n\boldsymbol{k}}(\boldsymbol{r}) = \mathrm{e}^{\mathrm{i}\boldsymbol{k}\cdot\boldsymbol{r}} u_{n\boldsymbol{k}}(\boldsymbol{r}) \tag{A.20}$$

where $u_{n\boldsymbol{k}}(\boldsymbol{r})$ is periodic in $\boldsymbol{r}$. It satisfies equation (A.18). At the point $\boldsymbol{k} = 0$, this equation for $u_{n0}(\boldsymbol{r})$ is

$$\left(-\frac{1}{2m_{\mathrm{e}}}\nabla^2 + V(\boldsymbol{r})\right) u_{n0}(\boldsymbol{r}) = E_{n0} u_{n0}(\boldsymbol{r}). \tag{A.21}$$

Hence, $u_{n0}(\boldsymbol{r})$ has the symmetry of the crystal space group. For a small $\boldsymbol{k}$, one can expand $u_{n\boldsymbol{k}}$ in the series,

$$u_{n\boldsymbol{k}}(\boldsymbol{r}) = \sum_{n'=s,x,y,z} a_{n'\boldsymbol{k}} u_{n'0}(\boldsymbol{r}) \tag{A.22}$$

to obtain a secular equation:

$$\det \begin{vmatrix} E_{\mathrm{g}} + \frac{k^2}{2m_{\mathrm{e}}} - E_{\boldsymbol{k}} & \frac{k_x p}{m_{\mathrm{e}}} & \frac{k_y p}{m_{\mathrm{e}}} & \frac{k_z p}{m_{\mathrm{e}}} \\ \frac{k_x p}{m_{\mathrm{e}}} & \frac{k^2}{2m_{\mathrm{e}}} - E_{\boldsymbol{k}} & 0 & 0 \\ \frac{k_y p}{m_{\mathrm{e}}} & 0 & \frac{k^2}{2m_{\mathrm{e}}} - E_{\boldsymbol{k}} & 0 \\ \frac{k_z p}{m_{\mathrm{e}}} & 0 & 0 & \frac{k^2}{2m_{\mathrm{e}}} - E_{\boldsymbol{k}} \end{vmatrix} = 0 \tag{A.23}$$

where $p \equiv \langle s| - \mathrm{i}\nabla_x |x\rangle = \langle s| - \mathrm{i}\nabla_y |y\rangle = \langle s| - \mathrm{i}\nabla_z |z\rangle$. Different bras and kets correspond to the four different Bloch functions $u_{n0}(\boldsymbol{r})$. There are four solutions to this equation. Two of them correspond to the conduction (c) and light hole (lh) valence band:

$$E_{\mathrm{c,lh}\boldsymbol{k}} = \frac{k^2}{2m_{\mathrm{e}}} + \frac{E_{\mathrm{g}}}{2} \pm \sqrt{\frac{E_{\mathrm{g}}^2}{4} + \frac{k^2 p^2}{m_{\mathrm{e}}^2}} \tag{A.24}$$

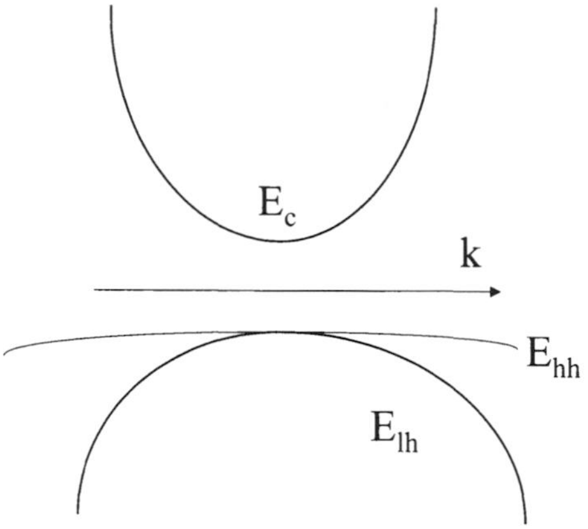

Figure A.1. '$\boldsymbol{k} \cdot \boldsymbol{p}$' energy bands near the Γ-point of the Brillouin zone in the cubic lattice.

and the other two, which are degenerate, correspond to heavy holes (hh):

$$E_{\mathrm{hh}\boldsymbol{k}} = \frac{k^2}{2m_{\mathrm{e}}} \tag{A.25}$$

One can split two degenerate heavy-hole bands and change the sign of their effective mass if one includes the states at the Γ-point split-off from those under consideration by an energy larger than E_{g} as shown by a thin line in figure A.1.

The effective mass approximation follows from equation (A.24) for small $k \ll m_{\mathrm{e}} E_{\mathrm{g}}/p$,

$$E_{\mathrm{c}\boldsymbol{k}} - E_{\mathrm{g}} \simeq \frac{k^2}{2m_{\mathrm{c}}} \tag{A.26}$$

and

$$E_{\mathrm{lh}\boldsymbol{k}} \simeq -\frac{k^2}{2m_{\mathrm{lh}}} \tag{A.27}$$

where the effective mass is

$$\frac{m_{\mathrm{c,lh}}}{m_{\mathrm{e}}} = \frac{1}{p^2/(m_{\mathrm{e}} E_{\mathrm{g}}) \pm 1}. \tag{A.28}$$

In many semiconductors, the interband dipole moment p is large and the bandgap is small, $p^2 \gg m_{\mathrm{e}} E_{\mathrm{g}}$. Therefore, the electron and hole masses can be significantly smaller than the free-electron mass:

$$\frac{m_{\mathrm{c}}}{m_{\mathrm{e}}} \simeq \frac{m_{\mathrm{lh}}}{m_{\mathrm{e}}} \ll 1. \tag{A.29}$$

As a result, when experiment does not show heavy carriers, it does not necessarily follow that the electron–phonon interaction is small and the renormalization of the *band* mass is absent. The effective mass approximation is applied if the external field varies slowly in time and smoothly in space. If the external or internal fields are strong and contain high-frequency and (or) short-wave Fourier components, they involve large momenta of the order of the reciprocal lattice constant making *all* states of the electron band relevant. In this case, *a tight binding approximation* is more appropriate.

A.3 Tight-binding approximation

For narrow-band semiconductors and metals, it is convenient to replace the Bloch states by the site (Wannier) wavefunctions, defined as

$$w_{\boldsymbol{m}}(\boldsymbol{r}) = \frac{1}{\sqrt{N}} \sum_{\boldsymbol{k}} \mathrm{e}^{-\mathrm{i}\boldsymbol{k}\cdot\boldsymbol{m}} \psi_{\boldsymbol{k}}(\boldsymbol{r}) \tag{A.30}$$

where $\boldsymbol{m} = \sum_j m_j \boldsymbol{a}_j$ is the lattice vector with the integer m_j and we drop the band quantum number n. The Wannier wavefunctions are orthogonal,

$$\int \mathrm{d}\boldsymbol{r}\; w^*_{\boldsymbol{m}}(\boldsymbol{r}) w_{\boldsymbol{n}}(\boldsymbol{r}) = \delta_{\boldsymbol{m},\boldsymbol{n}} \tag{A.31}$$

because the Bloch functions are orthogonal. Indeed calculating the integral in equation (A.31), one obtains

$$\int \mathrm{d}\boldsymbol{r}\; w^*_{\boldsymbol{m}}(\boldsymbol{r}) w_{\boldsymbol{n}}(\boldsymbol{r}) = \frac{1}{N} \sum_{\boldsymbol{k}} \mathrm{e}^{\mathrm{i}\boldsymbol{k}\cdot(\boldsymbol{m}-\boldsymbol{n})}. \tag{A.32}$$

The sum is

$$\begin{aligned}\sum_{\boldsymbol{k}} \mathrm{e}^{\mathrm{i}\boldsymbol{k}\cdot(\boldsymbol{m}-\boldsymbol{n})} = &\sum_{n_1=-N_1/2+1}^{N_1/2} \mathrm{e}^{2\pi \mathrm{i} n_1 l_1/N_1} \sum_{n_2=-N_2/2+1}^{N_2/2} \mathrm{e}^{2\pi \mathrm{i} n_2 l_2/N_2} \\ &\times \sum_{n_3=-N_3/2+1}^{N_3/2} \mathrm{e}^{2\pi \mathrm{i} n_3 l_3/N_3}\end{aligned} \tag{A.33}$$

where $l_j \leq N_j - 1$ are integers. Here each of the multipliers is the sum of a geometric progression:

$$\begin{aligned}\sum_{n_j=-N_j/2+1}^{N_j/2} \exp(2\mathrm{i}\pi n_j l_j/N_j) &= \mathrm{e}^{\mathrm{i}\pi l_i} \sum_{n_j=1}^{N_j} \exp(2\mathrm{i}\pi n_j l_j/N_j) \\ &= \mathrm{e}^{\mathrm{i}\pi l_i} \frac{(\mathrm{e}^{2\mathrm{i}\pi n_j} - 1)\exp(2\mathrm{i}\pi l_j/N_j)}{\mathrm{e}^{2\mathrm{i}\pi l_j/N_j} - 1}.\end{aligned} \tag{A.34}$$

The denominator of this expression is not zero for any allowed l_j, except $l_j = 0$. However, the numerator is zero because n_j is an integer. Hence, the sum equation (A.32) is non-zero only for $\boldsymbol{m} = \boldsymbol{n}$. In this case the sum is trivially equal to N which proves the orthogonality of the Wannier functions. In the same way, one can prove that

$$\frac{1}{N}\sum_{\boldsymbol{m}} \exp[\mathrm{i}(\boldsymbol{k} - \boldsymbol{k}') \cdot \boldsymbol{m}] = \delta_{\boldsymbol{k}\boldsymbol{k}'}. \tag{A.35}$$

Then multiplying equation (A.30) by $\exp(\mathrm{i}\boldsymbol{k} \cdot \boldsymbol{m})$ and performing the sum with respect to $\boldsymbol{m}$, we express the Bloch state as a linear combination of the Wannier states:

$$\psi_{\boldsymbol{k}}(\boldsymbol{r}) = \frac{1}{\sqrt{N}}\sum_{\boldsymbol{m}} \mathrm{e}^{\mathrm{i}\boldsymbol{k}\cdot\boldsymbol{m}} w_{\boldsymbol{m}}(\boldsymbol{r}). \tag{A.36}$$

If we substitute this sum into the Schrödinger equation and multiply it by $w_{\boldsymbol{n}}^*(\boldsymbol{r})$, then, after integrating with respect to the electron coordinate $\boldsymbol{r}$, the energy band dispersion, $E_{\boldsymbol{k}}$, is expressed via the hopping integrals $T(\boldsymbol{m} - \boldsymbol{n})$ as

$$E_{\boldsymbol{k}} = \frac{1}{N}\sum_{\boldsymbol{m}} T(\boldsymbol{m})\mathrm{e}^{\mathrm{i}\boldsymbol{k}\cdot\boldsymbol{m}} \tag{A.37}$$

where

$$T(\boldsymbol{m} - \boldsymbol{n}) = \int \mathrm{d}\boldsymbol{r}\, w_{\boldsymbol{n}}^*(\boldsymbol{r}) \left[-\frac{\nabla^2}{2m_{\mathrm{e}}} + V(\boldsymbol{r})\right] w_{\boldsymbol{m}}(\boldsymbol{r}). \tag{A.38}$$

The idea behind the tight-binding approximation is to fit the band dispersion calculated numerically with a finite number of hopping integrals. Many electronic structures, in particular perovskite ones, can be fitted with only the nearest-neighbour matrix elements between s-, p- and d-like orbitals. The hopping integrals could not be calculated by using tabulated atomic wavefunctions and potentials estimated for various solids. True atomic orbitals are not orthogonal for different sites and they do not provide a quantitative description of bands in solids. However, *atomic-like* Wannier orbitals (equation (A.30)) can provide a very good description already in a tight-binding nearest-neighbour approximation. For a non-degenerate band in a cubic lattice, the approximation yields

$$E_{\boldsymbol{k}} = \frac{1}{N}\sum_{|\boldsymbol{m}|=a} T(\boldsymbol{m})\mathrm{e}^{\mathrm{i}\boldsymbol{k}\cdot\boldsymbol{m}} = 2T(a)[\cos(k_x a) + \cos(k_y a) + \cos(k_z a)] \tag{A.39}$$

if the middle of the band is taken as zero ($T(0) = 0$). If the nearest-neighbour hopping integral, $T(a)$, is negative, the bottom of the band is found at $\boldsymbol{k} = 0$. Near the bottom the dispersion is parabolic with the effective *band* mass

$$m = \frac{1}{2a^2|T(a)|}. \tag{A.40}$$

Half the bandwidth is $D = z|T(a)|$, where $z = 6$ is the number of nearest neighbours. Many physical properties of solids depend on the density of states (DOS) per unit cell,

$$N(E) = \frac{1}{N}\sum_{\mathbf{k}} \delta(E - E_{\mathbf{k}}) \tag{A.41}$$

rather than on a particular dispersion $E_{\mathbf{k}}$. Replacing the sum by the integral, we obtain

$$N(E) = \frac{1}{4\pi^2|T(a)|}\sqrt{\frac{E}{|T(a)|}} \tag{A.42}$$

if the dispersion is parabolic, $E_{\mathbf{k}} \propto k^2$. If the dispersion is parabolic but two-dimensional, $E_{\mathbf{k}} \propto (k_x^2 + k_y^2)$, the DOS is energy independent:

$$N(E) = \frac{1}{4\pi|T(a)|}. \tag{A.43}$$

Finally, for a one-dimensional parabolic spectrum ($E_{\mathbf{k}} \propto k_x^2$), the DOS has a square-root singularity near the bottom of the band:

$$N(E) = \frac{1}{2\pi|T(a)|}\sqrt{\frac{|T(a)|}{E}}. \tag{A.44}$$

Appendix B

Quantum statistics and Boltzmann kinetics

B.1 Grand partition function

If the number of particles is large, the life history of a single particle is not so important. What really determines the physical properties of a macroscopic system is the *average* distribution of particles in real and momentum space. A statistical approach combined with quantum mechanics allow for a full description of the macroscopic system. Within this approach, we first assume that we know the exact eigenstates $|Q\rangle$ and energy levels U_Q of the many-particle system. The particles interact with the walls of their container (i.e. with a *thermostat*) so as to establish a thermal equilibrium but not to such an extent as to affect the whole set of quantum numbers Q and energy levels. Due to the interaction with the thermostat, the energy and the total number of particles (N) fluctuate. The system takes some time staying in every allowed quantum state Q. The statistical probability P of finding the system in a particular quantum state Q with the total number of particles N is proportional to this time. It depends on U_Q and N: $P = P(U_Q, N)$.

It is not hard to find the probability $P(U_Q, N)$ in thermal equilibrium. Let us consider a system containing two *independent* parts 1 and 2 with the energies $U_{1,2}$ and the number of particles $N_{1,2}$, respectively. If the two parts do not interact with each other, the probability $P_1(U_1, N_1)$ that system 1 should be in the state U_1, N_1 is independent of the probability $P_2(U_2, N_2)$. The probability for two independent events to occur is equal to the product of their separate probabilities. This means that the probability $P(U, N)$ of finding the whole system in the state U, N is the product of $P_1(U_1, N_1)$ and $P_2(U_2, N_2)$,

$$P(U, N) = P_1(U_1, N_1)P_2(U_2, N_2). \tag{B.1}$$

The thermal equilibrium is described by a *universal* probability, $P_1(U, N) = P_2(U, N) = P(U, N)$, which depends only on U and N but not on the particular system. Then taking the logarithm of both parts of equation (B.1), we obtain

$$\ln P(U, N) = \ln P(U_1, N_1) + \ln P(U_2, N_2). \tag{B.2}$$

The energy of non-interacting systems and the number of particles are additive:

$$U = U_1 + U_2 \tag{B.3}$$

$$N = N_1 + N_2 \tag{B.4}$$

and the only possibility to satisfy equation (B.2) is given by

$$\ln P(U, N) = -\ln Z - \beta(U - \mu N) \tag{B.5}$$

where Z, β and μ are independent of U and N. Hence, the equilibrium density matrix is given by

$$P(U_Q, N) = Z^{-1} e^{-\beta(U_Q - \mu N)}. \tag{B.6}$$

The constant Z (the *grand partition function)* depends on β and μ because the probability is normalized by the condition

$$\sum_{N=0}^{\infty} \sum_{Q} P(U_Q, N) = 1. \tag{B.7}$$

Hence, the grand partition function is obtained as

$$Z = \sum_{N=0}^{\infty} \sum_{Q} e^{-\beta(U_Q - \mu N)}. \tag{B.8}$$

The quantity $\beta = 1/T$ is the inverse temperature. It describes the interaction with the thermostat and determines the average energy of the macroscopic system. Finally, the constant μ is the chemical potential. It determines the average equilibrium number of particles in an open system.

B.2 Fermi–Dirac and Bose–Einstein distribution functions

The quantum statistics of non-interacting identical particles is readily derived using the grand partition function. The quantum state of an ideal Fermi or Bose gas is fully described by the numbers n_ν of identical particles in every single-particle quantum state $|\nu\rangle$. This means that proper quantum numbers of the whole gas are different sets of the occupation numbers of single-particle states, $Q = \{n_\nu\}$, where $n_\nu = 0, 1$ for fermions and $n_\nu = 0, 1, 2, \ldots, \infty$ for bosons. The energy levels are given by

$$U_Q = \sum_{\nu} E_\nu n_\nu \tag{B.9}$$

and the total number of particles is

$$N = \sum_{\nu} n_\nu. \tag{B.10}$$

Here E_ν are the single-particle energy levels. The average energy and the average number of particles of a gas depend on the average values $\bar{n}_\nu$ of the occupation numbers of each single-particle state. The *distribution function*

$$\bar{n}_\nu = \sum_Q n_\nu P(U_Q, N_Q) \tag{B.11}$$

can be expressed as

$$\bar{n}_\nu = -\frac{1}{Z\beta}\sum_{N=0}^{\infty}\sum_Q \frac{\partial}{\partial E_\nu}\exp\left[-\beta\sum_\nu n_\nu(E_\nu - \mu)\right] = -\frac{\partial \ln Z}{\partial E_\nu}. \tag{B.12}$$

The partition function is readily obtained by using

$$Z = \sum_{\{n_\nu\}}\prod_\nu \mathrm{e}^{-\beta n_\nu (E_\nu - \mu)} = \prod_\nu \sum_{n_\nu} \mathrm{e}^{-\beta n_\nu (E_\nu - \mu)}. \tag{B.13}$$

In the case of fermions, there are only two terms in the sum with $n_\nu = 0$ and $n_\nu = 1$, and

$$Z = \prod_\nu [1 + \mathrm{e}^{-\beta(E_\nu - \mu)}]. \tag{B.14}$$

Hence the (Fermi–Dirac) distribution of ideal fermions is

$$\bar{n}_\nu^{\mathrm{F}} = \frac{1}{\mathrm{e}^{\beta(E_\nu - \mu)} + 1}. \tag{B.15}$$

The partition function of ideal bosons is determined by the sum of the geometric progression as

$$\sum_{n_\nu = 0}^{\infty} \mathrm{e}^{-\beta n_\nu (E_\nu - \mu)} = \frac{1}{1 - \mathrm{e}^{-\beta(E_\nu - \mu)}} \tag{B.16}$$

and the (Bose–Einstein distribution) of ideal bosons is

$$\bar{n}_\nu^{\mathrm{B}} = \frac{1}{\mathrm{e}^{\beta(E_\nu - \mu)} - 1}.$$

B.3 Ideal Fermi gas

B.3.1 Fermi energy

The density of states (DOS) (per unit of volume) of a single particle in a macroscopic box is proportional to its kinetic energy $E = k^2/(2m)$ in a power depending on the dimensionality d of the box (appendix A.3):

$$N(E) \propto E^{(d-2)/2}. \tag{B.17}$$

If we place N_F non-interacting fermions with the spin-$\frac{1}{2}$ into a box of volume L^d, their total energy U and particle density n $(= N_F/L^d)$ are given by

$$U = 2\int dE\, N(E)\frac{E}{e^{\beta(E-\mu)}+1} \tag{B.18}$$

and

$$n = 2\int dE\, N(E)\frac{1}{e^{\beta(E-\mu)}+1}. \tag{B.19}$$

Here, 2 takes into account the degeneracy of every orbital state due to the spin. Let us calculate these integrals at zero temperature. The chemical potential should be positive at $T = 0$, otherwise the Fermi–Dirac distribution function would be zero for any positive E. For $\mu > 0$ and $\beta \to \infty$, the distribution is a step function

$$\frac{1}{e^{\beta(E-\mu)}+1} \approx \Theta(E_F - E) \tag{B.20}$$

where $E_F \equiv \mu(0)$ is the Fermi energy, and $\Theta(x) = 1$ for $x > 0$ and zero otherwise. Calculating the integrals with a step-like distribution yields

$$\begin{aligned} E_F &= \frac{(3\pi^2 n)^{2/3}}{2m} \\ E_F &= \frac{\pi n}{m} \\ E_F &= \frac{\pi^2 n^2}{8m} \end{aligned} \tag{B.21}$$

for a three-, two- and one-dimensional box, respectively, and the total energy $U \propto n^{(d+2)/d}$.

B.3.2 Specific heat

At almost all temperatures of interest, T remains much smaller than the Fermi temperature of ordinary metals, $T_F = E_F/k_B \gtrsim 1000$ K. Following Sommerfeld, we can expand the total energy in powers of T/T_F to calculate the electron contribution to the specific heat of a metal. Let us introduce a function of energy defined as

$$K_r(E) = \int_{-\infty}^{E} dE'\, N(E')E'^r. \tag{B.22}$$

In particular $K_0(E)$ represents the cumulative DOS, which is the number of states with an energy less than E. Then integrating by parts, the integral

$$\int_{-\infty}^{\infty} dE\, N(E)\frac{E^r}{e^{\beta(E-\mu)}+1}$$

becomes

$$\frac{\beta}{4}\int_{-\infty}^{\infty} dE\, \frac{K_r(E)}{\cosh^2[\beta(E-\mu)/2]}. \tag{B.23}$$

Since $\cosh^{-2}(x/2)$ falls exponentially for $|x| > 1$, only a narrow region of the order of a few T in the vicinity of the Fermi energy contributes to the integral. Provided that $K_r(E)$ is non-singular in the neighbourhood of $E = \mu$, we can expand it in a Taylor series around $E = \mu$ to obtain

$$\int_{-\infty}^{\infty} \mathrm{d}E\, N(E) \frac{E^r}{\mathrm{e}^{\beta(E-\mu)} + 1} = \sum_{p=0}^{\infty} T^{\mathrm{p}} a_p \frac{\partial^{\mathrm{p}} K_r(\mu)}{\mathrm{d}\mu^{\mathrm{p}}}. \tag{B.24}$$

Here the dimensionless coefficients a_p are given by

$$a_p = \frac{2^{p-1}}{p!} \int_{-\infty}^{\infty} \mathrm{d}x\, \frac{x^{\mathrm{p}}}{\cosh^2(x)}. \tag{B.25}$$

Only terms with even p contribute because $a_p = 0$ for odd p. The coefficients with $p \geqslant 2$ are usually written in terms of Riemann's zeta function as

$$a_p = [2 - 2^{2-p}]\zeta(p) \tag{B.26}$$

where

$$\begin{aligned} \zeta(z) &= \frac{1}{(1 - 2^{1-z})\Gamma(z)} \int_0^{\infty} \mathrm{d}t \frac{t^{z-1}}{\mathrm{e}^t + 1} \\ &= \frac{1}{\Gamma(z)} \int_0^{\infty} \mathrm{d}t\, \frac{t^{z-1}}{\mathrm{e}^t - 1} \end{aligned} \tag{B.27}$$

and $\Gamma(z)$ is the gamma function. The zero-order coefficient of the Taylor expansion is obtained by a straightforward integration:

$$a_0 = \int_0^{\infty} \mathrm{d}x\, \frac{1}{\cosh^2(x)} = 1. \tag{B.28}$$

At low temperatures, we can keep only this, $p = 0$, and the quadratic terms with $p = 2$ and $a_2 = \pi^2/6$. In particular, for the total energy ($r = 1$) in any dimensions, we obtain

$$U(T) \approx 2 \int_0^{\mu} \mathrm{d}E\, N(E) E + \frac{\pi^2}{3} T^2 [N(\mu) + \mu N'(\mu)] \tag{B.29}$$

where $N'(\mu) \equiv \mathrm{d}N(\mu)/\mathrm{d}\mu$ and for the particle density ($r = 0$)

$$n \approx 2 \int_0^{\mu} \mathrm{d}E\, N(E) + \frac{\pi^2}{3} T^2 N'(\mu). \tag{B.30}$$

The chemical potential differs from the Fermi energy by small terms of the order of T^2, so we write

$$\begin{aligned} U(T) &\approx U(0) + 2E_{\mathrm{F}} N(E_{\mathrm{F}})(\mu - E_{\mathrm{F}}) + \frac{\pi^2}{3} T^2 [N(E_{\mathrm{F}}) + E_{\mathrm{F}} N'(E_{\mathrm{F}})] \\ n &\approx n(0) + 2N(E_{\mathrm{F}})(\mu - E_{\mathrm{F}}) + \frac{\pi^2}{3} T^2 N'(E_{\mathrm{F}}) \end{aligned} \tag{B.31}$$

where $U(0)$ and $n(0)$ are the total energy and the particle density at zero temperature. To calculate the specific heat at constant density, we take n to be temperature independent ($n = n(0)$) and find that

$$\mu = E_{\mathrm{F}} - \frac{\pi^2}{6} T^2 N'(E_{\mathrm{F}})/N(E_{\mathrm{F}}). \tag{B.32}$$

Then the total energy becomes

$$U(T) \approx U(0) + \frac{\pi^2}{3} T^2 N(E_{\mathrm{F}}) \tag{B.33}$$

and the specific heat of a degenerate electron gas is, therefore,

$$C_{\mathrm{e}}(T) = \frac{2\pi^2}{3} T N(E_{\mathrm{F}}). \tag{B.34}$$

The lattice contribution to the specific heat falls off as the cube of the temperature and at very low temperatures, it drops below the electronic contribution, which is linear in T. In practical terms, one can separate these two contributions plotting $C(T)/T$ against T^2. Thus one can find $\gamma \equiv 2\pi^2 N(E_{\mathrm{F}})/3$ and the DOS at the Fermi level by extrapolating the $C(T)/T$ curve linearly down to $T^2 = 0$ and noting where it intercepts the $C(T)/T$ axis.

B.3.3 Pauli paramagnetism, Landau diamagnetism and de Haas–van Alphen quantum oscillations

Let us consider a two-dimensional ideal electron gas with the parabolic energy dispersion $E_{\mathbf{k}} = (k_x^2 + k_y^2)/2m$ on a lattice in an external magnetic field H directed along z. The energy levels of electrons are quantized (see section 1.6.4) as

$$E_n = \omega(n + 1/2) + \mu_{\mathrm{B}} \sigma H \tag{B.35}$$

where each level is g-fold degenerate with $g = L^2 eH/2\pi$ and the last term takes into account the splitting of the levels due to the spin with $\sigma = \pm 1$. Here $\omega = eH/m$ is the Larmour frequency and $\mu_{\mathrm{B}} = e/(2m_{\mathrm{e}})$ is the Bohr magneton. The band mass m might be different from the free-electron mass m_{e}. The DOS becomes a set of narrow peaks:

$$N(E) = d \sum_{n=0}^{\infty} \delta[E - \omega(n + 1/2) \mp \mu_{\mathrm{B}} H] \tag{B.36}$$

rather than a constant. This sharp oscillatory structure of the DOS imposed by the quantization of levels in a magnetic field combined with a step-like Fermi–Dirac distribution at low temperatures leads to the famous de Haas and van Alphen oscillations of magnetization of two- and three-dimensional metals with the magnetic field. We consider an open system (grand canonical ensemble) with a fixed chemical potential independent of the applied field. It is described by the thermodynamic potential $\Omega = F - \mu N = -T \ln Z$. The magnetization of the gas is

$$M(T, H) = -\frac{\partial \Omega}{\partial H}. \tag{B.37}$$

For non-interacting fermions, the partition function Z is the product equation (B.14) and

$$\Omega = -\mathrm{d}T \sum_{n=0}^{\infty} \sum_{\sigma=\pm 1} \ln \left\{ 1 + \exp \frac{\mu - \omega(n+1/2) - \mu_{\mathrm{B}} \sigma H}{T} \right\}. \tag{B.38}$$

We calculate this sum by applying the Poisson formula for an arbitrary real function $F(n)$:

$$\sum_{n=0}^{\infty} F(n) = \int_{-\delta}^{\infty} \mathrm{d}x\, F(x) + 2 \sum_{r=1}^{\infty} \mathrm{Re}\, F_r \tag{B.39}$$

where F_r is the transform of $F(x)$,

$$F_r = \int_{-\delta}^{\infty} \mathrm{d}x\, F(x) \mathrm{e}^{2\pi i r x} \tag{B.40}$$

and $\delta \to +0$. The Poisson formula is derived by Fourier transforming the sum of δ-functions

$$\Delta(x) = \sum_{n=-\infty}^{\infty} \delta(x - n).$$

This sum is periodic as a function of x with the period 1. Hence, we can expand it into the Fourier series

$$\Delta(x) = \sum_{r=-\infty}^{\infty} \Delta_r \mathrm{e}^{2\pi i r x}$$

where the Fourier coefficients are

$$\Delta_r = \int_{-1/2}^{1/2} \mathrm{d}x\, \mathrm{e}^{-2\pi i r x} \Delta(x) = 1.$$

Multiplying the Fourier expansion of $\Delta(x)$,

$$\sum_{n=-\infty}^{\infty} \delta(x - n) = \sum_{r=-\infty}^{\infty} \mathrm{e}^{2\pi i r x} \tag{B.41}$$

by $F(x)$ and integrating it with respect to x from $-\delta$ to $+\infty$ we obtain the Poisson formula (B.39). Applying the formula for $F(n) = -gT \sum_{\sigma=\pm 1} \ln\{1 + \exp[(\mu - \omega(n + 1/2) - \mu_{\mathrm{B}} \sigma H)/T]\}$ yields

$$\Omega = \Omega_{\mathrm{P}} + \Omega_{\mathrm{L}} \tag{B.42}$$

where

$$\Omega_P = -\frac{mL^2}{2\pi} T \sum_{\sigma=\pm 1} \int_0^\infty dE \ln\left\{1 + \exp \frac{\mu - E - \mu_B \sigma H}{T}\right\} \tag{B.43}$$

is the 'classical' part containing no orbital effect of the magnetic field but only spin splitting and

$$\Omega_L = -2\,dT\,\mathrm{Re} \sum_{r=1}^{\infty} \sum_{\sigma=\pm 1} \int_0^\infty dx\, e^{2\pi i r x} \ln\left\{1 + \exp \frac{\mu - \omega(x+1/2) - \mu_B \sigma H}{T}\right\} \tag{B.44}$$

is the 'quantum' part due to the quantization of the orbital motion. Taking the derivative of Ω_P, we obtain the Pauli paramagnetic contribution to the magnetization

$$M_P = \mu_B (N_\downarrow - N_\uparrow) \tag{B.45}$$

where

$$N_{\uparrow\downarrow} = \frac{L^2 m}{2\pi} \int_0^\infty dE \frac{1}{\exp\left(\frac{E \pm \mu_B H - \mu}{T}\right) + 1}$$

is the number of electrons with spin parallel ($\uparrow$) or antiparallel ($\downarrow$) to the magnetic field. Performing the integration for $T \ll T_F$ with the Fermi step-function, we find that

$$M_P = 2\mu_B^2 H N(E_F) \tag{B.46}$$

where $N(E_F) = L^2 m/(2\pi)$ is the two-dimensional zero-field DOS. The spin susceptibility $\chi_P = \partial M_P/\partial H$ is positive and essentially independent of temperature. It is proportional to the DOS at the Fermi level:

$$\chi_P = 2\mu_B^2 N(E_F). \tag{B.47}$$

This result is actually valid for any energy band spectrum of a metal. Now let us calculate the quantum part Ω_L. Integrating equation (B.44) twice by parts, we obtain

$$\Omega_L = \Omega_{LD} + \Omega_{dHvA} \tag{B.48}$$

where

$$\Omega_{LD} = \frac{d\omega}{2\pi^2} \mathrm{Re} \sum_{r=1}^{\infty} \frac{1}{r^2} \sum_{\sigma=\pm 1} \frac{1}{1 + \exp\left(\frac{\omega/2 + \mu_B \sigma H - \mu}{T}\right)} \tag{B.49}$$

and

$$\Omega_{dHvA} = \frac{d\omega^2}{8\pi^2 T} \mathrm{Re} \sum_{r=1}^{\infty} \frac{1}{r^2} \sum_{\sigma=\pm 1} \int_0^\infty dx \frac{e^{2\pi i r x}}{\cosh^2\left(\frac{\omega(x+1/2) + \mu_B \sigma H - \mu}{2T}\right)}. \tag{B.50}$$

At all temperatures and fields of interest, $\mu \simeq E_{\mathrm{F}} \gg T$ and $\mu \gg \omega$, so that

$$\Omega_{\mathrm{LD}} = \frac{L^2(eH)^2}{2m\pi^3}\zeta(2). \tag{B.51}$$

Here we applied the series representation of the Riemann's function $\zeta(2) = \sum_{r=1}^{\infty} r^{-2} = \pi^2/6$. Differentiating this expression twice with respect to H yields the diamagnetic (Landau) orbital susceptibility

$$\chi_{\mathrm{D}} = -\frac{\partial^2\Omega_{\mathrm{LD}}}{\partial H^2} = -\frac{1}{3}\left(\frac{m_{\mathrm{e}}}{m}\right)^2 \chi_{\mathrm{P}} \tag{B.52}$$

which might be larger or smaller than the paramagnetic Pauli term depending on the band mass. The remaining part Ω_{dHvA} is responsible for the de Haas–van Alphen (dHvA) effect. The main contribution to the integral in equation (B.50) yields the region of $x \simeq \mu/\omega \gg 1$, so that we can replace the lower limit in the integral by $-\infty$. Then using

$$\int_{-\infty}^{\infty} \mathrm{d}t \frac{\mathrm{e}^{\mathrm{i}\alpha t}}{\cosh^2(t/2)} = \frac{4\pi\alpha}{\sinh(\pi\alpha)}$$

we obtain

$$\Omega_{\mathrm{dHvA}} = \sum_{r=1}^{\infty} A_r \cos\left(\frac{rf}{H} + \pi r\right) \tag{B.53}$$

where the amplitudes of the Fourier harmonics are

$$A_r = \frac{L^2 eHT}{\pi r \sinh(2\pi^2 rT/\omega)} \cos\left(\frac{\pi m r}{m_{\mathrm{e}}}\right) \tag{B.54}$$

and $f = 2\pi m\mu/e$ is the frequency of oscillations with respect to the inverse magnetic field $1/H$. The oscillating part Ω_{dHvA} is small compared with the 'classical' part Ω_{P} as $\Omega_{\mathrm{dHvA}}/\Omega_{\mathrm{P}} \simeq (\omega/\mu)^2$ at any temperature. However, due to the high frequency of oscillations, the oscillating part of magnetization at zero temperature is larger than the monotonic part as μ/ω,

$$M_{\mathrm{dHvA}} = -\frac{\partial\Omega_{\mathrm{dHvA}}}{\partial H} \approx -\frac{f}{H^2}\sum_{r=1}^{\infty} rA_r \sin\left(\frac{rf}{H} + \pi r\right). \tag{B.55}$$

The period of oscillations with the inverse magnetic field of the first harmonic ($r = 1$),

$$\Delta(1/H) = \frac{2\pi e}{S}. \tag{B.56}$$

It directly measures the cross section of the Fermi surface $S = \pi k_{\mathrm{F}}^2$, where $k_{\mathrm{F}} = (2m\mu)^{1/2}$ is the Fermi momentum. With increasing temperature, the oscillating part of magnetization falls exponentially as

$$A_r \propto \mathrm{e}^{-2\pi^2 rT/\omega}. \tag{B.57}$$

The scattering of electrons by impurities also washes away dHvA oscillations as soon as the scattering rate is large enough compared with the Larmour frequency, $\omega\tau \lesssim 1$. The ratio equation (B.52) of the Landau diamagnetism and the Pauli paramagnetism does not depend on a particular distribution function and remains the same even for non-degenerate electrons. However, the dHvA effect can be observed only at low temperatures ($T \ll \omega$) in clean metals with a well-defined Fermi surface. The observation of dHvA oscillations is considered to be the ultimate evidence for a Fermi liquid.

B.4 Ideal Bose gas

B.4.1 Bose–Einstein condensation temperature

Let us consider N_B non-interacting Bose particles with zero spin in a cubic box of the volume L^3 under periodic boundary conditions. The one-particle energy spectrum is given by

$$E_{\boldsymbol{k}} = \frac{4\pi^2}{2mL^2} \sum_{j=1}^{3} n_j^2 \tag{B.58}$$

where n_j are positive integers including zero and $\boldsymbol{k} = (2\pi/L)\{n_1, n_2, n_3\}$ is the wavevector. The chemical potential of a gas $\mu(T, n)$ as a function of temperature and boson density $n = N_B/L^3$ is determined using the 'density sum rule':

$$n = \frac{1}{L^3} \sum_j \frac{1}{\exp[\beta(E_{\boldsymbol{k}} - \mu)] - 1} \tag{B.59}$$

where $\beta = 1/T$. This equation has a negative solution for μ at any temperature. If μ is negative, every term of the sum is finite. When $L, N_B \to \infty$ but n is finite (the so-called thermodynamic limit), the one-particle energy spectrum becomes continuous. In this limit the contribution of every individual term to the sum is negligible (as $1/N$) compared with n. Hence, we can replace the sum by the integral using the DOS. However, if the chemical potential approaches zero from below, a relative contribution of the term with $\boldsymbol{k} = 0$ and $E_{\boldsymbol{k}} = 0$ could be finite. Taking into account this singular contribution we replace the sum by the integral for all but the ground state. Let us take the number of bosons in the ground state per unit of volume as $n_s(T)$. Then equation (B.59) is written as

$$n = n_s(T) + \int_0^\infty \mathrm{d}E \, \frac{N(E)}{\exp[\beta(E - \mu)] - 1} \tag{B.60}$$

where $N(E) = m^{3/2}E^{1/2}/(\pi^2\sqrt{2})$. If the value of μ is of the order of level spacing $1/(mL^2) \to 0$, one can put $\mu = 0$ in this equation. Calculating the integral, we obtain the density of bosons in the ground state (i.e. the *condensate* density),

$$n_s(T) = n[1 - (T/T_c)^{3/2}] \tag{B.61}$$

where

$$T_c = 2\pi \left[\frac{n}{\zeta(3/2) m^{3/2}} \right]^{2/3} \tag{B.62}$$

is the Bose–Einstein condensation temperature. Here $\zeta(3/2) = 2.612$. In ordinary units, we have

$$k_B T_c = \frac{3.313 \hbar^2 n^{2/3}}{m}. \tag{B.63}$$

B.4.2 Third-order phase transition

To satisfy equation (B.60), the condensate density $n_s(T)$ should be zero at temperatures $T > T_c$, where the chemical potential is negative. In this *normal* state, the population of every individual level remains microscopic. At temperatures far above T_c, the gas is classical. Here the Bose–Einstein distribution is almost the same as the Maxwell–Boltzmann distribution,

$$\frac{1}{\exp[\beta(E-\mu)] - 1} \approx \exp[\beta(\mu - E)] \tag{B.64}$$

if $T \gg T_c$. The magnitude of the chemical potential is logarithmically large compared with the temperature in this limit,

$$\mu = -\frac{3}{2} T \ln \frac{T_c}{T}. \tag{B.65}$$

When the temperature approaches T_c from above, the chemical potential becomes small, and finally zero at and below T_c. To obtain its normal state value in the vicinity of T_c, we rewrite the density sum rule as

$$n[1 - (T/T_c)^{3/2}] = \int_0^\infty \mathrm{d}E\, N(E) \left[\frac{1}{\exp[\beta(E-\mu)] - 1} - \frac{1}{\exp[\beta E] - 1} \right]. \tag{B.66}$$

The function under the integral can be expanded for small E and μ, so that the integral becomes

$$n[1 - (T/T_c)^{3/2}] = T\mu \int_0^\infty \mathrm{d}E\, \frac{N(E)}{E(E-\mu)}. \tag{B.67}$$

The remaining integral yields

$$\mu = -\frac{9\pi^2 n^2}{2m^3 T_c^2} \left[\frac{T - T_c}{T_c} \right]^2 \tag{B.68}$$

for $0 < T - T_c \ll T_c$. Now we can calculate the free energy and the specific heat near the transition. The total energy of the gas per unit of volume is

$$U = \int_0^\infty \mathrm{d}E\, N(E) \frac{E}{\exp[\beta(E-\mu)] - 1}. \tag{B.69}$$

In the condensed state below T_c, the chemical potential is zero and

$$U = U_s(T) = \frac{3\zeta(5/2)}{2\zeta(3/2)} nT(T/T_c)^{3/2} \approx 0.128 m^{3/2} T^{5/2}. \tag{B.70}$$

Remarkably the energy density of the condensed phase does not depend on the particle density. This is because the macroscopic part of particles in the condensate does not contribute to the energy. The free energy density

$$F(T) = -T \int_0^T \mathrm{d}T \, \frac{U(T)}{T^2} \tag{B.71}$$

is found to be

$$F_s(T) = -\tfrac{2}{3} U_s(T). \tag{B.72}$$

The specific heat $C_s(T) = 5U_s(T)/(2T)$ is proportional to $T^{3/2}$ in the condensed phase. Differentiating equation (B.69) with respect to the chemical potential, one obtains

$$\frac{\partial U_n}{\partial \mu} = -\int_0^\infty \mathrm{d}E \, N(E) E \frac{\partial}{\partial E} \frac{1}{\exp[\beta(E-\mu)] - 1} \tag{B.73}$$

in the normal state. We can take $\mu = 0$ on the right-hand side of this equation because the chemical potential is small just above T_c. Then integrating by parts, we find that

$$\frac{\partial U_n}{\partial \mu} = \frac{3}{2} n \tag{B.74}$$

and

$$U_n(T) = U_s(T) + \frac{3}{2} n\mu = U_s(T) - \frac{27\pi^2 n^3}{4m^3 T_c^2} \left[\frac{T - T_c}{T_c}\right]^2. \tag{B.75}$$

Using equation (B.71), we arrive at the free-energy density in the normal state close to the transition as

$$F_n(T) = F_s(T) + \frac{9\pi^2 n^3}{4m^3 T_c^2} \left[\frac{T - T_c}{T_c}\right]^2 \tag{B.76}$$

and the normal state specific heat

$$C_n(T) = C_s(T) - \frac{27\pi^2 n^3}{2m^3 T_c^3} \left[\frac{T - T_c}{T_c}\right]. \tag{B.77}$$

While the specific heat of the ideal Bose gas is a continuous function (see figure 7.4), its derivative has a jump at $T = T_c$,

$$\left(\frac{\partial C_s}{\partial T} - \frac{\partial C_n}{\partial T}\right) = \frac{27\pi^2 n^3}{2m^3 T_c^4} \approx 3.66 \frac{n}{T_c}. \tag{B.78}$$

B.5 Boltzmann equation

An external perturbation can drive a macroscopic system out of the thermal equilibrium. The single-particle distribution function $f(\boldsymbol{r}, \boldsymbol{k}, t)$ in real $\boldsymbol{r}$ and momentum $\boldsymbol{k}$ space is no longer a Fermi–Dirac or a Bose–Einstein distribution. It satisfies the celebrated Boltzmann kinetic equation. The equation can be derived by considering a change in the number of particles in an elementary volume $\mathrm{d}\boldsymbol{r}\,\mathrm{d}\boldsymbol{k}$ of the phase space, which is given at the time t by $f(\boldsymbol{r}, \boldsymbol{k}, t)\,\mathrm{d}\boldsymbol{r}\,\mathrm{d}\boldsymbol{k}$. Particles with the momentum $\boldsymbol{k}$ move with the group velocity $\boldsymbol{v} = \mathrm{d}E_{\boldsymbol{k}}/\mathrm{d}\boldsymbol{k}$. Their number in the elementary volume $\mathrm{d}\boldsymbol{r}\,\mathrm{d}\boldsymbol{k}$ changes due to their motion. The change during an interval of time $\mathrm{d}t$ due to the motion in real space is given by

$$\begin{aligned}\delta N_{\boldsymbol{r}} &= [f(x, y, z, \boldsymbol{k}, t) - f(x + \mathrm{d}x, y, z, \boldsymbol{k}, t)]v_x\,\mathrm{d}t\,\mathrm{d}y\,\mathrm{d}z\,\mathrm{d}\boldsymbol{k}\\ &+ [f(x, y, z, \boldsymbol{k}, t) - f(x, y + \mathrm{d}y, z, \boldsymbol{k}, t)]v_y\,\mathrm{d}t\,\mathrm{d}x\,\mathrm{d}z\,\mathrm{d}\boldsymbol{k}\\ &+ [f(x, y, z, \boldsymbol{k}, t) - f(x, y, z + \mathrm{d}z, \boldsymbol{k}, t)]v_z\,\mathrm{d}t\,\mathrm{d}x\,\mathrm{d}y\,\mathrm{d}\boldsymbol{k}.\end{aligned}$$

Not only do the coordinates but also the momentum $\boldsymbol{k}$ change under the influence of a force $\boldsymbol{F}(\boldsymbol{r}, t)$. The change in the number of particles during the interval $\mathrm{d}t$ due to the 'motion' in the momentum space is

$$\begin{aligned}\delta N_{\boldsymbol{k}} &= [f(\boldsymbol{r}, k_x, k_y, k_z, t) - f(\boldsymbol{r}, k_x + \mathrm{d}k_x, k_y, k_z, t)]\dot{k}_x\,\mathrm{d}t\,\mathrm{d}k_y\,\mathrm{d}k_z\,\mathrm{d}\boldsymbol{r}\\ &+ [f(\boldsymbol{r}, k_x, k_y, k_z, t) - f(\boldsymbol{r}, k_x, k_y + \mathrm{d}k_y, k_z, t)]\dot{k}_y\,\mathrm{d}t\,\mathrm{d}k_x\,\mathrm{d}k_z\,\mathrm{d}\boldsymbol{r}\\ &+ [f(\boldsymbol{r}, k_x, k_y, k_z, t) - f(\boldsymbol{r}, k_x, k_y, k_z + \mathrm{d}k_z, t)]\dot{k}_z\,\mathrm{d}t\,\mathrm{d}k_x\,\mathrm{d}k_y\,\mathrm{d}\boldsymbol{r}\end{aligned}$$

where $\dot{\boldsymbol{k}} = \boldsymbol{F}(\boldsymbol{r}, t)$. As a result, the change in the number of particles in the elementary volume of the phase space is

$$\delta N = \delta N_{\boldsymbol{r}} + \delta N_{\boldsymbol{k}}. \tag{B.79}$$

This change can also be expressed via a time derivative of the distribution function as $\delta N = (\partial f/\partial t)\,\mathrm{d}t\,\mathrm{d}\boldsymbol{r}\,\mathrm{d}\boldsymbol{k}$, which finally yields

$$\frac{\partial f(\boldsymbol{r}, \boldsymbol{k}, t)}{\partial t} + \boldsymbol{v} \cdot \nabla_{\boldsymbol{r}} f(\boldsymbol{r}, \boldsymbol{k}, t) + \boldsymbol{F}(\boldsymbol{r}, t) \cdot \nabla_{\boldsymbol{k}} f(\boldsymbol{r}, \boldsymbol{k}, t) = 0. \tag{B.80}$$

The force acting upon a particle is the sum of the external force and the internal force of other particles of the system. Splitting the last term in equation (B.80) into the external and internal contributions, we obtain equation (1.4), where the collision integral describes the effect of the internal forces.

Appendix C

Second quantization

C.1 Slater determinant

Let us consider a quantum mechanical system of N identical particles with a mass m, a spin s and the interaction potential between any two of them $V(\boldsymbol{r}_i - \boldsymbol{r}_j)$. There is also an external field $V(\boldsymbol{r})$. Here $\boldsymbol{r}_i$ are three position coordinates $\{x_i, y_i, z_i\}$, $i, j = 1, 2, 3, \ldots, N$. The system is described by a many-particle wavefunction $\Phi(q_1, q_2, \ldots, q_N)$, which satisfies the stationary Schrödinger equation:

$$\left\{\sum_{i=1}^{N}\left[-\frac{1}{2m}\Delta_i + V(\boldsymbol{r}_i)\right] + \frac{1}{2}\sum_{i=1}^{N}\sum_{j\neq i} V(\boldsymbol{r}_i - \boldsymbol{r}_j)\right\}\Phi(q_1, q_2, \ldots, q_N)$$
$$= E\Phi(q_1, q_2, \ldots, q_N) \tag{C.1}$$

where $q_i \equiv (\boldsymbol{r}_i, \sigma_i)$ is a set of position $\boldsymbol{r}_i$ and spin σ_i coordinates. The spin coordinate (z-component of spin s_z) describes the 'internal' state of the particle, which might be a composite particle like an atom. We can readily solve this equation for non-interacting particles when $V(\boldsymbol{r}_i - \boldsymbol{r}_j) = 0$. In this case, many-particle eigenstates are products of one-particle wavefunctions,

$$\Phi_Q(q_1, q_2, \ldots q_i, \ldots q_j, \ldots q_N) = u_{k_1}(q_1)u_{k_2}(q_2) \times \cdots \times u_{k_N}(q_N) \tag{C.2}$$

where $u_k(q)$ is a solution of a one-particle Schrödinger equation

$$\left[-\frac{1}{2m}\Delta + V(\boldsymbol{r})\right] u_k(\boldsymbol{r}, \sigma) = \epsilon_k u_k(\boldsymbol{r}, \sigma) \tag{C.3}$$

k is a set of one-particle quantum numbers, for example the wavevector $\boldsymbol{k}$ and s_z, if the particles are free, or of n, $\boldsymbol{k}$ and s_z, if $V(\boldsymbol{r})$ is periodic (appendix A). Direct substitution of equation (C.2) into equation (C.1) yields the total energy of the system, which is the sum of one-particle energies,

$$E_Q = \sum_{i=1}^{N} \epsilon_{k_i}. \tag{C.4}$$

The quantum number Q of the whole system is the set of single-particle quantum numbers,

$$Q = \{k_1, k_2, \ldots k_i, \ldots k_j, \ldots k_N\}. \tag{C.5}$$

If we swap the coordinates of any two particles, the new state $\Phi_Q(q_1, q_2, \ldots q_j, \ldots q_i, \ldots q_N)$ will also be a solution of the Schrödinger equation with the same total energy. Any linear combination of such wavefunctions solves the problem as well. Importantly, only one of them is acceptable for identical particles. While in classical mechanics the existence of sharp trajectories makes it possible to distinguish identical particles by their paths, there is no way of keeping track of individual particles in quantum mechanics. This quantum 'indistinguishability' of identical particles puts severe constraints on our choice of a many-particle wavefunction. Let us introduce an operator $\hat{P}_{ij}$, which swaps the coordinates q_i and q_j,

$$\hat{P}_{ij}\Phi(q_1, q_2, \ldots q_i, \ldots q_j, \ldots q_N) = \Phi(q_1, q_2, \ldots q_j, \ldots q_i, \ldots q_N). \tag{C.6}$$

The Hamiltonian (C.1) is symmetric with respect to the permutation of any pair of the coordinates, that is $\hat{P}_{ij}$ and H commute:

$$[\hat{P}_{ij}, H] = 0. \tag{C.7}$$

Hence, the eigenfunctions of H are also the eigenfunctions of $\hat{P}_{ij}$, so that

$$\hat{P}_{ij}\Phi(q_1, q_2, \ldots q_i, \ldots q_j, \ldots q_N) = P\Phi(q_1, q_2, \ldots q_i, \ldots q_j, \ldots q_N). \tag{C.8}$$

Since two successive permutations of q_i and q_j bring back the original configuration, we have $P^2 = 1$ and $P = \pm 1$. The eigenstates with the eigenvalue $P = 1$ are called *symmetric* and the eigenstates with $P = -1$ *antisymmetric*. There is an extreme case of a *totally* symmetric (or totally antisymmetric) state which does not change (or changes its sign) under any permutation $\hat{P}_{ij}$. Only these extreme cases are realized for identical particles. Any state of *bosons* is symmetric and any state of *fermions* is antisymmetric. Relativistic quantum mechanics connects the symmetry of the many-particle wavefunction with the spin. Bosons have integer spins, $s = 0, 1, 2, 3, \ldots$ and fermions have half odd integer spins, $s = 1/2, 3/2, 5/2, \ldots$. A totally antisymmetric wavefunction Φ_A of non-interacting identical fermions is readily constructed as linear superpositions of $\Phi_Q(q_1, q_2, \ldots q_i, \ldots q_j, \ldots q_N)$ (equation (C.2)) with coefficients determined by the use of the Slater determinant:

$$\Phi_A = \frac{1}{\sqrt{N!}}\mathrm{Det}\begin{pmatrix} u_{k_1}(q_1) & u_{k_2}(q_1) & \ldots & u_{k_N}(q_1) \\ u_{k_1}(q_2) & u_{k_2}(q_2) & \ldots & u_{k_N}(q_2) \\ \ldots & \ldots & \ldots & \ldots \\ u_{k_1}(q_N) & u_{k_2}(q_N) & \ldots & u_{k_N}(q_N) \end{pmatrix}. \tag{C.9}$$

This function is totally antisymmetric because the permutation of the coordinates of any two particles (let us say q_1 and q_2) corresponds to the permutation of two rows in the matrix $N \times N$ which changes the sign of the whole determinant. The factor $1/\sqrt{N!}$ is introduced to make Φ_A properly normalized. The totally symmetric Φ_S of bosons is obtained as

$$\Phi_S = \sqrt{\frac{N_1!N_2!\ldots}{N!}} \sum \Phi_Q(q_1, q_2, \ldots q_i, \ldots q_j, \ldots q_N) \tag{C.10}$$

where the sum is taken over all permutations of *different* single-particle quantum numbers k_i in Q. $N_1, N_2, \ldots$ are the numbers of identical k_i in Q, so that $N_1 + N_2 + \cdots = N$. For example the system of two identical non-interacting fermions, which can be only in two one-particle states $u_k(q)$ and $u_p(q)$, is described by

$$\Phi_A(q_1, q_2) = \frac{1}{\sqrt{2}}\{u_k(q_1)u_p(q_2) - u_k(q_2)u_p(q_1)\} \tag{C.11}$$

while two bosons are described by

$$\Phi_S(q_1, q_2) = \frac{1}{\sqrt{2}}\{u_k(q_1)u_p(q_2) + u_k(q_2)u_p(q_1)\} \tag{C.12}$$

if $k \neq p$, and by

$$\Phi_S(q_1, q_2) = u_k(q_1)u_k(q_2) \tag{C.13}$$

if $k = p$. These functions are normalized,

$$\int dq_1 \int dq_2 \, |\Phi_{S,A}(q_1, q_2)|^2 = 1$$

if $u_{k,p}(q)$ are normalized and orthogonal,

$$\int dq \, u_k^*(q)u_p(q) = \delta_{kp}. \tag{C.14}$$

It is apparent that if two or more single-particle quantum numbers k_i are the same, the totally antisymmetric wavefunction vanishes, $\Phi_A = 0$. Only one fermion can occupy a given one-particle quantum state. This statement expresses the Pauli exclusion principle formulated in 1925.

C.2 Annihilation and creation operators

The collection of identical particles which do not mutually interact is a trivial problem. It is solved in the form of the Slater determinant for fermions (equation (C.9)) and of the sum (C.10) for bosons. However, the interaction term makes the problem hard. Exact many-particle wavefunctions can be expanded in a series of 'non-interacting' functions equation (C.9) or (C.10). Then a set of algebraic equations for the expansion coefficients may be treated perturbatively, if all matrix elements of the many-particle Hamiltonian are known. These coefficients and matrices perfectly replace the original wavefunctions and operators, respectively. The non-interacting eigenstates (C.9) and (C.10) are fully identified by a set $\{n_{k_1}, n_{k_2}, \ldots n_{k_i}, \ldots\}$ of the occupation numbers n_{ki} of each single-particle quantum state. Differing from the original wavefunctions, which are functions of the position and spin coordinates of particles q_i, the expansion coefficients are functions of the occupation numbers $\{n_{k_1}, n_{k_2}, \ldots n_{k_i}, \ldots\}$. A representation, where the initial coordinates $\{q_1, q_2, \ldots q_N\}$ are replaced by new 'coordinates' $\{n_{k_1}, n_{k_2}, \ldots n_{k_i}, \ldots\}$, is known as the *second quantization*. The wavefunction of non-interacting particles (equations (C.9) and (C.10)) is written in the second quantization using the Dirac notations as a *ket*,

$$\Phi_{\{n_i\}} = |n_{k_1}, n_{k_2}, \ldots n_{k_i}, \ldots\rangle \tag{C.15}$$

and its complex conjugate as a *bra*,

$$\Phi^*_{\{n_i\}} = \langle n_{k_1}, n_{k_2}, \ldots n_{k_i}, \ldots | \tag{C.16}$$

where $n_{k_i} = 0, 1, 2, 3, \ldots, \infty$ for bosons and $n_{k_i} = 0, 1$ for fermions. The matrix elements of any operator $\hat{A}(q_1, q_2, \ldots, q_N)$ are written as

$$\langle m_{k_1}, m_{k_2}, \ldots m_{k_i}, \ldots |\hat{A}| n_{k_1}, n_{k_2}, \ldots n_{k_i}, \ldots\rangle$$
$$\equiv \int dq_1 \int dq_2 \ldots \int dq_N \Phi^*_{\{m_i\}}(q_1, q_2, \ldots, q_N) \hat{A} \Phi_{\{n_i\}}(q_1, q_2, \ldots, q_N). \tag{C.17}$$

There are two major types of many-particle symmetric Hermitian operators:

$$\hat{A}^{(1)} \equiv \sum_{i=1}^{N} \hat{h}(q_i) \tag{C.18}$$
$$\hat{A}^{(2)} \equiv \sum_{i,j=1}^{N} \hat{V}(q_i, q_j).$$

While $\hat{A}^{(1)}$ is the sum of identical one-particle operators (like the kinetic energy), $\hat{A}^{(2)}$ is the sum of identical two-particle operators (like the potential energy). Their matrix elements can be readily calculated by the use of annihilation $b_k(a_k)$ and creation $b_k^\dagger(a_k^\dagger)$ operators, whose matrices are defined in the second quantization as

$$\langle m_{k_1}, m_{k_2}, \ldots m_{k_i}, \ldots |b_{k_i}| n_{k_1}, n_{k_2}, \ldots n_{k_i}, \ldots\rangle$$
$$= \sqrt{n_{k_i}} \delta_{m_{k_1}, n_{k_1}} \delta_{m_{k_2}, n_{k_2}} \times \cdots \times \delta_{m_{k_i}, n_{k_i}-1} \times \cdots \times \delta_{m_{k_j}, n_{k_j}} \times \cdots \tag{C.19}$$

$$\langle m_{k_1}, m_{k_2}, \ldots m_{k_i}, \ldots |b^\dagger_{k_i}| n_{k_1}, n_{k_2}, \ldots n_{k_i}, \ldots\rangle$$
$$= \sqrt{n_{k_i} + 1} \delta_{m_{k_1}, n_{k_1}} \delta_{m_{k_2}, n_{k_2}} \times \cdots \times \delta_{m_{k_i}, n_{k_i}+1} \times \cdots \times \delta_{m_{k_j}, n_{k_j}} \times \cdots$$

for bosons and

$$\langle m_{k_1}, m_{k_2}, \dots m_{k_i}, \dots | a_{k_i} | n_{k_1}, n_{k_2}, \dots n_{k_i}, \dots \rangle$$
$$= \pm\sqrt{n_{k_i}} \delta_{m_{k_1}, n_{k_1}} \delta_{m_{k_2}, n_{k_2}} \times \dots \times \delta_{m_{k_i}, n_{k_i}-1} \times \dots \times \delta_{m_{k_j}, n_{k_j}} \times \dots \tag{C.20}$$
$$\langle m_{k_1}, m_{k_2}, \dots m_{k_i}, \dots | a^{\dagger}_{k_i} | n_{k_1}, n_{k_2}, \dots n_{k_i}, \dots \rangle$$
$$= \pm\sqrt{n_{k_i} + 1} \delta_{m_{k_1}, n_{k_1}} \delta_{m_{k_2}, n_{k_2}} \times \dots \times \delta_{m_{k_i}, n_{k_i}+1} \times \dots \times \delta_{m_{k_j}, n_{k_j}} \times \dots$$

for fermions. Here $+$ or $-$ depend on the evenness or oddness, respectively, of the number of one-particle occupied states, which precede the state k_i. It follows from this definition that $b^{\dagger}_k$ is the Hermitian conjugate of b_k and $a^{\dagger}_k$ is the Hermitian conjugate of a_k. We see that $b_k(a_k)$ decreases the number of particles in the state k by one and $b^{\dagger}_k(a^{\dagger}_k)$ increases this number by one. The products $b^{\dagger}_k b_k$ and $a^{\dagger}_k a_k$ have diagonal matrix elements only, which are n_k. These are the occupation number operators. The operators $b_k b^{\dagger}_k$ and $a_k a^{\dagger}_k$ are also diagonal and their eigenvalues are $1 + n_k$ and $1 - n_k$, respectively. As a result, the commutation rules for these operators are:

$$b_k b^{\dagger}_k - b^{\dagger}_k b_k \equiv [b_k, b^{\dagger}_k] = 1 \tag{C.21}$$

and

$$a_k a^{\dagger}_k + a^{\dagger}_k a_k \equiv \{a_k a^{\dagger}_k\} = 1. \tag{C.22}$$

The bosonic annihilation and (or) creation operators for different single-particle states commute and the fermionic annihilation and (or) creation operators anticommute:

$$[b_k, b^{\dagger}_p] = [b_k, b_p] = 0 \tag{C.23}$$
$$\{a_k, a^{\dagger}_p\} = \{a_k, a_p\} = 0$$

if $k \neq p$. For example, the matrix elements of fermionic annihilation and creation operators are:

$$\langle 0_k | a_k | 1_k \rangle = \langle 1_k | a^{\dagger}_k | 0_k \rangle = (-1)^{N(1,k-1)} \tag{C.24}$$

where $N(1, k-1)$ is the number of *occupied* states, which precede the state k in the adopted ordering of one-particle states and the occupation numbers of all other single-particle states are the same in both *bra* and *ket*. Then multiplying $a^{\dagger}_p$ and $a^{\dagger}_k$, we obtain a non-zero matrix element

$$\langle 1_p, 0_k | a^{\dagger}_p a_k | 0_p, 1_k \rangle = \langle 1_p, 0_k | a^{\dagger}_p | 0_p, 0_k \rangle \langle 0_p, 0_k | a_k | 0_p, 1_k \rangle$$
$$= (-1)^{N(1,p-1)} (-1)^{N(1,k-1)} = (-1)^{N(p+1,k-1)}. \tag{C.25}$$

Multiplying these operators in the inverse order yields

$$\begin{aligned}\langle 1_p, 0_k | a_k a_p^\dagger | 0_p, 1_k\rangle &= \langle 1_p, 0_k | a_k | 1_p, 1_k\rangle \langle 1_p, 1_k | a_p^\dagger | 0_p, 1_k\rangle \\ &= (-1)^{N(1,k-1)}(-1)^{N(1,p-1)} = (-1)^{N(p+1,k-1)+1}.\end{aligned} \tag{C.26}$$

Equations (C.25) and (C.26) have opposite signs because the p-state of the intermediate *bra* and *ket* in equation (C.26) is occupied, while it is empty in the intermediate *bra* and *ket* in equation (C.25). Hence, we obtain $a_k a_p^\dagger + a_p^\dagger a_k = 0$ for $k \neq p$.

Any operator of the first type is expressed in terms of the annihilation and creation operators as

$$\begin{aligned}\hat{A}^{(1)} &\equiv \sum_{k,p} h_{k'k} b_{k'}^\dagger b_k \\ \hat{A}^{(1)} &\equiv \sum_{k,p} h_{k'k} a_{k'}^\dagger a_k\end{aligned} \tag{C.27}$$

for bosons and fermions, respectively. And an operator of the second type is expressed as

$$\begin{aligned}\hat{A}^{(2)} &\equiv \sum_{k,p,k',p'} V_{pk}^{p'k'} b_{k'}^\dagger b_{p'}^\dagger b_p b_k \\ \hat{A}^{(2)} &\equiv \sum_{k,p,k',p'} V_{pk}^{p'k'} a_{k'}^\dagger a_{p'}^\dagger a_p a_k.\end{aligned} \tag{C.28}$$

Here, h_{pk} and $V_{pk}^{p'k'}$ are the matrix elements of $\hat{h}$ and $\hat{V}$ in the basis of one-particle wavefunctions:

$$\begin{aligned}h_{k'k} &\equiv \int \mathrm{d}q\, u_{k'}^*(q)\hat{h}(q)u_k(q) \\ V_{pk}^{p'k'} &\equiv \int \mathrm{d}q \int \mathrm{d}q'\, u_{k'}^*(q)u_{p'}^*(q')\hat{V}(q,q')u_p(q')u_k(q).\end{aligned} \tag{C.29}$$

For example, let us consider a system of two bosons, $N = 2$, which can be in the two different one-particle states $u_m(q)$ and $u_l(q)$. If bosons do not interact, the eigenstates of the whole system are

$$\begin{aligned}|2_m, 0_l\rangle &= u_m(q_1)u_m(q_2) \\ |0_m, 2_l\rangle &= u_l(q_1)u_l(q_2) \\ |1_m, 1_l\rangle &= \frac{1}{\sqrt{2}}[u_l(q_1)u_m(q_2) + u_l(q_2)u_m(q_1)].\end{aligned} \tag{C.30}$$

The diagonal matrix elements of $\hat{A}^{(1)}$, calculated directly by the use of the orthogonality of $u_m(q)$ and $u_l(q)$ for $m \neq l$, are

$$\begin{aligned}
\langle 2_m, 0_l|\hat{A}^{(1)}|2_m, 0_l\rangle &= \int\int \mathrm{d}q_1\,\mathrm{d}q_2\,u_m^*(q_1)u_m^*(q_2)[\hat{h}(q_1)+\hat{h}(q_2)] \\
&\quad \times u_m(q_1)u_m(q_2) = 2h_{mm} \qquad \text{(C.31)} \\
\langle 0_m, 2_l|\hat{A}^{(1)}|0_m, 2_l\rangle &= 2h_{ll} \\
\langle 1_m, 1_l|\hat{A}^{(1)}|1_m, 1_l\rangle &= h_{mm} + h_{ll}.
\end{aligned}$$

The off-diagonal elements are:

$$\begin{aligned}
\langle 0_m, 2_l|\hat{A}^{(1)}|2_m, 0_l\rangle &= 0 \qquad \text{(C.32)} \\
\langle 0_m, 2_l|\hat{A}^{(1)}|1_m, 1_l\rangle &= \sqrt{2}h_{lm} \\
\langle 2_m, 0_l|\hat{A}^{(1)}|1_m, 1_l\rangle &= \sqrt{2}h_{ml}.
\end{aligned}$$

However, the matrix elements of $b_{k'}^\dagger b_k$, calculated by the use of the matrix elements of the annihilation and creation operators (equation (C.19)), are:

$$\begin{aligned}
\langle 2_m, 0_l|b_{k'}^\dagger b_k|2_m, 0_l\rangle &= 2\delta_{km}\delta_{k'm} \qquad \text{(C.33)} \\
\langle 0_m, 2_l|b_{k'}^\dagger b_k|0_m, 2_l\rangle &= 2\delta_{kl}\delta_{k'l} \\
\langle 1_m, 1_l|b_{k'}^\dagger b_k|1_m, 1_l\rangle &= \delta_{km}\delta_{k'm} + \delta_{kl}\delta_{k'l} \\
\langle 0_m, 2_l|b_{k'}^\dagger b_k|2_m, 0_l\rangle &= 0 \\
\langle 0_m, 2_l|b_{k'}^\dagger b_k|1_m, 1_l\rangle &= \sqrt{2}\delta_{km}\delta_{k'l} \\
\langle 2_m, 0_l|b_{k'}^\dagger b_k|1_m, 1_l\rangle &= \sqrt{2}\delta_{kl}\delta_{k'm}.
\end{aligned}$$

Substituting these matrix elements into equation (C.27), we obtain the result of direct calculations (equations (C.31) and (C.32)). The expressions (C.27) and (C.28) allow us to replace the first (coordinate) representation by the second quantization (occupation numbers) representation.

C.3 Ψ-operators

The second quantization representation depends on our choice of a complete set of one-particle wavefunctions. In many cases we do not know the eigenstates of the one-particle Hamiltonian $\hat{h}(q)$, and a representation in terms of the field Ψ-operators is more convenient. These operators are defined as

$$\begin{aligned}
\Psi(q) &= \sum_k u_k(q)b_k \qquad \text{(C.34)} \\
\Psi^\dagger(q) &= \sum_k u_k^*(q)b_k^\dagger
\end{aligned}$$

for bosons and in the same way for fermions,

$$\Psi(q) = \sum_k u_k(q)a_k \qquad \text{(C.35)}$$
$$\Psi^\dagger(q) = \sum_k u_k^*(q)a_k^\dagger.$$

Differing from the annihilation and creation operators, Ψ-operators do not depend on a particular choice of one-particle states. Both $\hat{A}^{(1)}$ and $\hat{A}^{(2)}$ many-particle operators are readily expressed in terms of Ψ-operators:

$$\hat{A}^{(1)} = \int dq\, \Psi^\dagger(q)\hat{h}(q)\Psi(q) \qquad \text{(C.36)}$$
$$\hat{A}^{(2)} = \int dq \int dq'\, \Psi^\dagger(q)\Psi^\dagger(q')\hat{V}(q,q')\Psi(q')\Psi(q).$$

Indeed, if we substitute equations (C.34) or (C.35) into equations (C.36) we obtain equations (C.27) and (C.28). Here the operators $\hat{h}(q)$ and $\hat{V}(q,q')$ act on the coordinates, which are parameters (i.e. *c-numbers*), while the 'true' operators are the field operators acting on the occupation numbers. The commutation rules for the field operators are derived using the commutators of the annihilation and creation operators,

$$[\Psi(q)\Psi^\dagger(q')] = \sum_{k,p} u_k(q)u_p^*(q')[a_k a_p^\dagger]$$
$$= \sum_k u_k(q)u_k^*(q') = \delta(\mathbf{r}-\mathbf{r}')\delta_{\sigma\sigma'} \qquad \text{(C.37)}$$
$$[\Psi(q)\Psi(q')] = 0 \qquad \text{(C.38)}$$

for bosons and anticommutators

$$\{\Psi(q)\Psi^\dagger(q')\} = \delta(\mathbf{r}-\mathbf{r}')\delta_{\sigma\sigma'} \qquad \text{(C.39)}$$
$$\{\Psi(q)\Psi(q')\} = 0$$

for fermions. The Hamiltonian of identical interacting particles can be expressed in terms of Ψ-operators:

$$H = \int dq\, \Psi^\dagger(q)\hat{h}(q)\Psi(q) + \int dq \int dq'\, \Psi^\dagger(q)\Psi^\dagger(q')\hat{V}(q,q')\Psi(q')\Psi(q). \qquad \text{(C.40)}$$

Now we can use any complete set of one-particle states to express it in terms of the annihilation and creation operators. For example, for fermions on a lattice a convenient set is the set of Bloch functions (appendix A),

$$\Psi(q) = \sum_{n,\mathbf{k},s} \psi_{n\mathbf{k}}(\mathbf{r})\chi_s(\sigma)a_{n\mathbf{k}s} \qquad \text{(C.41)}$$

where we also include the spin component $\chi_s(\sigma)$. If we drop the band index n in the framework of the single-band approximation, the Hamiltonian takes the following form

$$H = \sum_{k,s} \xi_k a_{ks}^\dagger a_{ks} + \sum_{k,k',s} \sum_{p,p',s'} V_{pk}^{p'k'} a_{k's}^\dagger a_{p's'}^\dagger a_{ps'} a_{ks}. \tag{C.42}$$

Here $V_{pk}^{p'k'}$ is the matrix element of the particle–particle interaction $V(\boldsymbol{r}, \boldsymbol{r}')$ calculated by the use of the Bloch states. We assume that $V(\boldsymbol{r}, \boldsymbol{r}')$ does not depend on the spin. One can equally use the Wannier functions (A.30) as another complete set,

$$\Psi(q) = \sum_{m,s} w_m(\boldsymbol{r}) \chi_s(\sigma) a_{ms} \tag{C.43}$$

to express the same Hamiltonian in the site representation:

$$H = \sum_{m,n,s} [T(\boldsymbol{m} - \boldsymbol{n}) - \mu \delta_{m,n}] a_{ms}^\dagger a_{ns} + \sum_{m,m',s} \sum_{n,n',s'} V_{nm}^{n'm'} a_{m's}^\dagger a_{n's'}^\dagger a_{ns'} a_{ms}. \tag{C.44}$$

Here $T(\boldsymbol{m} - \boldsymbol{n})$ is the hopping integral (equation (A.38)) and $V_{nm}^{n'm'}$ are the matrix elements of the interaction potential, calculated by the use of the Wannier functions.

Appendix D

Analytical properties of one-particle Green's functions

Green's functions were originally introduced to solve differential equations and link them to the corresponding integral equations. Later on, their applications became remarkably broad in theoretical physics. Nowadays Green's functions (GFs) serve as a very powerful tool in many-body theory. Regarded as a description of the time development of a system, a *one-particle* GF is applied to any interacting system of identical particles. We use the occupation number representation and suppose that the quantum state is described by $|n\rangle$. If at time $t = 0$ we add a particle with the quantum number ν, the system is described immediately after this addition by $c_\nu^\dagger|n\rangle$, where $c_\nu^\dagger$ is the creation operator for the state $|\nu\rangle$. Now the development of the system proceeds according to $\mathrm{e}^{-\mathrm{i}\hat{H}t}c_\nu^\dagger|n\rangle$. However, ν is not usually an eigenstate of $\hat{H}$, so the particle in the state ν gets scattered. Thus, when at a later time we measure to see how much of a probability amplitude is left in the state ν, the measurement provides information about the interaction in the system. If we require the probability amplitude for the persistence of the added particle in the quantum state ν, we must take the scalar product of $\mathrm{e}^{-\mathrm{i}\hat{H}t}c_\nu^\dagger|n\rangle$ with a function describing the quantum state plus the particle added at the time t. At this time, the state is described by $\mathrm{e}^{-\mathrm{i}\hat{H}t}|n\rangle$, and immediately after the addition of the particle it is described by $c_\nu^\dagger\mathrm{e}^{-\mathrm{i}\hat{H}t}|n\rangle$. The quantum number ν can be anything depending on the problem of interest. Usually, it is assigned as the quantum number of a free particle, i.e. $\nu = (\boldsymbol{k}, \sigma)$ representing both the momentum and spin. Such is the idea behind the following definition of the one-particle GF at any temperature:

$$G(\boldsymbol{r}, \boldsymbol{r}', t - t') = -\mathrm{i}\,\mathrm{Tr}\{\mathrm{e}^{(\Omega - H)/T}T_t\psi(\boldsymbol{r}, t)\psi^\dagger(\boldsymbol{r}', t')\} \tag{D.1}$$

$$\equiv -\mathrm{i}\langle\langle T_t\psi(\boldsymbol{r}, t)\psi^\dagger(\boldsymbol{r}', t')\rangle\rangle \tag{D.2}$$

where

$$\psi(\boldsymbol{r}, t) = \mathrm{e}^{\mathrm{i}Ht}\Psi(\boldsymbol{r})\mathrm{e}^{-\mathrm{i}Ht}$$

is the time-dependent (Heisenberg) field operator and

$$T_t \hat{A}(t)\hat{B}(t) \equiv \Theta(t-t')\hat{A}(t)\hat{B}(t') \pm \Theta(t'-t)\hat{B}(t')\hat{A}(t)$$

with $+$ for bosons and $-$ for fermions. In the following we consider fermions with only one component of their spin and drop the spin index. The GF allows us to calculate the single-particle excitation spectrum of a many-particle system. However, the perturbation theory is easier to formulate for the Matsubara GF defined as

$$\mathcal{G}(\boldsymbol{r}, \boldsymbol{r}', \tau - \tau') = -\langle\langle T_\tau \psi(\boldsymbol{r}, \tau)\psi^\dagger(\boldsymbol{r}', \tau')\rangle\rangle. \tag{D.3}$$

Here $\psi(\boldsymbol{r}, \tau) = \exp(H\tau)\Psi(\boldsymbol{r})\exp(-H\tau)$ and the thermodynamic 'time' τ is defined within the interval $0 \leqslant \tau \leqslant 1/T$. If the system is homogeneous, GFs depend only on the difference $(\boldsymbol{r} - \boldsymbol{r}')$ and their Fourier transforms depend on a single wavevector $\boldsymbol{k}$:

$$\begin{aligned} G(\boldsymbol{k}, t) &= -\,\mathrm{i}\langle\langle T_t c_{\boldsymbol{k}}(t) c^\dagger_{\boldsymbol{k}}\rangle\rangle \\ \mathcal{G}(\boldsymbol{k}, \tau) &= -\,\langle\langle T_\tau c_{\boldsymbol{k}}(\tau) c^\dagger_{\boldsymbol{k}}\rangle\rangle. \end{aligned} \tag{D.4}$$

Let us connect the GF and the Matsubara GF. By the use of the exact eigenstates $|n\rangle$ and eigenvalues E_n of the Hamiltonian, we can write

$$G(\boldsymbol{k}, t) = -\mathrm{i}\mathrm{e}^{\Omega/T} \sum_{n,m} \mathrm{e}^{-E_n/T} \mathrm{e}^{\mathrm{i}\omega_{nm} t} |\langle n|c_{\boldsymbol{k}}|m\rangle|^2 \tag{D.5}$$

for $t > 0$ and

$$G(\boldsymbol{k}, t) = \mathrm{i}\mathrm{e}^{\Omega/T} \sum_{n,m} \mathrm{e}^{-E_m/T} \mathrm{e}^{\mathrm{i}\omega_{nm} t} |\langle n|c_{\boldsymbol{k}}|m\rangle|^2 \tag{D.6}$$

for $t < 0$, where $\omega_{nm} \equiv E_n - E_m$. The Fourier transform with respect to time yields for the Fourier component

$$\begin{aligned} G(\boldsymbol{k}, \omega) = &-\mathrm{i}\mathrm{e}^{\Omega/T} \sum_{n,m} \mathrm{e}^{-E_n/T} |\langle n|c_{\boldsymbol{k}}|m\rangle|^2 \int_0^\infty \mathrm{d}t\, \mathrm{e}^{\mathrm{i}(\omega_{nm}+\omega)t} \\ &+ \mathrm{i}\mathrm{e}^{\Omega/T} \sum_{n,m} \mathrm{e}^{-E_m/T} |\langle n|c_{\boldsymbol{k}}|m\rangle|^2 \int_{-\infty}^0 \mathrm{d}t\, \mathrm{e}^{\mathrm{i}(\omega_{nm}+\omega)t}. \end{aligned} \tag{D.7}$$

Using the formula

$$\begin{aligned} \int_0^\infty \mathrm{d}t\, \mathrm{e}^{\mathrm{i}zt} &= \lim_{\delta \to +0} \int_0^\infty \mathrm{d}t\, \mathrm{e}^{\mathrm{i}zt - \delta t} \\ &= \frac{\mathrm{i}}{z + \mathrm{i}\delta} \\ &= \mathrm{i}P\frac{1}{z} + \pi\delta(z) \end{aligned} \tag{D.8}$$

we obtain

$$G(\boldsymbol{k},\omega) = \mathrm{e}^{\Omega/T}\sum_{n,m}\mathrm{e}^{-E_n/T}|\langle n|c_{\boldsymbol{k}}|m\rangle|^2 \times \left\{P\frac{1+\mathrm{e}^{\omega_{nm}/T}}{\omega_{nm}+\omega} - \mathrm{i}\pi[1-\mathrm{e}^{\omega_{nm}/T}]\delta(\omega_{nm}+\omega)\right\}. \tag{D.9}$$

Here P means the principal value of the integral, when integrating over ω. It is convenient to introduce the spectral function $A(\boldsymbol{k},\omega)$:

$$A(\boldsymbol{k},\omega) \equiv \pi(1+\mathrm{e}^{-\omega/T})\mathrm{e}^{\Omega/T}\sum_{n,m}\mathrm{e}^{-E_n/T}|\langle n|c_{\boldsymbol{k}}|m\rangle|^2\delta(\omega_{nm}+\omega) \tag{D.10}$$

which is real and strictly positive ($A(\boldsymbol{k},\omega) > 0$). It obeys an important sum rule:

$$\frac{1}{\pi}\int_{-\infty}^{\infty}\mathrm{d}\omega\, A(\boldsymbol{k},\omega) = 1. \tag{D.11}$$

Indeed, integrating equation (D.10) yields

$$\frac{1}{\pi}\int_{-\infty}^{\infty}\mathrm{d}\omega A(\boldsymbol{k},\omega) = \mathrm{e}^{\Omega/T}\sum_{n,m}\mathrm{e}^{-E_n/T} \times (\langle n|c_{\boldsymbol{k}}|m\rangle\langle m|c_{\boldsymbol{k}}^{\dagger}|n\rangle + \langle n|c_{\boldsymbol{k}}^{\dagger}|m\rangle\langle m|c_{\boldsymbol{k}}|n\rangle). \tag{D.12}$$

The set of exact eigenstates is complete, i.e.

$$\sum_m |m\rangle\langle m| = 1$$

so that we can eliminate the summation over m in equation (D.12),

$$\frac{1}{\pi}\int_{-\infty}^{\infty}\mathrm{d}\omega\, A(\boldsymbol{k},\omega) = \mathrm{e}^{\Omega/T}\sum_{n,m}\mathrm{e}^{-E_n/T}\langle n|\{c_{\boldsymbol{k}}c_{\boldsymbol{k}}^{\dagger}\}|n\rangle = \mathrm{e}^{\Omega/T}\sum_{n}\mathrm{e}^{-E_n/T} = 1. \tag{D.13}$$

The spectral function plays an important role in the theoretical description of the photoemission spectra (part 2). Using equation (D.9) we can connect the imaginary part of the GF with the spectral function as

$$A(\boldsymbol{k},\omega) = -\coth\left(\frac{\omega}{2T}\right)\operatorname{Im} G(\boldsymbol{k},\omega). \tag{D.14}$$

The real part of the GF can also be expressed via $A(\boldsymbol{k},\omega)$ as an integral,

$$\operatorname{Re} G(\boldsymbol{k},\omega) = \frac{1}{\pi}P\int_{-\infty}^{\infty}\mathrm{d}\omega'\,\frac{A(\boldsymbol{k},\omega')}{\omega-\omega'}. \tag{D.15}$$

Let us also introduce retarded and advanced GFs defined as

$$G^{\mathrm{R}}(\boldsymbol{k}, t) = -\mathrm{i}\Theta(t)\langle\langle\{c_{\boldsymbol{k}}(t)c^{\dagger}_{\boldsymbol{k}}\}\rangle\rangle$$
$$G^{\mathrm{A}}(\boldsymbol{k}, t) = \mathrm{i}\Theta(-t)\langle\langle\{c_{\boldsymbol{k}}(t)c^{\dagger}_{\boldsymbol{k}}\}\rangle\rangle \tag{D.16}$$

respectively. By the use of the exact eigenstates and eigenvalues of the many-particle Hamiltonian, we obtain, as before,

$$G^{\mathrm{R}}(\boldsymbol{k}, \omega) = \mathrm{e}^{\Omega/T}\sum_{n,m} e^{-E_n/T}|\langle n|c_{\boldsymbol{k}}|m\rangle|^2(1+\mathrm{e}^{\omega_{nm}/T}) \times\left\{P\frac{1}{\omega_{nm}+\omega} - \mathrm{i}\pi\delta(\omega_{nm}+\omega)\right\}. \tag{D.17}$$

We see that

$$\mathrm{Im}\, G^{\mathrm{R}}(\boldsymbol{k}, \omega) = -A(\boldsymbol{k}, \omega) \tag{D.18}$$

and

$$\mathrm{Re}\, G^{\mathrm{R}}(\boldsymbol{k}, \omega) = \frac{1}{\pi}P\int_{-\infty}^{\infty}\mathrm{d}\omega'\,\frac{A(\boldsymbol{k}, \omega')}{\omega-\omega'}. \tag{D.19}$$

As a result, we connect the GF with the retarded (or advanced) GF:

$$G^{\mathrm{R}}(\boldsymbol{k}, \omega) = \mathrm{Re}\, G(\boldsymbol{k}, \omega) + \mathrm{i}\coth\left(\frac{\omega}{2T}\right)\mathrm{Im}\, G(\boldsymbol{k}, \omega) \tag{D.20}$$

and

$$G^{\mathrm{A}}(\boldsymbol{k}, \omega) = \mathrm{Re}\, G(\boldsymbol{k}, \omega) - \mathrm{i}\coth\left(\frac{\omega}{2T}\right)\mathrm{Im}\, G(\boldsymbol{k}, \omega). \tag{D.21}$$

From equations (D.18) and (D.19), we also have

$$G^{\mathrm{R}}(\boldsymbol{k}, \omega) = \frac{1}{\pi}\int_{-\infty}^{\infty}\mathrm{d}\omega'\,\frac{A(\boldsymbol{k}, \omega')}{\omega-\omega'+\mathrm{i}\delta} \tag{D.22}$$

and, in a similar way,

$$G^{\mathrm{A}}(\boldsymbol{k}, \omega) = \frac{1}{\pi}\int_{-\infty}^{\infty}\mathrm{d}\omega'\,\frac{A(\boldsymbol{k}, \omega')}{\omega-\omega'-\mathrm{i}\delta}. \tag{D.23}$$

It becomes clear that the retarded GF is analytical in the upper half-plane of ω and the advanced GF is analytical in the lower half-plane. Finally, using the definition (equation (D.4)) of the Matsubara GF, we can write, for $\tau > 0$,

$$\mathcal{G}(\boldsymbol{k}, \tau) = -\mathrm{e}^{\Omega/T}\sum_{n,m}\mathrm{e}^{-E_n/T}|\langle n|c_{\boldsymbol{k}}|m\rangle|^2\mathrm{e}^{\omega_{nm}\tau}. \tag{D.24}$$

Transforming this function into a Fourier series yields

$$\mathcal{G}(\boldsymbol{k}, \omega_k) = \int_0^{1/T}\mathrm{d}\tau\,\mathcal{G}(\boldsymbol{k}, \tau)\exp(\mathrm{i}\omega_k\tau) = \mathrm{e}^{\Omega/T}\sum_{n,m}\mathrm{e}^{-E_n/T}|\langle n|c_{\boldsymbol{k}}|m\rangle|^2\frac{1+\mathrm{e}^{\omega_{nm}/T}}{\omega_{nm}+\mathrm{i}\omega_k} \tag{D.25}$$

where $\omega_k = \pi T(2k+1)$, $k = 0, \pm 1, \pm 2, \ldots$. This Fourier transform can be also expressed via the spectral function:

$$\mathcal{G}(\boldsymbol{k}, \omega_k) = \frac{1}{\pi} \int_{-\infty}^{\infty} \mathrm{d}\omega' \, \frac{A(\boldsymbol{k}, \omega')}{\mathrm{i}\omega_k - \omega'}. \tag{D.26}$$

By comparison of equations (D.26) and (D.22), we obtain, for $\omega_k > 0$,

$$G^{\mathrm{R}}(\boldsymbol{k}, i\omega_k) = \mathcal{G}(\boldsymbol{k}, \omega_k) \tag{D.27}$$

and, from equation (D.26),

$$\mathcal{G}(\boldsymbol{k}, -\omega_k) = \mathcal{G}^*(\boldsymbol{k}, \omega_k). \tag{D.28}$$

These relations allow us to construct $G^{\mathrm{R}}(\boldsymbol{k}, \omega)$ for real frequencies continuing $\mathcal{G}(\boldsymbol{k}, \omega_k)$ analytically from a discrete set of points to the upper half-plane. Then using equation (D.20), one can obtain $G(\boldsymbol{k}, \omega)$.

Appendix E

Canonical transformation

Various canonical transformations serve as a powerful tool in the polaron theory [124]. Using an appropriate transformation, we can approximately diagonalize the Hamiltonian of strongly coupled electrons and phonons and then apply the perturbation theory with respect to residual off-diagonal terms. The philosophy of the method is simple. We are looking for a complete orthonormal set of multi-particle eigenstates $|n\rangle$, which obey the steady-state Schrödinger equation

$$H|n\rangle = E_n|n\rangle. \tag{E.1}$$

There exists a unitary transformation U such that the eigenstates $|n\rangle$ may be generated from another arbitrary complete orthonormal set $|\tilde{n}\rangle$ such that

$$|n\rangle = U|\tilde{n}\rangle. \tag{E.2}$$

The requirement that the states generated by equation (E.2) form an eigenbase of the given Hamiltonian H is equivalent to the condition that the transformed Hamiltonian

$$\tilde{H} = U^{\dagger}HU \tag{E.3}$$

is diagonal with respect to $|\tilde{n}\rangle$:

$$\tilde{H}|\tilde{n}\rangle = E_n|\tilde{n}\rangle. \tag{E.4}$$

The orthogonality should conserve ($\langle n \mid n'\rangle = \langle\tilde{n}|U^{\dagger}U|\tilde{n}'\rangle = \delta_{nn'}$) so that

$$U^{\dagger} = U^{-1}. \tag{E.5}$$

A frequently imployed transformation in polaron theory is the displacement transformation introduced by Lee *et al* [125] for a single polaron and by Lang and Firsov [71] for a multi-polaron system. It displays ions to new equilibrium positions depending on the electron coordinates,

$$U = \mathrm{e}^{-S} \tag{E.6}$$

where

$$S = \sum_{q,\nu,i} \hat{n}_i [u_i^*(\mathbf{q}, \nu) d_{\mathbf{q}\nu}^\dagger - H.c.] \tag{E.7}$$

is such that $S^\dagger = S^{-1} = -S$. The electron and phonon operators are transformed as

$$\begin{aligned} \tilde{c}_i &= \mathrm{e}^S c_i \mathrm{e}^{-S} \\ \tilde{d}_{\mathbf{q}\nu} &= \mathrm{e}^S d_{\mathbf{q}\nu} \mathrm{e}^{-S}. \end{aligned} \tag{E.8}$$

We can simplify equations (E8) scaling all matrix elements by the same amount, $u_i(\mathbf{q}, \nu) \to \eta u_i(\mathbf{q}, \nu)$, and differentiating the transformed operators with respect to the scaling parameter η as

$$\frac{\partial \tilde{c}_i}{\partial \eta} = \sum_{\mathbf{q},\nu} \mathrm{e}^S [\hat{n}_i, c_i](u_i^*(\mathbf{q}, \nu) d_{\mathbf{q}\nu}^\dagger - u_i(\mathbf{q}, \nu) d_{\mathbf{q}\nu}) \mathrm{e}^{-S} \tag{E.9}$$

and

$$\frac{\partial \tilde{d}_{\mathbf{q}\nu}}{\partial \eta} = \sum_i \mathrm{e}^S \hat{n}_i u_i^*(\mathbf{q}, \nu)[d_{\mathbf{q}\nu}^\dagger, d_{\mathbf{q}\nu}] \mathrm{e}^{-S}. \tag{E.10}$$

Using commutators $[\hat{n}_i, c_i] = -c_i$, $[d_{\mathbf{q}\nu}^\dagger, d_{\mathbf{q}\nu}] = -1$ and $[\hat{n}_i, S] = 0$, we find

$$\begin{aligned} \frac{\partial \tilde{c}_i}{\partial \eta} &= -\tilde{c}_i \sum_{\mathbf{q},\nu} (u_i^*(\mathbf{q}, \nu) \tilde{d}_{\mathbf{q}\nu}^\dagger - u_i(\mathbf{q}, \nu) \tilde{d}_{\mathbf{q}\nu}) \\ \frac{\partial \tilde{d}_{\mathbf{q}\nu}}{\partial \eta} &= -\sum_i \hat{n}_i u_i^*(\mathbf{q}, \nu). \end{aligned} \tag{E.11}$$

The solutions of these differential equations, which respects the 'boundary' conditions $\tilde{d}_{\mathbf{q}\nu} = d_{\mathbf{q}\nu}$ and $\tilde{c}_i = c_i$ when $\eta = 0$, are

$$\tilde{c}_i = c_i \exp\left[\eta \sum_{\mathbf{q},\nu} (u_i(\mathbf{q}, \nu) d_{\mathbf{q}\nu} - u_i^*(\mathbf{q}, \nu) d_{\mathbf{q}\nu}^\dagger)\right] \tag{E.12}$$

and

$$\tilde{d}_{\mathbf{q}\nu} = d_{\mathbf{q}\nu} - \eta \sum_i \hat{n}_i u_i^*(\mathbf{q}, \nu). \tag{E.13}$$

By taking $\eta = 1$ in equations (E.12) and (E.13), we obtain equations (4.28) and (4.29), respectively.

Let us now calculate the statistical average of the multi-phonon operator, which determines the polaron bandwidth in equations (4.38) and (4.42):

$$\langle\langle \hat{X}_i^\dagger \hat{X}_j \rangle\rangle \equiv \prod_{\mathbf{q},\nu} \langle\langle \exp[u_i^*(\mathbf{q}, \nu) d_{\mathbf{q}\nu}^\dagger - H.c.] \exp[u_j(\mathbf{q}, \nu) d_{\mathbf{q}\nu} - H.c.] \rangle\rangle. \tag{E.14}$$

Here an operator identity

$$e^{\hat{A}+\hat{B}} = e^{\hat{A}}e^{\hat{B}}e^{-[\hat{A},\hat{B}]/2} \tag{E.15}$$

is instrumental. It is applied when the commutator $[\hat{A}, \hat{B}]$ is a number. To prove the identity, we use the parameter differentiation [126] as before. Let us assume that

$$e^{\eta(\hat{A}+\hat{B})} = e^{\eta\hat{A}}e^{\eta\hat{B}}e^{\eta^2 C} \tag{E.16}$$

where C is a number. Differentiating both sides of equation (E.16) with respect to the parameter η and multiplying it from the right by

$$e^{-\eta(\hat{A}+\hat{B})} = e^{-\eta^2 C}e^{-\eta\hat{B}}e^{-\eta\hat{A}}$$

one obtains

$$\hat{A} + \hat{B} = \hat{A} + e^{\eta\hat{A}}\hat{B}e^{-\eta\hat{A}} + 2\eta C. \tag{E.17}$$

The quantity $e^{\eta\hat{A}}\hat{B}e^{-\eta\hat{A}}$ is expanded by means of

$$\begin{aligned} e^{\eta\hat{A}}\hat{B}e^{-\eta\hat{A}} &= \sum_{r,q=0}^{\infty} \frac{(-1)^q \eta^{r+q}}{r!q!} \hat{A}^r \hat{B} \hat{A}^q \\ &= \hat{B} + \eta[\hat{A}, \hat{B}] + \frac{\eta^2}{2}[\hat{A}, [\hat{A}, \hat{B}]] + \cdots. \end{aligned} \tag{E.18}$$

Because $[\hat{A}, \hat{B}]$ is a number, all terms, starting from the third one, vanish in the right-hand side of equation (E.18). Then using equation (E.17), one obtains

$$C = -\tfrac{1}{2}[\hat{A}, \hat{B}]. \tag{E.19}$$

The identity (E.15) allows us to write

$$\begin{aligned} e^{[u_i^*(\boldsymbol{q},\nu)d_{\boldsymbol{q}\nu}^\dagger - H.c.]}e^{[u_j(\boldsymbol{q},\nu)d_{\boldsymbol{q}\nu} - H.c.]} &= e^{(\alpha^* d_{\boldsymbol{q}\nu}^\dagger - \alpha d_{\boldsymbol{q}\nu})} \\ &\quad \times e^{[u_i(\boldsymbol{q},\nu)u_j^*(\boldsymbol{q},\nu) - u_i^*(\boldsymbol{q},\nu)u_j(\boldsymbol{q},\nu)]/2} \end{aligned}$$

where $\alpha \equiv u_i(\boldsymbol{q}, \nu) - u_j(\boldsymbol{q}, \nu)$. Applying once again the same identity yields

$$\begin{aligned} e^{[u_i^*(\boldsymbol{q},\nu)d_{\boldsymbol{q}\nu}^\dagger - H.c.]}e^{[u_j(\boldsymbol{q},\nu)d_{\boldsymbol{q}\nu} - H.c.]} &= e^{\alpha^* d_{\boldsymbol{q}}^\dagger}e^{-\alpha d_{\boldsymbol{q}}}e^{-|\alpha|^2/2} \\ &\quad \times e^{[u_i(\boldsymbol{q},\nu)u_j^*(\boldsymbol{q},\nu) - u_i^*(\boldsymbol{q},\nu)u_j(\boldsymbol{q},\nu)]/2}. \end{aligned} \tag{E.20}$$

Quantum and statistical averages are calculated by expanding the exponents in the trace as

$$\langle\langle e^{\alpha^* d^\dagger}e^{-\alpha d}\rangle\rangle = (1-p)\sum_{N=0}^{\infty}\sum_{n=0}^{N} p^N (-1)^n \frac{|\alpha|^{2n}}{(n!)^2} N(N-1) \times \cdots \times (N-n+1) \tag{E.21}$$

where we have dropped the phonon and site quantum numbers for transparency. Here $p = \exp(-\omega_{\mathbf{q}\nu}/T)$, so that a single-mode phonon partition function (appendix B) is

$$Z_{\mathrm{ph}} = \frac{1}{1-p}.$$

Equation (E.21) can be written in the form [71]

$$\langle\langle e^{\alpha^* d^\dagger} e^{-\alpha d}\rangle\rangle = (1-p)\sum_{n=0}^{N}(-1)^n \frac{|\alpha|^{2n}}{(n!)^2} p^n \frac{d^n}{dp^n}\sum_{N=0}^{\infty} p^N. \tag{E.22}$$

Taking the sum over N,

$$\sum_{N=0}^{\infty} p^N = \frac{1}{1-p}$$

and differentiating it n times yields $n!$ in the numerator, after which the series over n turns out equal to

$$\langle\langle e^{\alpha^* d^\dagger} e^{-\alpha d}\rangle\rangle = e^{-|\alpha|^2 n_\omega} \tag{E.23}$$

where $n_\omega = [\exp(\omega_{\mathbf{q}}/T) - 1]^{-1}$ is the Bose–Einstein distribution function of phonons. Now collecting all multipliers we obtain equations (4.38) and (4.42).

References

[1] Bednorz J G and Müller K A 1986 *Z. Phys.* B **64** 189
[2] Bardeen J, Cooper L N. and Schrieffer J R 1957 *Phys. Rev.* **108** 1175
[3] London F 1938 *Phys. Rev.* **54** 947
[4] Bogoliubov N N 1947 *Izv. Acad. Sci. (USSR)* **11** 77
[5] Landau L D 1947 *J. Phys. (USSR)* **11** 91
[6] Ogg Jr. R A 1946 *Phys. Rev.* **69** 243
[7] Schafroth M R 1955 *Phys. Rev.* **100** 463
[8] Blatt J M and Butler S T 1955 *Phys. Rev.* **100** 476
[9] Fröhlich H 1950 *Phys. Rev.* **79** 845
[10] Alexandrov A S and Ranninger J 1981 *Phys. Rev.* B **23** 1796
[11] Alexandrov A S 1983 *Zh. Fiz. Khim.* **57** 273
Alexandrov A S 1983 *Russ. J. Phys. Chem.* **57** 167
Alexandrov A S 1998 *Models and Phenomenology for Conventional and High-temperature Superconductivity (Course CXXXVI of the Intenational School of Physics 'Enrico Fermi')* ed G Iadonisi, J R Schrieffer and M L Chiofalo (Amsterdam: IOS) p 309
[12] Alexandrov A S and Mott N F 1994 *High Temperature Superconductors and Other Superfluids* (London: Taylor and Francis)
[13] Alexandrov A S and Mott N F 1995 *Polarons and Bipolarons* (Singapore: World Scientific)
[14] Anderson P W 1975 *Phys. Rev. Lett.* **34** 953
[15] Chakraverty B K 1979 *J. Physique Lett.* **40** L99
Chakraverty B K 1981 *J. Physique Lett.* **42** 1351
[16] Bednorz J G and Müller K A 1988 *Angew. Chem. Int. Ed. Engl.* **27** 735
[17] Meissner W and Ochsenfeld R 1933 *Naturw.* **21** 787
[18] London F and London H 1935 *Proc. R. Soc.* A**149** 71
[19] Deaver Jr. B S and Fairbank W M 1961 *Phys. Rev. Lett.* **7** 43
Doll R and Näbauer M 1961 *Phys. Rev. Lett.* **7** 51
[20] Pippard A B 1953 *Proc. R. Soc.* A **216** 547
[21] Ginzburg V L and Landau L D 1950 *Zh. Eksp. Teor. Fiz.* **20** 1064
[22] Levanyuk A P 1959 *Zh. Eksp. Teor. Fiz.* **36** 810
Pippard A B 1950 *Proc. R. Soc.* A **203** 210
[23] Landau L D and Lifshitz E M 1980 *Statistical Mechanics* 3rd edn (Oxford: Pergamon)
[24] Abrikosov A A 1957 *Zh. Eksp. Teor. Fiz.* **32** 1442
[25] Fock V 1928 *Z. Phys.* **47** 446

Landau L D 1930 *Z. Phys.* **64** 629
[26] Clogston A M 1962 *Phys. Rev. Lett.* **9** 266
Chandrasekhar B S 1962 *Appl. Phys. Lett.* **1** 7
[27] Werthamer N R, Helfand E and Hohenberg P C 1966 *Phys. Rev.* **147** 295
[28] Schmidt V V 1997 *Physics of Superconductors* ed P Muller and A V Ustinov (Berlin: Springer)
[29] Josephson B D 1962 *Phys. Lett.* **1** 251
[30] Feynman R P, Leighton R B and Sands M 1989 *Feynman Lectures on Physics, The Commemorative Issue* (San Francisco, CA: Benjamin Cummings)
[31] Schrieffer J R 1964 *Theory of Superconductivity* (Reading, MA: Benjamin)
[32] Abrikosov A A, Gor'kov L P and Dzaloshinski I E 1963 *Methods of Quantum Field Theory in Statistical Physics* (Englewood Cliffs, NJ: Prentice-Hall)
[33] de Gennes P G 1989 *Superconductivity of Metals and Alloys* (New York: Addison-Wesley)
[34] Bardeen J and Pines D 1955 *Phys. Rev.* **99** 1140
[35] Cooper L N 1957 *Phys. Rev.* **104** 1189
[36] Eliashberg G M 1960 *Zh. Eksp. Teor. Fiz.* **38** 966
Eliashberg G M 1960 *Zh. Eksp. Teor. Fiz.* **39** 1437
Eliashberg G M 1960 *Sov. Phys.–JETP* **11** 696
Eliashberg G M 1960 *Sov. Phys.–JETP* **12** 1000
[37] Giaever I 1960 *Phys. Rev. Lett.* **5** 147, 464
[38] Hebel L C and Slichter C P 1959 *Phys. Rev.* **113** 1504
[39] Geilikman B T 1958 *Zh. Eksp. Teor. Fiz.* **34** 1042
[40] Lifshits E M and Pitayevski 1978 *Statistical Physics Part 2 Theory of Condensed State* (Oxford: Butterworth–Heinemann)
[41] Svidzinskii A V 1982 *Space Inhomogeneous Problems in the Theory of Superconductivity* (Moscow: Nauka)
[42] Abrikosov A A and Gor'kov L P 1960 *Zh. Eksp. Teor. Fiz.* **38** 178
[43] Gor'kov L P 1958 *Zh. Eksp. Teor. Fiz.* **34** 735
Gor'kov L P 1958 *Sov. Phys.–JETP* **7** 505
[44] Matsubara T 1955 *Prog. Theor. Phys.* **14** 351
[45] Gor'kov L P 1959 *Zh. Eksp. Teor. Fiz.* **36** 1918
Gor'kov L P 1959 *Sov. Phys.–JETP* **9** 1364
[46] Andreev A F 1964 *Zh. Eksp. Teor. Fiz.* **46** 1823
Andreev A F 1964 *Sov. Phys.–JETP* **19** 1228
[47] Blonder G E, Tinkham M and Klapwijk T M 1982 *Phys. Rev.* B **25** 4515
[48] Migdal A B 1958 *Zh. Eksp. Teor. Fiz.* **34** 1438
Migdal A B 1958 *Sov. Phys.–JETP* **7** 996
[49] Nambu Y 1960 *Phys. Rev.* **117** 648
[50] Scalapino D J 1969 *Superconductivity* ed R D Parks (New York: Marcel Dekker)
[51] Allen P B and Mitrovic B 1982 *Solid State Physics* vol 37, ed H Ehrenreich, F Seitz and D Turnbull (New York: Academic)
[52] Ginzburg V L and Kirzhnits D A (ed) 1977 *The Problem of High-Temperature Superconductivity* (Moscow: Nauka)
[53] Tolmachev V V 1959 *A new Method in the Theory of Superconductivity* ed N N Bogoliubov *et al* (New York: Consultant Bureau)
[54] McMillan W J 1968 *Phys. Rev.* B **167** 331
[55] Klein B M, Boyer L L and Papaconstantopoulos D A 1978 *Phys. Rev.* B **18** 6411

[56] Allen P B and Dynes R C 1975 *Phys. Rev.* B **12** 905
[57] Kohn W and Luttinger J M 1965 *Phys. Rev. Lett.* **15** 524
[58] Friedel H 1954 *Adv. Phys.* **3** 446
[59] Fay D and Layzer A 1968 *Phys. Rev. Lett.* **20** 187
[60] Landau L D 1933 *J. Phys. (USSR)* **3** 664
[61] Pekar S I 1946 *Zh. Eksp. Teor. Fiz.* **16** 335
[62] Fröhlich H 1954 *Adv. Phys.* **3** 325
[63] Feynman R P 1955 *Phys. Rev.* **97** 660
[64] Devreese J T 1996 *Encyclopedia of Applied Physics* vol 14 (Weinheim: Wiley–VCH) p 383 and references therein
[65] Tjablikov S V 1952 *Zh. Eksp. Teor. Fiz.* **23** 381
[66] Yamashita J and Kurosawa T 1958 *J. Phys. Chem. Solids* **5** 34
[67] Sewell G L 1958 *Phil. Mag.* **3** 1361
[68] Holstein T 1959 *Ann. Phys.* **8** 325
Holstein T 1959 *Ann. Phys.* **8** 343
[69] Friedman L and Holstein T 1963 *Ann. Phys.* **21** 494
[70] Emin D and Holstein T 1969 *Ann. Phys.* **53** 439
[71] Lang I G and Firsov Yu A 1962 *Zh. Eksp. Teor. Fiz.* **43** 1843
Lang I G and Firsov Yu A 1963 *Sov. Phys.–JETP* **16** 1301
[72] Eagles D M 1963 *Phys. Rev.* **130** 1381
Eagles D M 1969 *Phys. Rev.* **181** 1278
Eagles D M 1969 *Phys. Rev.* **186** 456
[73] Appel J 1968 *Solid State Physics* vol 21, ed F Seitz, D Turnbull and H Ehrenreich (New York: Academic)
[74] Firsov Yu A (ed) 1975 *Polarons* (Moscow: Nauka)
[75] Böttger H and Bryksin V V 1985 *Hopping Conduction in Solids* (Berlin: Academie)
[76] Mahan G D 1990 *Many Particle Physics* (New York: Plenum)
[77] Geilikman B T 1975 *Usp. Fiz. Nauk.* **115** 403
Geilikman B T 1975 *Sov. Phys.–Usp.* **18** 190
[78] Alexandrov A S and Mazur E A 1989 *Zh. Eksp. Teor. Fiz.* **96** 1773
[79] Rashba E I 1957 *Opt. Spectrosc.* **2** 75
Rashba E I 1985 *Excitons* ed E I Rashba and D M Struge (Moscow: Nauka)
[80] Kabanov V V and Mashtakov O Yu 1993 *Phys. Rev.* B **47** 6060
[81] Aubry S 1995 *Polarons and Bipolarons in High-T_c Superconductors and Related Materials* ed E K H Salje, A S Alexandrov and W Y Liang (Cambridge: Cambridge University Press) p 271
[82] Alexandrov A S 1992 *Phys. Rev.* B **46** 14 932
[83] Alexandrov A S, Kabanov V V and Ray D K 1994 *Phys. Rev.* B **49** 9915
[84] Bishop A R and Salkola M 1995 *Polarons and Bipolarons in High-T_c Superconductors and Related Materials* ed E K H Salje, A S Alexandrov and W Y Liang (Cambridge: Cambridge University Press) p 353
[85] Marsiglio F 1995 *Physica* C **244** 21
[86] Takada Y and Higuchi T 1995 *Phys. Rev.* B **52** 12 720
[87] Fehske H, Loos J and Wellein G 1997 *Z. Phys.* B **104** 619
[88] Hotta T and Takada Y 1997 *Phys. Rev.* B **56** 13 916
[89] Romero A H, Brown D W and Lindenberg K 1998 *J. Chem. Phys.* **109** 6504
[90] La Magna A and Pucci R 1996 *Phys. Rev.* B **53** 8449
[91] Benedetti P and Zeyher R 1998 *Phys. Rev.* B **58** 14 320

[92] Frank T and Wagner M 1999 *Phys. Rev.* B **60** 3252
[93] Bonca J, Trugman S A and Batistic I 1999 *Phys. Rev.* B **60** 1633
[94] Alexandrov A S and Kornilovich P E 1999 *Phys. Rev. Lett.* **82** 807
[95] Proville L and Aubry S 1999 *Eur. Phys. J.* B **11** 41
[96] Alexandrov A S 2000 *Phys. Rev.* B **61** 12 315
[97] Hiramoto H and Toyozawa Y 1985 *J. Phys. Soc. Japan* **54** 245
[98] Alexandrov A S 1992 *Phys. Rev.* B **46** 2838
[99] Gogolin A A 1982 *Phys. Status Solidi* B **109** 95
[100] Alexandrov A S 1996 *Phys. Rev.* B **53** 2863
[101] Alexandrov A S and Sricheewin C 2000 *Europhys. Lett.* **51** 188
[102] Alexandrov A S and Ranninger J 1992 *Phys. Rev.* B **45** 13 109
Alexandrov A S and Ranninger J 1992 *Physica* C **198** 360
[103] Vinetskii V L and Giterman M Sh 1957 *Zh. Eksp. Teor. Fiz.* **33** 730
Vinetskii V L and Giterman M Sh 1958 *Sov. Phys.–JETP* **6** 560
Vinetskii V L and Pashitskii E A 1983 *Fiz. Tverd. Tela (Leningrad)* **25** 1744
Vinetskii V L and Pashitskii E A 1983 *Sov. Phys. Solid State* **25** 1005
[104] Müller K A 2002 *Phys. Scr.* T **102** 39 and references therein
[105] Alexandrov(Aleksandrov) A S and Kabanov V V 1986 *Fiz. Tverd. Tela* **28** 1129
Alexandrov(Aleksandrov) A S and Kabanov V V 1986 *Sov. Phys. Solid State* **28** 631
[106] Bryksin V V and Gol'tsev A V 1988 *Fiz. Tverd. Tela* **30** 1476
Bryksin V V and Gol'tsev A V 1988 *Sov. Phys. Solid State* **30** 851
[107] Bonča J and Trugman S A 2001 *Phys. Rev.* B **64** 094507
[108] Alexandrov A S and Kornilovitch P E 2002 *J. Phys.: Condens. Matter* **14** 5337
[109] Alexandrov A S, Ranninger J and Robaszkiewicz S 1986 *Phys. Rev.* B **33** 4526
[110] Alexandrov A S, Samarchenko D A and Traven S V 1987 *Zh. Eksp. Teor. Fiz.* **93** 1007
Alexandrov A S, Samarchenko D A and Traven S V 1987 *Sov. Phys.–JETP* **66** 567
[111] Peierls R E 1933 *Z. Phys.* **80** 763
[112] Alexandrov A S and Beere W H 1995 *Phys. Rev.* B **51** 5887
[113] Gross E P 1961 *Lett. Nuovo Cimento* **20** 454
Pitaevskii L P 1961 *Sov. Phys.–JETP* **13** 451
[114] Foldy L L 1961 *Phys. Rev.* **124** 649
[115] Alexandrov A S and Kabanov V V 2002 *J. Phys.: Condens. Matter* **14** L327
[116] Tonks L and Langmuir I 1929 *Phys. Rev.* **33** 195
[117] Pines D and Schrieffer J R 1961 *Phys. Rev.* **124** 1387
[118] Fetter A L 1971 *Ann. Phys.* **64** 1
[119] Bishop R F 1974 *J. Low Temp. Phys.* **15** 601
[120] Alexandrov A S 1999 *Phys. Rev.* B **60** 14 573
[121] Alexandrov A S 1984 *Doctoral Thesis* MIFI (Moscow)
Alexandrov A S 1993 *Phys. Rev.* B **48** 10 571
[122] Alexandrov A S, Beere W H and Kabanov V V 1996 *Phys. Rev.* B **54** 15 363
[123] Alexandrov A S 1998 *Physica* C **305** 46
[124] Wagner M 1986 *Unitary Transformations in Solid State Physics* (Amsterdam: North-Holland)
[125] Lee T D, Low F and Pines D 1953 *Phys. Rev.* **90** 297
[126] Wilcox R M 1967 *J. Math. Phys.* **8** 962

[127] Wu M K, Ashburn J R, Torng C J, Hor P H, Meng R L, Gao L, Huang Z J, Wang Y Q and Chu C W 1987 *Phys. Rev. Lett.* **58** 908

[128] Hebard A F, Rosseinsky M J, Haddon R C, Murphy D W, Glarum S H, Palstra T T, Ramirez A P, and Kortan A R 1991 *Nature* **350** 600

[129] Holczer K, Klein O, Huang S M, Kaner R B, Fu K J, Whetten R L and Diederich F 1991 *Science* **252** 1154

[130] Nagamatsu J, Nakagawa N, Muranaka T, Zenitani Y and Akimitsu J 2001 *Nature* **410** 63

[131] Alexandrov A S and Edwards P P 2000 *Physica* C **331** 97

[132] Simon A 1997 *Angew: Chem. Int. Ed. Engl.* **36** 1788

[133] Ginzburg V L 1968 *Contemp. Phys.* **9** 355

[134] Little W A 1964 *Phys. Rev.* A **134** 1416

[135] Fröhlich H 1968 *Phys. Lett.* A **26** 169

[136] Pashitskii E A 1968 *Zh. Eksp. Teor. Fiz.* **55** 2387
Pashitskii E A 1969 *Sov. Phys.–JETP* **28** 1267

[137] Schrieffer J R, Wen X G and Zhang S C 1989 *Phys. Rev.* B **39** 11 663

[138] Millis A J, Monien H and Pines D 1990 *Phys. Rev.* B **42** 167

[139] Mott N F 1990 *Adv. Phys.* **39** 55

[140] Anderson P W 1997 *The Theory of Superconductivity in the Cuprates* (Princeton NJ: Princeton University Press)

[141] Carlson E W, Emery V J, Kivelson S A and Orgad D 2002 Concepts in high temperature superconductivity *Preprint* cond-mat/0206217 and references therein

[142] Gabovich A M, Voitenko A I, Annett J F and Ausloos M 2001 *Supercond. Sci. Technol.* **14** R1

[143] Mott N F 1990 *Metal–Insulator Transitions* 2nd edn (London: Taylor and Francis)

[144] Wheatley J M, Hsu T C and Anderson P W 1988 *Phys. Rev.* B **37** 5897

[145] Schützmann J, Somal H S, Tsvetkov A A, van der Marel D, Koops G E J, Koleshnikov N, Ren Z F, Wang J H, Brück E and Menovsky A A 1997 *Phys. Rev.* B **55** 11 118

[146] Nagaosa N and Lee P A 2000 *Phys. Rev.* B **61** 9166

[147] Hubbard J 1963 *Proc. R. Soc.* A **276** 238

[148] For a review of the Hubbard and t–J models see Plakida N M 2001 *Lectures on the Physics of Highly Correlated Electron Systems V (AIP Conf. Proc. 580)* ed F Mancini (Melville, NY: American Institute of Physics) p 121

[149] Scalapino D J, Loh J E and Hirsch J E 1986 *Phys. Rev.* B **34** 8190

[150] Scherman A and Scrieber M 1995 *Phys. Rev.* B **52** 10 621

[151] Laughlin R B 2002 *Preprint* cond-mat/0209269

[152] Xiang X D, Vareka W A, Zettl A, Corkill J L, Cohen M L, Kijima N and Gronsky R 1992 *Phys. Rev. Lett.* **68** 530

[153] Forro L, Ilakovac V, Cooper J R, Ayache C and Henry J Y 1992 *Phys. Rev.* B **46** 6626

[154] Bourges P, Sidis Y, Fong H F, Regnault L P, Bossy J, Ivanov A and Keimer B 2000 *Science* **288** 1234
Bourges P, Keimer B, Regnault L P and Sidis Y 2000 *J. Supercond.* **13** 735

[155] Abbamonte P, Venema L, Rusydi A, Sawatzky G A, Logvenov G and Bozovic I 2002 *Science* **297** 581
Cho A 2002 *Science* **297** 499 and references therein

[156] Hirsch J E 2002 *Science* **295** 2226
Hirsch J E 2000 *Phys. Rev.* B **62** 14 487 and references therein
[157] Molegraaf H J A, Presura C, van der Marel D, Kes P H and Li M 2002 *Science* **295** 2239
[158] Labbe J and Bok J 1987 *Europhys. Lett.* **3** 1225
Newns D M, Pattnaik P C and Tsuei C C 1991 *Phys. Rev.* B **43** 3075
Abrikosov A A 2000 *Physica* C **341** 97
[159] Castellani C, DiCastro C and Grilli M 1995 *Phys. Rev. Lett.* **75** 4650
[160] Sachdev S, Chubukov A V and Ye J 1994 *Phys. Rev.* B **49** 11 919
[161] Smith T J, Andersen K H, Beck U, Capellmann H, Kremer R K, Neumann K U, Scharpf O, Simon A and Ziebeck K R A 1998 *J. Magnetism Magn. Mater.* **177** 543
[162] Radtke R J, Levin K, Schüttler H B and Norman M R 1993 *Phys. Rev.* B **48** 15 957
[163] Varma C M, Littlewood P B, Schmitt-Rink S, Abrahams E and Ruckenstein A E 1989 *Phys. Rev. Lett.* **63** 1996
[164] Emery V J and Kivelson S A *Nature* **374** 434
[165] Leggett A J 1973 *Physica Fennica* **8** 125
Leggett A J 1998 *J. Stat. Phys.* **93** 927
Popov V N 1987 *Functional Integrals and Collective Excitations* (Cambridge: Cambridge University Press)
[166] Alexandrov A S 1987 *Pis'ma Zh. Eksp. Teor. Fiz. (Prilozh.)* **46** 128
Alexandrov A S 1987 *JETP Lett. Suppl.* **46** 107
[167] Prelovsek P, Rice T M and Zhang F C 1987 *J. Phys. C: Solid State Phys.* **20** L229
[168] Emin D 1989 *Phys. Rev. Lett.* **62** 1544
[169] See different contributions in 1995 *Polarons and Bipolarons in High-T_c Superconductors and Related Materials* ed E K H Salje, A S Alexandrov and W Y Liang (Cambridge: Cambridge University Press) and in 1995 *Anharmonic properties of High Tc cuprates* ed D Mihailovic *et al* (Singapore: World Scientific)
[170] Mihailovic D, Foster C M, Voss K and Heeger A J 1990 *Phys. Rev.* B **42** 7989
[171] Calvani P, Capizzi M, Lupi S, Maselli P, Paolone A, Roy P, Cheong S-W, Sadowski W and Walker E 1994 *Solid State Commun.* **91** 113
[172] Timusk T, Homes C C and Reichardt W 1995 *Anharmonic Properties of High T_c Cuprates* ed D Mihailovic D *et al* (Singapore: World Scientific) p 171
[173] Zhao G, Hunt M B, Keller H and Müller K A 1997 *Nature* **385** 236
[174] Lanzara A, Bogdanov P V, Zhou X J, Kellar S A, Feng D L, Lu E D, Yoshida T, Eisaki H, Fujimori A, Kishio K, Shimoyama J I, Noda T, Uchida S, Hussain Z and Shen Z X 2001 *Nature* **412** 510
[175] Egami T 1996 *J. Low Temp. Phys.* **105** 791
[176] Temprano D R, Mesot J, Janssen S, Conder K, Furrer A, Mutka H and Müller K A 2000 *Phys. Rev. Lett.* **84** 1990
[177] Shen Z X, Lanzara A, Ishihara S and Nagaosa N 2002 *Phil. Mag.* B **82** 1349
[178] Phillips J C 1989 *Physics of high T_c Superconductors* (New York: Academic)
[179] Alexandrov A S and Mott N F 1993 *Supercond. Sci. Technol.* **6** 215
Alexandrov A S and Mott N F 1994 *J. Supercond.* **7** 599
Alexandrov A S and Mott N F 1994 *Rep. Prog. Phys.* **57** 1197
[180] Allen P B 2001 *Nature* **412** 494
[181] Gor'kov L P 1999 *J. Supercond.* **12** 9

[182] Tempere J, Fomin V M and Devreese J T 1997 *Solid State Commun.* **101** 661
[183] Alexandrov A S 1998 *Phil. Trans. R. Soc.* A **356** 197
[184] Merz M, Nucker N, Schweiss P, Schuppler S, Chen C T, Chakarian V, Freeland J, Idzerda Y U, Klaser M, Muller-Vogt G and Wolf Th 1998 *Phys. Rev. Lett.* **80** 5192
Dow J D, Howard U, Blackstead A 1998 *Bull. Am. Phys. Soc.* **43** 877
[185] Alexandrov A S 1992 *J. Low Temp. Phys.* **87** 721
[186] Alexandrov A S, Kabanov V V and Mott N F 1996 *Phys. Rev. Lett.* **77** 4796
[187] Alexandrov A S and Bratkovsky A M 2000 *Phys. Rev. Lett.* **84** 2043
[188] Kim J H, Feenstra B J, Somal H S, van der Marel D, Lee W Y, Gerrits A M and Wittlin A 1994 *Phys. Rev.* B **49** 13 065
[189] Alexandrov A S 2001 *Physica* C **363** 231
[190] Alexandrov A S and Kabanov V V 1996 *Phys. Rev.* B **54** 3655
[191] Pietronero L 1992 *Europhys. Lett.* **17** 365
Pietronero L and Strässler S 1992 *Europhys. Lett.* **18** 627
Grimaldi C, Pietronero L and Strässler S 1995 *Phys. Rev. Lett.* **75** 1158
Pietronero L, Strässler S and Grimaldi C 1995 *Phys. Rev.* B **52** 10 516
[192] Catlow C R A, Islam M S and Zhang X 1998 *J. Phys.: Condens. Matter* **10** L49
[193] Anselm A 1981 *Introduction of Semiconductor Theory* (Englewood Cliffs, NJ: Prentice-Hall)
[194] Alexandrov A S, Bratkovsky A M and Mott N F 1994 *Phys. Rev. Lett.* **72** 1734
[195] Mihailovic D, Kabanov V V, Zagar K, and Demsar J 1999 *Phys. Rev.* B **60** 6995 and references therein
[196] Zhang Y, Ong N P, Xu Z A, Krishana K, Gagnon R and Taillefer L 2000 *Phys. Rev. Lett.* **84** 2219
[197] Chen X H, Yu M, Ruan K Q, Li S Y, Gui Z, Zhang G C and Cao L Z 1998 *Phys. Rev.* B **58** 14 219
Chen W M, Franck J P and Jung J 2000 *Physica* C **341** 1875
[198] Hwang H Y, Batlogg B, Takagi H, Kao H L, Kwo J, Cava R J, Krajewski J J and Peck W F 1994 *Phys. Rev. Lett.* **72** 2636
Batlogg B, Hwang H Y, Takagi H, Cava R J, Kao H L and Kwo J 1994 *Physica* C **235** 130
[199] Bucher B, Steiner P, Karpinski J, Kaldis E and Wachter P 1993 *Phys. Rev. Lett.* **70** 2012
[200] Mott N F 1993 *Phil. Mag. Lett.* **68** 245
[201] Boebinger G S, Ando Y, Passner A, Kimura T, Okuya M, Shimoyama J, Kishio K, Tamasaku K, Ichikawa N and Uchida S 1996 *Phys. Rev. Lett.* **77** 5417
Ando Y, Boebinger G S, Passner A, Kimura T and Kishio K 1995 *Phys. Rev. Lett.* **75** 4662
[202] Alexandrov A S 1997 *Phys. Lett.* A **236** 132
[203] Alexandrov A S and Mott N F 1993 *Phys. Rev. Lett.* **71** 1075
[204] Takenaka K, Fukuzumi Y, Mizuhashi K, Uchida S, Asaoka H and Takei H 1997 *Phys. Rev.* B **56** 5654
[205] Yu R C, Salamon M B, Lu J P and Lee W C 1992 *Phys. Rev. Lett.* **69** 1431
[206] Hill R W, Proust C, Taillefer L, Fournier P and Greene R L 2001 *Nature* **414** 711
[207] Takeya J, Ando Y, Komiya S and Sun X F 2002 *Phys. Rev. Lett.* **88** 077001
[208] Proust C, Boakin E, Hill R W, Taillefer L and Mackenzie A P 2002 *Preprint* cond-mat/0202101

[209] Machi T, Tomeno I, Miyatake T, and Tanaka S 1991 *Physica* C **173** 32

[210] Rossat-Mignod J, Regnault L P, Bourges P, Vettier C, Burlet P and Henry J Y 1992 *Phys. Scr.* **45** 74

[211] Mook H A, Yethiraj M, Aeppli G, Mason T E and Armstrong T 1993 *Phys. Rev. Lett.* **70** 3490

[212] Hofer J, Karpinski J, Willemin M, Meijer G I, Kopnin E M, Molinski R, Schwer H, Rossel C and Keller H 1998 *Physica* C **297** 103

[213] Zverev V N and Shovkun D V 2000 *JETP Lett.* **72** 73

[214] Nakamura Y and Uchida S 1993 *Phys. Rev.* B **47** 8369

[215] Nakano T, Oda M, Manabe C, Momono N, Miura Y and Ido M 1994 *Phys. Rev.* B **49** 16 000

[216] Fehske H, Röder H, Wellein G and Mistriotis A 1995 *Phys. Rev.* B **51** 16 582

[217] Alexandrov A S 1999 *Phys. Rev. Lett.* **82** 2620
Alexandrov A S and Kabanov V V 1999 *Phys. Rev.* B **59** 13 628

[218] Uemura Y J 1995 *Polarons and Bipolarons in High-T_c Superconductors and Related Materials* ed E K H Salje, A S Alexandrov and W Y Liang (Cambridge: Cambridge University Press) p 453

[219] Alder B J and Peters D S 1989 *Europhys. Lett.* **10** 1

[220] Pokrovsky V L 1988 *Pis'ma Zh. Teor. Fiz.* **47** 5395

[221] Schneider T and Singer J M 2000 *Phase Transition Approach To High Temperature Superconductivity* (London: Imperial College Press)

[222] Legget A J 1980 *J. Physique* **41** C7

[223] Nozieres P and Schmitt-Rink S 1985 *J. Low Temp. Phys.* **59** 195

[224] Micnas R and Kostyrko T 1995 *Recent developments in High Temperature Superconductivity* ed P W Klamut and M Kazimierski (Berlin: Springer) p 221 and references therein

[225] Gyorffy B L, Staunton J B and Stocks G M 1991 *Phys. Rev.* B **44** 5190
Gorbar E V, Loktev V M and Sharapov S G 1996 *Physica* C **257** 355
Pistolesi F and Strinati G C 1996 *Phys. Rev.* B **53** 15 168
Marini M, Pistolesi F and Strinati G C 1998 *Eur. Phys. J.* B **1** 158
Maly J, Levin K and Liu D Z 1996 *Phys. Rev.* B **54** 15 657

[226] Alexandrov A S and Rubin S G 1993 *Phys. Rev.* B **47** 5141

[227] Tiverdi N and Randeria M 1995 *Phys. Rev. Lett.* **75** 312
Singer J M, Pedersen M H and Schneider T 1995 *Physica* B **230** 955

[228] Schmitt Rink S, Varma C M and Ruckenstein A E 1989 *Phys. Rev. Lett.* **63** 445

[229] Traven S V 1994 *Phys. Rev.* B **51** 3242

[230] Crawford M K, Farneth W E, McCarron E M III, Harlow R L and Moudden A H 1990 *Science* **250** 1309

[231] Bornemann H J, Morris D E, Liu H B, Sinha A P, Narwankar P and Chandrachood M 1991 *Physica* C **185–189** 1359

[232] Franck J P, Jung J, Salomons G J, Miner W A, Mohamed M A K, Chrzanowski J, Gygax S, Irwin J C, Mitchell D F and Sproule I 1989 *Physica* C **162** 753
Franck J P, Jung J, Mohamed M A K, Gygax S, and Sproule G I 1991 *Phys. Rev.* B **44** 5318

[233] Fisher R A, Kim S, Lacy S E, Phillips N E, Morris D E, Markelz A G, Wei J Y T and Ginley D S 1988 *Phys. Rev.* B **38** 11 942

[234] Loram J W, Cooper J R, Wheatley J M, Mirza K A and Liu R S 1992 *Phil. Mag.* B **65** 1405

[235] Inderhees S E, Salamon M B, Goldenfeld N, Rice J P, Pazol B G and Ginzberg D M 1988 *Phys. Rev. Lett.* **60** 1178

[236] Junod A, Eckert D, Triscone G, Lee V Y and Muller J 1989 *Physica* C **159** 215

[237] Schnelle W, Braun E, Broicher H, Dömel R, Ruppel S, Braunisch W, Harnischmacher J and Wohlleben D 1990 *Physica* C **168** 465

[238] Salamon M B, Inderhees S E, Rice J P and Ginsberg D M 1990 *Physica* A **168** 283

[239] Junod A, 1996 *Studies of High Temperature Superconductors* vol 19, ed A Narlikar (New York: Nova Science, Commack) p 1

Revaz B, Junod A and Erb A 1998 *Phys. Rev.* B **58** 11 153

Roulin M, Junod A and Walker E 1998 *Physica* C **296** 137

[240] Alexandrov A S, Beere W H, Kabanov V V and Liang W Y 1997 *Phys. Rev. Lett.* **79** 1551

[241] Zavaritsky V N, Kabanov V V and Alexandrov A S 2002 *Europhys. Lett.* **60** 127

[242] Bucher B, Karpinski J, Kaldis E and Wachter P 1990 *Physica* C **167** 324

[243] Mackenzie A P, Julian S R, Lonzarich G G, Carrington A, Hughes S D, Liu R S and Sinclair D C 1993 *Phys. Rev. Lett.* **71** 1238

[244] Osofsky M S, Soulen R J, Wolf S A, Broto J M, Rakoto H, Ousset J C, Coffe G, Askenazy S, Pari P, Bozovic I, Eckstein J N and Virshup G F 1993 *Phys. Rev. Lett.* **71** 2315

Osofsky M S, Soulen R J, Wolf S A, Broto J M, Rakoto H, Ousset J C, Coffe G, Askenazy S, Pari P, Bozovic I, Eckstein J N and Virshup G F 1994 *Phys. Rev. Lett.* **72** 3292

[245] Alexandrov A S, Zavaritsky V N, Liang W Y and Nevsky P L 1996 *Phys. Rev. Lett.* **76** 983

[246] Lawrie D D, Franck J P, Beamish J R, Molz E B, Chen W M and Graf M J 1997 *J. Low Temp. Phys.* **107** 491

[247] Gantmakher V F, Tsydynzhapov G E, Kozeeva L P and Lavrov A N *Zh. Eksp. Teor. Fiz.* **88** 148

Gantmakher V F, Emel'chenko G A, Naumenko I G, and Tsydynzhapov G E 2000 *JETP Lett.* **72** 21

[248] Nakanishi T, Motoyama N, Mitamura H, Takeshita N, Takahashi H, Eisaki H, Uchida S and Môri N 2000 *Int. J. Mod. Phys.* **14** 3617

[249] Lee I J, Chaikin P M and Naughton M J 2000 *Phys. Rev.* B **62** R14 669

[250] Wen H H, Li S L and Zhao Z X 2000 *Phys. Rev.* B **62** 716

[251] Annett J, Goldenfeld N and Legget A J 1996 *Physical Properties of High Temperature Superconductors* vol 5, ed D M Ginsberg (Singapore: World Scientific) p 375

[252] Tsuei C C and Kirtley J R 1997 *Physica* C **282** 4 and references therein

[253] Bonn D A, Kamal S, Zhang K, Liang R X, Baar D J, Klein E and Hardy W N 1994 *Phys. Rev.* B **50** 4051

[254] Xiang T, Panagopoulos C and Cooper J R 1998 *Int. J. Mod. Phys.* B **12** 1007

[255] Li Q, Tsay Y N, Suenaga M, Klemm R A, Gu G D and Koshizuka N 1999 *Phys. Rev. Lett.* **83** 4160

[256] Müller K A 2002 *Phil. Mag. Lett.* **82** 279 and references therein

[257] Walter H, Prusseit W, Semerad R, Kinder H, Assmann W, Huber H, Burkhardt H, Rainer D and Sauls J A 1998 *Phys. Rev. Lett.* **80** 3598

[258] Alexandrov A S and Giles R T 1999 *Physica* C **325** 35

[259] Alexandrov A S 1998 *Physica* C **305** 46

[260] Alexandrov A S and Andreev A F 2001 *Europhys. Lett.* **54** 373
[261] Alexandrov A S and Sricheewin C 2002 *Europhys. Lett.* **58** 576
[262] Alexandrov A S and Dent C J 1999 *Phys. Rev.* B **60** 15 414
[263] Alexandrov A S 2002 *Phil. Mag.* B **81** 1397
[264] For a review see Shen Z X 1998 *Models and Phenomenology for Conventional and High-temperature Superconductivity (Course CXXXVI of the Intenational School of Physics 'Enrico Fermi')* ed G Iadonisi, J R Schrieffer and M L Chiofalo (Amsterdam: IOS) p 141
[265] Renner Ch, Revaz B, Genoud J Y, Kadowaki K and Fischer Ø 1998 *Phys. Rev. Lett.* **80** 149
[266] Mott N F and Davis E 1979 *Electronic Processes in Non-crystalline Materials* 2nd edn (Oxford: Oxford University Press)
[267] Müller K A, Zhao G M, Conder K and Keller H 1998 *J. Phys.: Condens. Matter* **10** L291
[268] Hancotte H, Deltour R, Davydov D N, Jansen A G M and Wyder P 1997 *Phys. Rev.* B **55** R3410
[269] Mourachkine A 2000 *Europhys. Lett.* **49** 86
[270] Ding H, Yokoya T, Campuzano J C, Takahashi T, Randeria M, Norman M R, Mochiku T, Kadowaki K and Giapintzakis J 1996 *Nature* **382** 51
[271] Shen Z X and Schrieffer J R 1997 *Phys. Rev. Lett.* **78** 1771 and references therein
[272] Loram J W, Mirza K A, Wade J M, Cooper J R and Liang W Y 1994 *Physica* C **235** 134
[273] Deutscher G 1999 *Nature* **397** 410
Yagil Y, Hass N, Desgardin G and Monot I 1995 *Physica* C **250** 59
[274] Krasnov V M, Yurgens A, Winkler D, Delsing P and Claeson T 2000 *Phys. Rev. Lett.* **84** 5860
[275] Suzuki M and Watanabe T 2000 *Phys. Rev. Lett.* **85** 4787
[276] Tinkham M 1996 *Introduction to Superconductivity* (Singapore: McGraw-Hill) p 20
[277] Demuth J E, Persson B N J, Holtzberg F and Chandrasekhar C V 1990 *Phys. Rev. Lett.* **64** 603
[278] Gofron K, Campuzano J C, Abrikosov A A, Lindroos M, Bansil A, Ding H, Koelling D and Dabrowski B 1994 *Phys. Rev. Lett.* **73** 3302
[279] Schabel M C, Park C H, Matsuura A, Shen Z X, Bonn D A, Liang R and Hardy W N 1998 *Phys. Rev.* B**57** 6090
[280] Park C H, Shen Z X, Loeser A G, Dessau D S, Mandrus D G, Migliori A, Sarrao J and Fisk Z 1995 *Phys. Rev.* B **52** R16 981
[281] Halperin B I 1965 *Phys. Rev.* **139** A104
[282] Frisch H L and Lloyd S P 1960 *Phys. Rev.* **120** 1175
[283] Randeria M and Campuzano J C 1998 *Models and Phenomenology for Conventional and High-temperature Superconductivity (Course CXXXVI of the Intenational School of Physics 'Enrico Fermi')* ed G Iadonisi, J R Schrieffer and M L Chiofalo (Amsterdam: IOS) p 115
[284] Kaminski A, Mesot J, Fretwell H, Campuzano J C, Norman M R, Randeria M, Ding H, Sato T, Takahashi T, Mochiku T, Kadowaki K and Hoechst H 2000 *Phys. Rev. Lett.* **84** 1788
[285] Fedorov A V, Valla T, and Johnson P D, Li Q, Gu G D and Koshizuka N 1999 *Phys. Rev. Lett.* **82** 2179

[286] Zhang Y, Ong N P, Anderson P W, Bonn D A, Liang R and Hardy W N 2001 *Phys. Rev. Lett.* **86** 890

[287] Alexandrov A S and Dent C J 2001 *J. Phys.: Condens. Matter* **13** L417

[288] Landau L D and Lifshitz E N 1977 *Quantum Mechanics (Non-relativistic Theory)* 3rd edn (Oxford: Butterworth–Heinemann)

[289] Calogero F 1967 *Variable Phase Approach to Potential Scattering* (New York: Academic)

[290] Hore S R and Frankel N E 1975 *Phys. Rev.* B **12** 2619

[291] Zaanen J and Gunnarsson O 1989 *Phys. Rev.* B **40** 7391

[292] Tranquada J M, Sternlieb B J, Axe J D, Nakamura Y and Uchida 1996 *Nature* **375** 561

[293] Bianconi A 1999 *J. Physique* IV **9** 325 and references therein

[294] Alexandrov A S and Kabanov V V 2000 *JETP Lett.* **72** 569

[295] Hoffman J E, Hudson E W, Lang K M, Madhavan V, Eisaki H, Uchida S and Davis J C 2002 *Science* **295** 466

[296] Howald C, Eisaki H, Kaneko N and Kapitulnik A 2002 *Preprint* cond-mat/0201546

Index

Printed and bound by CPI Group (UK) Ltd, Croydon, CR0 4YY
23/10/2024
01778238-0005